Advancements in Colorectal Cancer

This book provides a comprehensive exploration of colorectal cancer (CRC), covering its development, cellular and molecular pathologies, diagnostic methods, treatment modalities, challenges, and future research directions. The introductory chapters provide foundational aspects of CRC, its historical evolution, prevalent symptoms, and underlying causes. This book presents the pathology of CRC, delving into microscopic analysis techniques, and covers the classification and various stages of CRC. Additionally, this book examines the cellular and molecular aspects of the disease, unraveling the intricate signaling pathways involved, and shedding light on the genetic and epigenetic factors that contribute to its development and progression. This book further explores tools and techniques utilized in clinical practice, including cellular, molecular, and genetic testing methods. It examines various treatment modalities for CRC, including chemotherapy, radiation therapy, and immunotherapy, highlighting their associated advantages and disadvantages. Toward the end, this book highlights innovative trends and advancements in CRC diagnosis and treatment, including the transformative role of machine learning, artificial intelligence, stem cells, and nano drug delivery systems. This book serves as a vital resource for the latest information for researchers, cancer research students, and oncologists.

Key Features

- Reviews cellular and molecular pathologies, diagnostic methods, and treatment modalities of CRC.
- Discusses diagnostic methods and treatment strategies, including chemotherapy, radiation therapy, and immunotherapy.
- Highlights the role of machine learning, artificial intelligence, stem cells, and nano drug delivery systems in CRC diagnosis and treatment.
- Examines the genetic and epigenetic factors that contribute to the development and progression of CRC.
- Delves into the cellular and molecular aspects of CRC, elucidating complex signaling pathways involved in disease initiation and progression.

Advancements in Colorectal Cancer

Edited by
Firdos Alam Khan

CRC Press
Taylor & Francis Group
Boca Raton London New York

CRC Press is an imprint of the
Taylor & Francis Group, an **informa** business

Designed cover image: Firdos Alam Khan

First edition published 2026
by CRC Press
2385 NW Executive Center Drive, Suite 320, Boca Raton FL 33431

and by CRC Press
4 Park Square, Milton Park, Abingdon, Oxon, OX14 4RN

CRC Press is an imprint of Taylor & Francis Group, LLC

ISBN: 978-1-032-82277-8 (hbk)
ISBN: 978-1-032-82280-8 (pbk)
ISBN: 978-1-003-50384-2 (ebk)

DOI: 10.1201/9781003503842

Typeset in Times
by codeMantra

Contents

Preface

Colorectal cancer (CRC) is a major global cancer, and an estimated more than 1.9 million new cases of CRC in 2020 and more than 930,000 deaths due to CRC were estimated to have occurred worldwide. It has been reported that in 2023, approximately 153,020 individuals were diagnosed with CRC. The course of therapy for CRC is determined by screening or diagnosing the unique expression of certain markers in the tumor. The treatment for CRC involves surgical removal of the tumor, which is the most common approach across all stages of colon cancer. Other treatments include radiofrequency ablation, cryosurgery, chemotherapy, targeted therapy, and immunotherapy. While there are a few books written about cancer and its treatment, they don't cover all aspects of CRC pathology, screening, diagnosis, and treatment. The title of this book is likely to attract readers such as medical students, oncologists, and cancer researchers who are interested in learning comprehensive information about CRC, its pathology, diagnosis, and treatment modalities, all in a single scan.

In this book, we have discussed CRC development, its cellular and molecular pathologies, stages of CRC, methods of CRC screening and diagnosis, various treatment modalities, issues and challenges of CRC diagnosis and treatments, and future research in these areas. This book has been divided into ten chapters, covering topics such as CRC development and causes, cellular and molecular pathologies of CRC, stages of CRC, methods of CRC screening and diagnosis, various available treatment modalities, issues and challenges in CRC diagnosis and treatments, and future research in CRC diagnosis and treatments. This book aids medical students, biomedical and CRC researchers, and medical teachers in understanding all aspects of CRC diagnosis, treatments, and current research trends in a single scan concisely.

Prof. Firdos Alam Khan
Editor

Acknowledgments

I am grateful to the Almighty Allah, who gave me the strength to complete the editing of this book on time. I am thankful to many people, especially Dr. Bhavik Sawhney, Editor-Biomedicine at CRC Press, who supported me in completing the task, as well as to all the members of the Taylor & Francis production team. I am also grateful to all the corresponding authors, namely, Dr. Sara Mohammad Ibrahim, Prof. Alaa A. A. Aljabali, Dr. Khalid Umar Fakhri, Dr. Anil Kumar Mavi, Dr. Nazia Afroze, Dr. Mohammad Raghibul Hasan, Dr. Nasrullah Jan, Prof. Syed Arman Rabbani, and Dr. Gayatri Sharma, for their immense contributions and timely completion of the work.

I am thankful to the Dean of the Institute for Research & Medical Consultations (IRMC), Imam Abdulrahman Bin Faisal University, Dammam, Saudi Arabia, for her support and encouragement. I am also thankful to the President of Imam Abdulrahman Bin Faisal University, Dammam, Saudi Arabia, and the Vice President of Higher Education and Scientific Research, Imam Abdulrahman Bin Faisal University, Dammam, Saudi Arabia, for their encouragement and appreciation. I am grateful to all my teachers and mentors, especially Professor Nishikant Subhedar and the late Professor Obaid Siddiqi FRS, for their immense contributions to shaping my research career. I extend my gratitude to all my friends, well-wishers, and colleagues for their support and cooperation.

I am grateful to my entire family, especially to my late father, Nayeemuddin Khan, and my late mother, Sarwari Begum; my brothers (Aftab Alam Khan, Javed Alam Khan, Intekhab Alam Khan, and Sarfaraz Alam Khan); my sisters (Late Sayeeda Khanum, Faheemida Khanum, Kahkashan Khanum, and Ayesha Khanum); my wife, Samina Khan; my sons (Zuhayr Ahmad Khan, Zaid Ahmad Khan, and Zahid Ahmad Khan); my daughter (Azraa Khan); my father-in-law (Abdul Qayyum Siddiqi); my mother-in-law (Late Uzma Siddiqi); and my brothers-in-law (Rehan Ahmed Siddiqi, Haroon Ahmed Siddiqi, and Noman Ahmed Siddiqi) for their prayers and best wishes.

Enjoy reading!

Prof. Firdos Alam Khan
Department of Stem Cell Research
Institute for Research & Medical Consultations,
Imam Abdulrahman Bin Faisal University
Dammam
Saudi Arabia

About the Editor

Firdos Alam Khan is a Professor at the Department of Stem Cell Research, Institute for Research and Medical Consultations (IRMC), Imam Abdulrahman Bin Faisal University, Dammam, Saudi Arabia. He received his Ph.D. in zoology with a specialization in neuroscience from Nagpur University, India. Professor Khan has been involved in teaching various courses, including Biochemistry and Cell Physiology, Stem Cell Biology and Regenerative Medicine, Cell Biology, Pharmacology, Business of Biotechnology, Biomedicine, Cell & Tissue Engineering, and Bioethics & IPR, to both undergraduate and postgraduate students. Professor Khan was associated with the Manipal Academy of Higher Education, Dubai Campus, United Arab Emirates, where he was a Professor and Chairperson of the School of Life Sciences, and served as the Chairman of the Research & Development Program. Professor Khan completed his first postdoctoral fellowship at the National Centre for Biological Sciences (NCBS), Bangalore, India, and a second postdoctoral fellowship at the Massachusetts Institute of Technology (MIT), Cambridge, USA. Professor Khan has also worked with Reliance Life Sciences as a Research Scientist in the Stem Cell Research Laboratory. Professor Khan has been granted 17 US patents and has authored and edited nine books published by international publishers. Professor Khan is a recognized reviewer of *Scientific Reports, Nanomedicine, Pharmaceutics, Pharmaceuticals, Nanomedicine: Nanomedicine, Biology, and Medicine, International Journal of Nanomedicine, Microscopy and Microanalysis, IET Nanobiotechnology, Polymers, Drug Design, Development and Therapy, Current Pharmaceutical Biotechnology, Tissue Engineering Part A, Journal of Biomolecular Structure and Dynamics, Nanotechnology Reviews, 3-Biotechnology, Molecules, Biotechnology Letters, Journal of Nanostructure Chemistry, Pharmaceutical Patent Analysis, World Journal of Surgical Oncology, NanoExpress, Advances in Natural Sciences: Nanoscience and Nanotechnology,* and *Cancers.* He has published more than 145 research publications. Professor Khan is a Fellow of the Royal Society of Biology (RSB), UK, and a member of the Sigma Xi Honor Society, USA (MIT Chapter). He is currently studying the role of different types of nanomaterials and biomaterials in regenerative medicine and cancer treatment.

1 Colorectal Cancer
Definition, History, Symptoms, Causes, Worldwide Scenario, and Global Statistics

Firdos Alam Khan
Department of Stem Cell Research, Institute for Research and Medical Consultations (IRMC), Imam Abdulrahman Bin Faisal University, Dammam, Saudi Arabia

1.1 COLORECTAL CANCER

Colorectal cancer (CRC) is cancer of the colon and rectum, and it is sometimes called colon cancer or rectal cancer. As per the anatomical location in the body, both colon and rectal cancer are often assembled together because they have several topographies in common (American Cancer Society, 2024a, b). The human adult intestine mainly consists of the colon, rectum, and anal canal. Based on the anatomical location and position, the colon is divided into the right-side colon, which consists of the cecum, ascending colon, and part of the transverse colon, and the left-side colon, which consists of the remaining transverse colon, descending colon, and sigmoid colon (Sakorafas et al., 2006; Houston et al., 2016) (Figure 1.1). The physiological function of the intestine is to facilitate food digestion and food particle absorption to provide the human body energy. Due to genetic or environmental influence, the colon and rectum may produce abnormal tissues or lumps, which may later on develop into cancerous tumors if not treated properly. Approximately 49.66% of patients develop rectal cancer, whereas 49.09% of patients develop colon cancer, and the remaining 1.25% of patients develop both colon and rectal cancers (Wang, 2019). Based on the studies, it was shown that

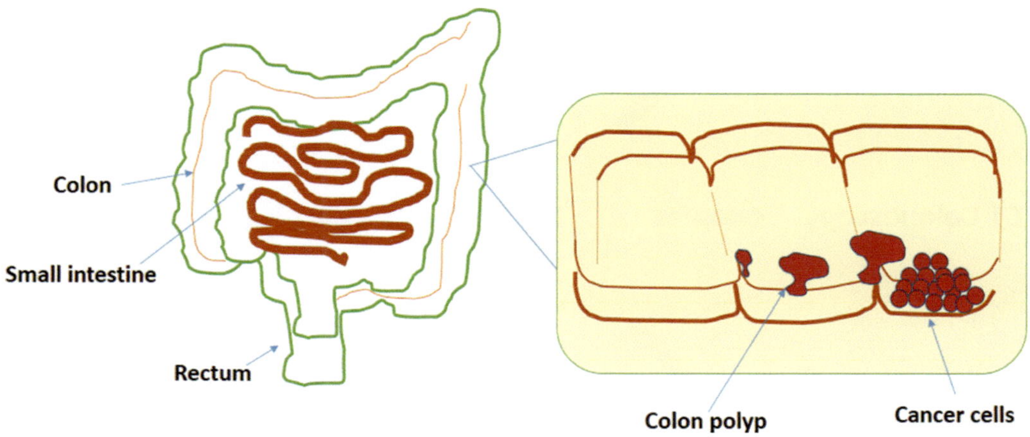

FIGURE 1.1 Diagrammatic illustration of the colon and rectum and the development of polyps leading to cancer development.

DOI: 10.1201/9781003503842-1

the sigmoid colon accounts for 55%, followed by the ascending colon (23.3%), the transverse colon (8.5%), the descending colon (8.1%), and the cecum (8.0%), respectively (Wang, 2019).

1.2 HISTORICAL PERSPECTIVES

CRC has a long history, with numerous discoveries made at different time intervals to understand how it is caused and develops in humans. In this section, we have briefly discussed the historical perspectives of breast cancer and how breast cancer research and clinical investigation have evolved. There are three types of ragged CRC polyps accepted in the past 60 years. The hyperplastic polyp was first identified, and the serrated polyp is the most commonly encountered form of CRC. The second type of CRC is the carcinoma-associated sessile serrated adenoma or polyp, which has been documented, but its diagnosis can be difficult due to overlapping histologic topographies with hyperplastic polyps. The third type of CRC is serrated adenoma, which is a less common type of cancer compared to hyperplastic polyp and carcinoma-associated sessile serrated adenoma cancers (Figure 1.2).

The topography of the serrated cancer in the colon's mucosa appears as sawtooth-like indentations along the luminal edges of the basements. The serrations or grooves are formed by small clumps of epithelial cells and are also referred to as micro-papillary processes, which are detached by single cells along the sides of vaults and mucosal surfaces (William et al., 1980). The development of serrations is due to the formation of small crypt-like buds lengthwise of the main tubules and is also referred to as ectopic crypt formations (Torlakovic et al., 2008). With advanced diagnostic techniques, it's possible to identify serrated polyps in colorectal biopsies, but in the old days, almost 60 years ago, it was not possible to identify the serrated polyps in the patients.

The term hyperplastic polyp was first coined by two German researchers in the early 19th century (Feyrter, 1929; Westhues, 1934). The use of histochemical techniques made it possible to measure oxidative enzymes in colonic lesions, which are also called as focal mucosal hyperplasia in cancer-producing areas. Hyperplastic polyps were identified using photomicrographs (Wattenberg, 1959). The pioneering work of British scientist Basil Morson led to the coining of the name "metaplastic polyp" in 1962, which now is called as hyperplastic polyp to differentiate it from adenoma-based cell proliferation (Morson, 1962). He coined the name "metaplasia" because of a change in the appearance of the mucous membrane compared with the normal tissue. However, the researcher was not comfortable using the name "hyperplasia" as he found that there was little evidence of any increase in cellular activity in these polyps. Nevertheless, the term

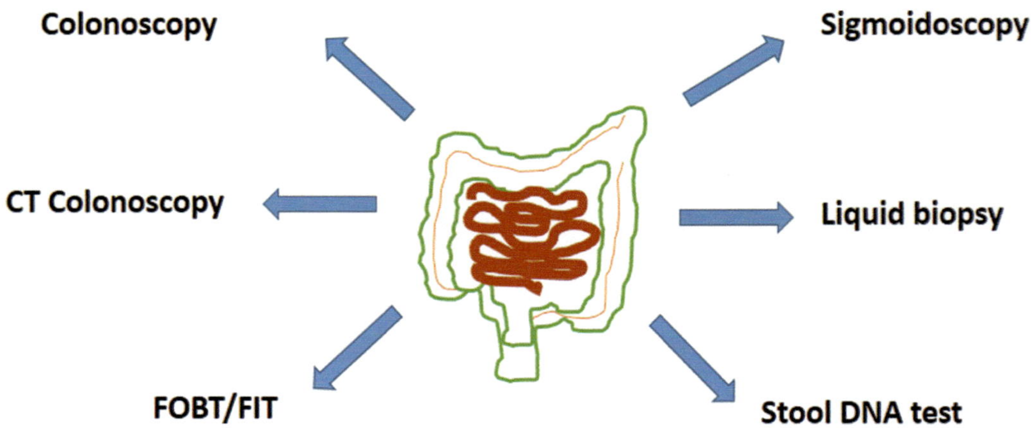

FIGURE 1.2 Illustration of different methods of diagnosis of colorectal cancer: FOBT/FIT (fecal immunochemical test). The FOBT test is a screening test for colorectal cancer before patients have symptoms. FOBT, fecal occult blood test.

metaplastic polyp has been used in the United Kingdom until 2006, after which it was replaced by hyperplastic polyp (Carey et al., 2007).

In the United States, the term polyp's hyperplastic was first used to differentiate it from adenoma cancer cells (Lane and Lev, 1963). During the anatomical examination of the serial section, they found tiny polyps in the patients with familial adenomatous polyposis; nonetheless, they also refused to use the name polyps. They coined a new term for the polyps based on the topographies, which they called sawtooth architecture hyperplastic polyps. This modified name, nevertheless, did not accomplish common acceptance in the medical fraternity until the late 1960s. It was not clear why these serrated polyps were called polyps, and some researchers argued that they could actually be adenomas. The argument was based on the observation that all polyps were considered adenomas because the proliferative zone in hyperplastic polyps occasionally resembles that of adenoma epithelium, which could be a reason for calling them adenomas.

With a view to better understand the physiology of hyperplastic polyps, researchers have applied cell kinetics to study the cell kinetics of hyperplastic polyps (Kaye et al., 1973). The study compared the normal maturation of cells in colonic crypts with that in the hyperplastic polyps and found that hyperplastic epithelium exhibits a progression similar to that of normal colon cells. The main variance was in the formation of crypts in the colon (Kaye et al., 1973). In another study, using electron microscopy, the researchers studied the surface of the hyperplastic polyps, and they found that hyperplastic polyps appeared hypermature and verified a reduced rate of cell migration from base to surface as examined by autoradiography (Hayashi et al., 1974). It was suggested that the migration time of cells in the mucosa was delayed in hyperplastic polyps compared to normal mucosa. This study was further investigated by using advanced techniques to study the hyperplastic polyps, and it was found that hyperplastic polyps are caused due to the inhibition of programmed cell death or apoptosis because normal cells undergo programmed cell death and apoptosis (Higuchi and Jass, 2004). The formation of hyperplastic polyps is due to continuous cell growth or proliferation of the cells in the mucosa membrane. Hypermaturation of cells in the mucosa membrane resulted in hyperplastic polyps due to inhibition of programmed cell death, or apoptosis (Higuchi and Jass, 2004).

Ultrastructural analysis of hyperplastic polyps using electron microscopy revealed that they contain a large number of cells with vacuolated cytoplasm, known as absorptive cells (Fenoglio et al., 1975; Goldman et al., 1970). The ratio of these absorptive cells to goblet cells differs in hyperplastic polyps. In another study, it was found that hyperplastic polyps possessed micro-vesicular cell types, which are the most often come across type with an abundance of absorptive cells and goblet cell type cells. The goblet type of cells possessed fewer absorptive cells (Torlakovic et al., 2003).

The anatomical feature of hyperplastic polyps showed remarkable peculiarities in the rectum region of the intestine, as they have large packs of smooth muscle spreading into different layers of mucosa. The hyperplastic polyps have a peculiar epithelium in the basal region that is highly proliferative in nature. These hyperplastic polyps consist of multinucleated giant cells with large nuclei and nucleoli (Kambham et al., 2005). The functional attributes of hyperplastic polyps were also examined using immunohistochemistry, and it was found that they do not possess any viral proteins (Kambham et al., 2005). The absence of viral protein in the hyperplastic polyps suggests that CRC is not caused by any viral infection and may be either a genetic mutation or environmentally induced cancer.

After examining the cell kinetics and cell types of hyperplastic polyps, it was found that only one kind of CRC serrated polyp was identified, and the "hyperplastic polyp" was not classified as precancerous. This was observed in six patients, where they found numerous serrated polyps in both colonic adenocarcinoma and adenomatous tissue (Torlakovic and Snover, 1996). The average size of the polyps is 1.2 cm, and they were dispersed all over the colon, principally left-sided in 2 cm and right-sided in 1 cm. The researchers assigned the name serrated adenomatous polyposis to the condition observed in this group of patients, and it has become the first record of carcinoma-associated serrated polyps (Torlakovic and Snover, 1996; Choi and Appelman, 2016).

1.3 COLORECTAL CANCER: SYMPTOMS

The major signs and symptoms of CRC are a change in bowel habits, and patients get frequent diarrhea, constipation, or tapering of the stool, which persists for a few more days. Patients also feel a continuous bowel movement and rectal bleeding with bright red blood. They also have blood in their stool, which causes it to appear dark brown or black; abdominal cramps and pain; weakness and exhaustion; and unintentional weight loss. Sometimes, CRC can cause bleeding in the digestive tract, and sometimes blood is observed in the stool. Then later on, continuous release of blood in the stool may cause a reduction in the red blood cells, and patients become anemic.

1.3.1 Signs of Colorectal Cancer

It has been reported that in some patients, CRC can spread to other parts of the body and can cause liver enlargement, resulting in yellowing of the skin. The CRC can affect the lungs, where patients have problems in breathing. Patients with CRC do not always show any symptoms, but in a few patients, CRC showed bleeding in the rectum and also in the stool, which caused the stool to appear red or black. Another issue is that bowel movements cause constipation and diarrhea. The patients also have low red blood cell counts due to iron deficiency, which develops into anemia. These symptoms could also be due to drug side effects or other medical conditions.

1.4 COLORECTAL CANCER: CAUSES

The etiology of CRC in patients is not very clear, but several factors may be responsible for causing CRC.

1.4.1 Genetic Influence

An estimated 20% of CRCs are mostly related to genetic reasons, as it has been found that there was a threefold increase in the first-generation family members who have CRC. It was reported that familial adenomatous polyposis, which is a genetic syndrome, caused the CRC, and the mismatch repair genes have been associated with inheriting CRC (Maida et al., 2017).

1.4.2 Dietary Contribution

It has been reported that there is an increased incidence of CRC in people who consume food with high animal protein, high fat, and low cellulose. The excessive use of fat causes an increase in bile secretion, bile acid decomposition, and increase the secretion of intestine. It is currently believed that high-fat, high-animal-protein, and low-cellulose diets are related to the incidence of CRC. Excessive fat intake will promote bile secretion, bile acid decomposition, and augmented intestinal carcinogens while also increasing the activity of intestinal anaerobic bacteria (Murphy et al., 2019).

1.4.3 Other Factors

There are other factors that also cause CRC in many patients. For example, colorectal polyps, ulcerative colitis, and Crohn's colorectal adenoma disease can cause the development of CRC. It has been reported that an estimated 3%–5% of ulcerative colitis patients develop CRC, and the cancer rate increases by 10% in patients who have had ulcerative colitis for more than 20 years. In addition, an estimated increase from 15% to 40% has been observed in patients with colonic polyps for 2–5 years. Moreover, adenomas with a size of less than 1 cm and a diameter of less than 1 cm can induce CRC, whereas polyps larger than 3 cm have greater chances of inducing CRC in the patients (Sun et al., 2014; Roncucci and Mariani, 2015). Other factors that also induce CRC include exposure to cancer-causing drugs or chemicals and lifestyle factors, such as obesity, which can greatly increase the risk of CRC. Exposure to radiation therapy also increases the chances of developing CRC (Sun, 2014).

1.5 CLINICAL INDICATOR FOR CRC

The development of CRC shows different symptoms (Simon, 2016), and the disease progressively develops in patients as these symptoms appear. With the progress of the disease, the following symptoms will generally occur. The fecal occultation test in CRC patients shows the presence of blood clots or drops in the stool (Sun, 2014). In some CRC patients, they experience abdominal pain, abdominal distention, and feelings of exhaustion, nausea, and vomiting. In some cases, patients will have trouble passing on the stool as the stool mass becomes big (Sun, 2014).

There are other factors, such as the age of the patients, family history, and environmental factors, which also can cause CRC. There are a few genetic disorders, such as Lynch syndrome (hereditary non-polyposis CRC), familial adenomatous polyposis, and MUTYH-associated polyposis, which are known to cause CRC. An estimated 5% increase in CRC was reported in patients with Lynch syndrome and familial adenomatous polyposis (Finlay et al., 2018). There is a 20% increase in the development of CRC in patients with degree relatives who have a family history of CRC. The risk of CRC development is double compared to patients who don't have any hereditary or genetic disease in the family members.

There are other factors that play an important role in the development of CRC; for example, the chances of having CRC are increased in patients who are African American, male, or have inflammatory bowel disease, such as ulcerative colitis and Crohn's disease. The chances of developing CRC increased in people who are overweight, consume red meat or processed meat, use tobacco, or drink alcohol. Patients who have undergone radiation therapy, used immunosuppressive drugs, have diabetes, or received androgen deprivation therapy have a higher risk of developing CRC.

There are also factors that help in reducing the occurrence of CRC in people who do regular physical exercise and eat fruits, vegetables, fish, high-fiber diets, folate-rich diets, calcium, and dairy products. The eating of vitamin D, vitamin B6, magnesium intake, garlic, regular use of aspirin, and non-steroidal anti-inflammatories are also known to reduce the risk of CRC (Finlay et al., 2018). It has been reported that the risk of CRC occurrence increased in patients who ate foods with a Dietary Inflammatory Index (DII) (Shivappa et al., 2017), as a higher score of DII is known to be linked with pro-inflammation of the cells that caused the CRC. Hence, the lower DII score may result in a decrease in the chances of getting CRC. The application of anti-inflammatory foods such as fiber, monounsaturated fatty acids, polyunsaturated fatty acids, different types of vitamins, folic acid, and some types of vegetables and spices are known to reduce the risk of CRC (Shivappa et al., 2017).

There are studies performed that demonstrated a relationship between CRC and alcohol consumption. For example, a person who consumes more than 50 g/day alcohol has a higher chance of dying from CRC than a non-alcohol-consuming person (Cai et al., 2014). It has been found that this direct relationship between alcohol consumption and CRC is higher in the Asian population than in the European population. The Asian population has a different genetic makeup that processes the alcohol metabolism than the European population. The link between alcohol consumption and CRC-induced death was found to be similar in both the colon and rectum. In another study, a direct link between alcohol consumption and CRC-induced death was reported (Dashti et al., 2022). Chemicals such as acetaldehyde, which is a metabolite of ethanol, are well-known carcinogenic agents that affect the cell DNA, alter the structure and function of glutathione, and increase colonic mucosal proliferation. The link between obesity and CRC has been reported to increase by 20%–30% per 5 kg/m^2 increase in men and approximately increase by 10% per 5 kg/m^2 in women (Thrift et al., 2015), and there was a higher degree of association between obesity and CRC in females than in males (Thrift et al., 2015).

In contrast, the consumption of vitamin D decreases the risk of CRC, as the consumption of free and total 25-hydroxyvitamin D was found to decrease the occurrence of CRC. The presence of 10 ng of vitamin D in the blood circulation decreased the risk of CRC by 26% (Cross et al., 2011). The protective effect of vitamin D is mediated through vitamin D receptors present in the cells. In one more study, the higher amount of vitamin D circulating in the bloodstream in several patients caused a drastic decrease in the risk of CRC, preferably in females (McCullough et al., 2013). Based on this study, the concentration of vitamin D required for lowering the risk of CRC is determined to be 75–100 nmol/L (McCullough et al., 2018). Interestingly, vitamin D supplementation was recommended for bone strength and stability by the Institute of Medicine.

1.6 GLOBAL SCENARIO AND STATISTICS

Globally, CRC is the third most diagnosed cancer, where the occurrence of CRC in males is the third rank, and in females, it is the second rank (Sung et al., 2021). With regard to mortality, CRC is the second most common reason for cancer-related deaths globally (Sung et al., 2021). With regard to China, there is a dramatic increase in CRC prevalence in both males and females, with a ratio estimated to be 24.7/100,000. The male ratio is 28.1/100,000, and the female ratio is 19.4/100,000 (Chen et al., 2016) (Figure 1.3). There is a link between environmental factors, lifestyle, and diet and CRC, as cancer incidence increases with the migration of the cells. Factors such as religion and ethnicity have not been linked to an increase in CRC, but geographical position or location plays a role in the development of CRC, as the incidence of CRC varies by region. The highest number of CRC cases is reported in North America, Western Europe, and Oceania, whereas the lowest number of CRC cases is reported in Africa, Asia, and South America. In the case of China, the rate of CRC incidence increases from east to west China (Chen et al., 2016). Economics plays an important role in the development of CRC, as the rate of CRC is more prevalent in the economically developed eastern region than in the eastern region, which is relatively economically backward (Chen et al., 2016). There is no difference between Eastern and Western regions with regard to second-generation Chinese migrants who migrated from low-incidence to high-incidence areas (Zheng et al., 2019).

Another factor that influences CRC is the age of the patients, as the risk of CRC grows with age. It has been reported that people less than 45 years old are less likely to get CRC than those who are 45 and above. In the case of China, only 10%–20% of people get CRC under the age

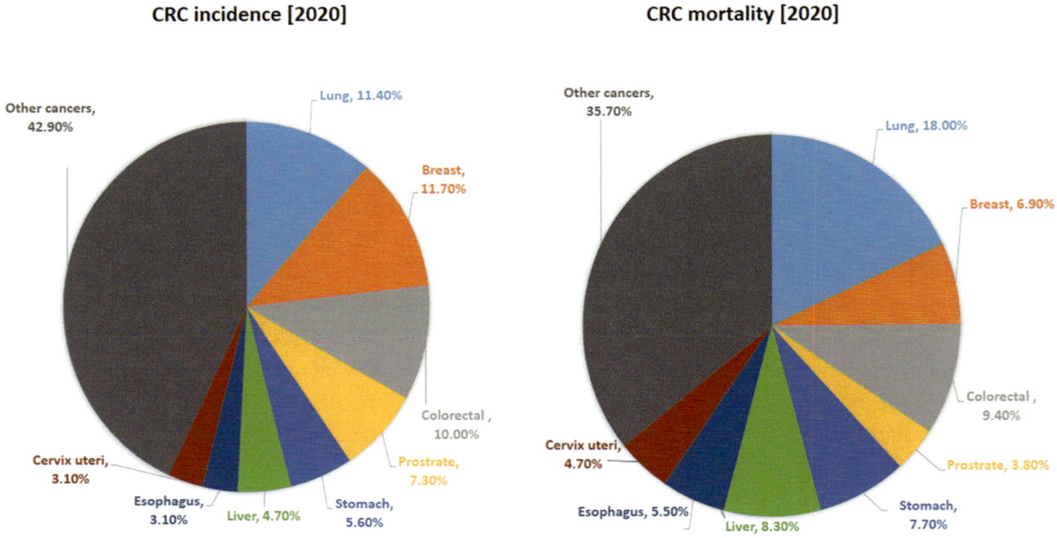

FIGURE 1.3 The rate of incidence and the rate of mortality of different types of cancer at the global level.

of 30 years (Wang, 2019; Ye et al., 2016; Thanikachalam and Khan, 2019). As per the American Cancer Society, there were about 106K new cases diagnosed for colon cancer in the year 2024 and about 46K new cases diagnosed for rectal cancer in the year 2024. Interestingly, the rate of colon or rectum diagnosis has dropped since the 1980s due to the availability of advanced and accurate diagnostic methods. Awareness about the early detection of CRC has been created on both social media and advertisements, which is one of the main reasons for the decline in CRC cases. Another factor that is contributing to the reduction in CRC rates is early screening, along with lifestyle changes and the consumption of foods that directly or indirectly lower the risk of CRC. It has been found that the occurrence rate decreased by 1% between 2011 and 2019 among older people. However, in people over 55 years of age, the rate of decline was between 1% and 2%. In the United States, CRC is the third leading cause of cancer-related death in males, whereas it is the second leading cause of death in both males and females. Approximately 53,010 deaths were reported in 2024. The rate of CRC has been declining in the United States, especially in older age patients, and one of the main reasons for this is the early diagnosis of cancer and a change of lifestyle, which includes the eating of balanced foods. Another reason is the advancement of cancer treatment technology, which improves the precise location and early removal of the CRC polyps from the site. Early screening of CRC also contributes to a decline in CRC-related deaths (American Cancer Society, 2024a, 2024b).

REFERENCES

American Cancer Society (29 January 2024a). https://www.cancer.org/cancer/types/colon-rectal-cancer/about/what-is-colorectal-cancer.html

American Cancer Society (29 January 2024b). https://www.cancer.org/cancer/types/colon-rectal-cancer/about/key-

Cai Z, Ke J, He X, Yuan R, Chen Y, Wu X, Wang L, Wang J, Lan P, Wu X. Significance of mTOR signaling and its inhibitor against cancer stem-like cells in colorectal cancer. *Ann Surg Oncol.* 2014 Jan;21(1):179–88. doi: 10.1245/s10434-013-3146-8. Epub 2013 Aug 2. PMID: 23907312.

Carey F, Newbold M, Quirke P, Shepherd N, Warren B, Williams G. *Reporting Lesions in the NHS Bowel Cancer Screening Programme: Guidelines from the Bowel Cancer Screening Programme Pathology Group: Publication No. 1.* Sheffield, United Kingdom: NHS Cancer Screening Programmes. 2007.

Chen W, Zheng R, Baade PD, Zhang S, Zeng H, Bray F, et al. Cancer statistics in China, 2015. *CA Cancer J Clin.* 2016;66(2):115–132. https://doi.org/10.3322/caac.21338

Choi EY, Appelman HD. A historical perspective and exposé on serrated polyps of the colorectum. *Arch Pathol Lab Med.* 2016;140(10), 1079–1084. https://doi.org/10.5858/arpa.2016-0278-RA

Cross HS, Nittke T, Kallay E. Colonic vitamin D metabolism: implications for the pathogenesis of inflammatory bowel disease and colorectal cancer. *Mol Cell Endocrinol.* 2011 Dec 5;347(1-2):70–9. doi: 10.1016/j.mce.2011.07.022. Epub 2011 Jul 27. PMID: 21801808.

Dashti H, Dehzangi I, Bayati M, Breen J, Beheshti A, Lovell N, Rabiee HR, Alinejad-Rokny H. Integrative analysis of mutated genes and mutational processes reveals novel mutational biomarkers in colorectal cancer. *BMC Bioinformatics.* 2022 Apr 19;23(1):138. doi: 10.1186/s12859-022-04652-8. PMID: 35439935; PMCID: PMC9017053.

Fenoglio CM, Richart RM, Kaye GI. Comparative electron-microscopic features of normal, hyperplastic, and adenomatous human colonic epithelium, II: variations in surface architecture found by scanning electron microscopy. *Gastroenterology.* 1975;69(1):100–109.

Feyrter F. Zur Lehre von der Polypenbildung im menschlichen Darm. *Wiener Med Wochenschr.* 1929;79:338–342.

Finlay A, Macrae RMG, Seres D, Savarese DMF. *Colorectal Cancer: Epidemiology, Risk Factors, and Protective Factors.* Accessed on 24 December 2018; Available online: https://www.uptodate.com/contents/colorectal-cancer-epidemiology-risk-factors-and-protective-factors

Goldman H, Ming S, Hickock DF. Nature and significance of hyperplastic polyps of the human colon. *Arch Pathol.* 1970;89(4):349–354.

Hayashi T, Yatani R, Apostol J, Stemmermann GN. Pathogenesis of hyperplastic polyps of the colon: a hypothesis based on ultrastructure and in vitro cell kinetics. *Gastroenterology.* 1974;66(3):347–356.

Higuchi T, Jass JR. My approach to serrated polyps of the colorectum. *J Clin Pathol.* 2004;57(7):682–686.

Houston SA, Cerovic V, Thomson C, Brewer J, Mowat AM, Milling S. The lymph nodes draining the small intestine and colon are anatomically separate and immunologically distinct. *Mucosal Immunol.* 2016;9(2):468–478. https://doi.org/10.1038/mi.2015.77

Kambham N, Troxell M, Longacre TA. Multinucleated epithelial giant cells in colorectal polyps: a potential mimic of viropathic and/or dysplastic changes. *Am J Surg Pathol.* 2005;29(7):912–919.

Kaye GI, Fenoglio CM, Pascal RR, Lane N. Comparative electron microscopic features of normal, hyperplastic, and adenomatous human colonic epithelium: variations in cellular structure relative to the process of epithelial differentiation. *Gastroenterology.* 1973;64(5):926–945.

Lane N, Lev R. Observations on the origin of adenomatous epithelium of the colon: serial section of minute polyps in familial polyposis. *Cancer.* 1963;16:751–764.

Maida M, Macaluso FS, Ianiro G, Mangiola F, Sinagra E, Hold G, et al. Screening of colorectal cancer: present and future. *Expert Rev Anticancer Ther.* 2017;17(12):1131–1146. https://doi.org/10.1080/14737140.2017.1392243

McCullough ML, Gapstur SM, Shah R, Jacobs EJ, Campbell PT. Association between red and processed meat intake and mortality among colorectal cancer survivors. *J Clin Oncol.* 2013 Aug 1;31(22):2773-82. doi: 10.1200/JCO.2013.49.1126. Epub 2013 Jul 1. PMID: 23816965; PMCID: PMC3718877.

Morson BC. Some peculiarities in the histology of intestinal polyps. *Dis Colon Rectum.* 1962;5:337–344.

Murphy N, Moreno V, Hughes DJ, Vodicka L, Vodicka P, Aglago EK, et al. Lifestyle and dietary environmental factors in colorectal cancer susceptibility. *Mol Aspects Med.* 2019;69:2–9. https://doi.org/10.1016/j.mam.2019.06.005

Roncucci L, Mariani F. Prevention of colorectal cancer: how many tools do we have in our basket? *Eur J Intern Med.* 2015;26(10):752–6. https://doi.org/10.1016/j.ejim.2015.08.019

Sakorafas GH, Zouros E, Peros G. Applied vascular anatomy of the colon and rectum: clinical implications for the surgical oncologist. *Surg Oncol.* 2006;15(4):243–255. https://doi.org/10.1016/j.suronc.2007.03.002

Simon K. Colorectal cancer development and advances in screening. *Clin Interv Aging.* 2016;11:967–976. https://doi.org/10.2147/CIA.S109285

Shivappa N, Wirth MD, Hurley TG, Hébert JR. Association between the dietary inflammatory index (DII) and telomere length and C-reactive protein from the National Health and Nutrition Examination Survey-1999-2002. *Mol Nutr Food Res.* 2017 Apr;61(4). doi: 10.1002/mnfr.201600630. Epub 2017 Feb 1. PMID: 27878970; PMCID: PMC5380547.

Sun Y. *Advanced Course in Clinical Oncology.1.* Beijing: People's Military Medical Press; 2014. p. 889.

Sung H, Ferlay J, Siegel RL, Laversanne M, Soerjomataram I, Jemal A, et al. Global cancer statistics 2020: GLOBOCAN estimates of incidence and mortality worldwide for 36 cancers in 185 countries. *CA Cancer J Clin.* 2021;71(3):209–49. https://doi.org/10.3322/caac.21660

Thanikachalam K, Khan G. Colorectal cancer and nutrition. *Nutrients.* 2019 Jan 14;11(1):164. https://doi.org/10.3390/nu11010164

Thrift AP, Gong J, Peters U, Chang-Claude J, Rudolph A, Slattery ML, Chan AT, Esko T, Wood AR, Yang J, Vedantam S, Gustafsson S, Pers TH; GIANT Consortium; Baron JA, Bezieau S, Küry S, Ogino S, Berndt SI, Casey G, Haile RW, Du M, Harrison TA, Thornquist M, Duggan DJ, Le Marchand L, Lemire M, Lindor NM, Seminara D, Song M, Thibodeau SN, Cotterchio M, Win AK, Jenkins MA, Hopper JL, Ulrich CM, Potter JD, Newcomb PA, Schoen RE, Hoffmeister M, Brenner H, White E, Hsu L, Campbell PT. Mendelian randomization study of height and risk of colorectal cancer. *Int J Epidemiol.* 2015 Apr;44(2):662-72. doi: 10.1093/ije/dyv082. Epub 2015 May 20. PMID: 25997436; PMCID: PMC4481609.

Torlakovic E, Skovlund E, Snover DC, Torlakovic G, Nesland JM. Morphologic reappraisal of serrated colorectal polyps. *Am J Surg Pathol.* 2003;27(1):65–81.

Torlakovic E, Snover DC. Serrated adenomatous polyposis in humans. *Gastroenterology.* 1996;110(3):748–755.

Torlakovic EE, Gomez JD, Driman DK, Parfitt JR, Wang C, Benerjee T, et al. Sessile serrated adenoma (SSA) vs. traditional serrated adenoma (TSA). *Am J Surg Pathol.* 2008;32(1):21–29.

Wang X. Epidemiological characteristics and prevention and control strategies of colorectal cancer in China and American. *Chin J Colorectal Dis.* 2019;8(1):1–5.

Wattenberg LW. A histochemical study of five oxidative enzymes in carcinoma of the large intestine in man. *Am J Pathol.* 1959;35(1):113–137.

Westhues H. *Die pathologisch – anatomischen Grundlagen der Chirurgie des Rektumkarzinoms.* Leipzig, Germany: Georg Thieme Verlag; 1934.

Williams GT, Arthur JF, Bussey HJ, Morson BC. Metaplastic polyps and polyposis of the colorectum. *Histopathology.* 1980;4(2):155–170.

Ye D, Liu C, Zhang Y. Epidemiology and research status of colorectal cancer. *World Lat Med Inf.* 2016;16(A2):34–35.

Zheng RS, Sun KX, Zhang SW, Zeng HM, Zou XN, Chen R, et al. Report of cancer epidemiology in China, 2015. *Zhonghua Zhong Liu Za Zhi.* 2019;41(1): 19–28.

2 Pathology of Colorectal Cancer

Microscopic Analysis, Staging, and Classification

Sara Mohammed Ibrahim
Faculty of Science, Department of Biochemistry, King Abdulaziz University, Jeddah, Saudi Arabia

2.1 INTRODUCTION

Colorectal cancer (CRC) is the third most commonly diagnosed cancer worldwide and is the second leading cause of cancer-related mortality in both men and women. The incidence is further expected to increase by 63%, with a staggering 73% rise in mortality by 2040 (Rawla et al., 2019). Despite tremendous advancements in understanding the molecular landscape of cancers and the emergence of therapies targeting specific genetic abnormalities, there has been limited improvement in the clinical outcome of CRC due to the extensive complexity of the disease.

CRC is considered the most preventable cancer, as almost 95% of CRC inadvertently arise from premalignant lesions or polyps, which gradually metamorphose into adenocarcinoma over time, thereby giving the opportunity for their removal before they undergo neoplastic transformation.

Screening is looking for the diseases in the absence or before the onset of symptoms. Screening allows the removal of polyps (polypectomy) before they mature into cancer and facilitates the early detection of the disease, significantly affecting the prognosis and outcome. The clinical utility of screening is evident by the 5-year overall survival (OS) of tumors detected in early stages (stages I and II), being 90% in comparison to 13% for tumors detected later (stage IV) (Cervantes et al., 2023).

Screening and diagnosis of CRC are essentially carried out by (i) stool-based assays such as the fecal occult blood test, which looks for the presence of precancerous polyps/blood in the stool, or (ii) endoscopic examination of a part or the entire colon, i.e., colonoscopy or flexible sigmoidoscopy, which are invasive and can simultaneously detect and remove polyps, and (iii) imaging techniques such as CT colonography and ultrasound that enable virtual visualization of any colonic abnormalities or distant metastases and are non-invasive.

Major risk factors for CRC include age, family history, certain hereditary conditions (familial adenomatous polyposis – FAP, Lynch syndrome), pathogenic conditions [inflammatory bowel diseases (IBDs) like Crohn's disease and ulcerative colitis], environmental factors such as diet, lifestyle, and genetic epigenetic dispositions. Age is the most important risk factor implicated in the development of CRC, with the risk escalating substantially after 50 years and screening recommended every 5 years after the age of 45.

Over the past three decades, the role of pathologists has evolved from providing mere histopathological diagnoses to assessing prognostic markers that may potentially affect response to neoadjuvant and adjuvant therapies, in addition to looking for pathological features suggestive of genetic abnormalities and selecting appropriate tissue sections for mutational analysis and its interpretation. Pathologic examination essentially involves the analysis of biopsied tissues retrieved during screening or surgery to render a disease diagnosis, which is vital in assessing prognosis and guiding

DOI: 10.1201/9781003503842-2

treatment. Microscopic analysis involves a detailed characterization of the histologic features of the tumor cells. The primary objective of microscopic examination is to ascertain the presence/absence of invasion, the diagnosis of which is made when the tumor cells have at least invaded the submucosa of the colorectal wall. Apart from histopathological examination, resected tissues may be additionally tested for molecular and immunohistochemical (IHC) markers of prognostic significance.

Treatment for CRC depends upon the stage, specific molecular profile, and patient compliance. Surgery is the mainstay of CRC treatment, complemented by neoadjuvant and adjuvant chemotherapy and radiation. Targeted therapies that have been positively shown to improve survival are limited and include anti-epidermal growth factor receptor (anti-EGFR) that inhibits the commonly activated *KRAS* and *BRAF* oncogenes in CRC, anti-vascular endothelial growth factor receptor (anti-VEGFR), and antiangiogenic drugs (Conlin et al., 2005). An enhanced understanding of mechanisms underlying CRC is crucial to further refine diagnosis and develop and design efficient treatment modalities for improved clinical outcomes.

2.2 INTESTINAL POLYPS

CRC is the cancer of the large intestine confined to the colon/rectum, which is a muscular passage that connects the large intestine to the anus. Anatomically, the large intestine can be divided into the proximal colon, distal colon, and rectum. The proximal colon includes the cecum, the ascending colon, and the transverse colon, while the distal colon comprises the descending colon and the sigmoid colon (Figure 2.1a). Tumors of the colon and rectum are collectively referred to as CRC due to their shared anatomy and physiologic characteristics.

The inner lining of the colon, called the mucosa is associated with a thin muscle layer known as muscularis mucosa. The submucosa, a fibrous tissue, separates the muscularis mucosa from the muscularis propria. The colon is surrounded by thin connective tissues called the serosa and subserosa, forming the outermost layers of the colon that are absent in the rectum (Figure 2.1b).

CRCs are characterized by abnormal growths in the inner lining of the colorectal wall known as polyps, which may eventually lead to cancer. A polyp is simply an epithelial protrusion of abnormal cell aggregates within the mucosal lining into the intestinal lumen. They are generally benign and are considered premalignant lesions that can develop anywhere along the colon and rectum. Polyps may be pedunculated with thin stalks or sessile. Only a small fraction of polyps (<10%) progress to CRC, with the risk increasing with larger polyp size and the presence of multiple polyps.

Based on their histomorphology, polyps are classified as follows.

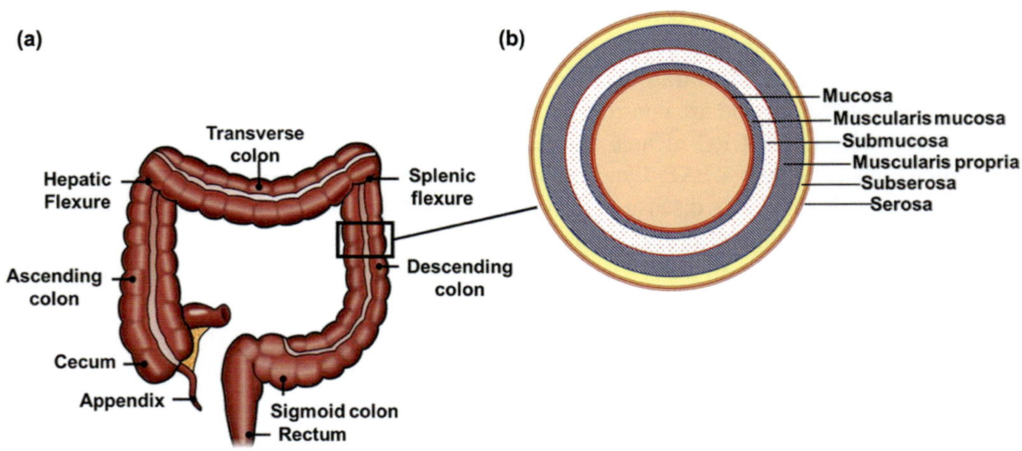

FIGURE 2.1 (a) Anatomy of the large intestine. (b) Different layers of the intestinal wall.

2.2.1 Adenomatous

About 90% of sporadic CRC classically evolve from adenomatous polyps or adenomas, which are precancerous lesions demonstrating low-grade dysplasia, i.e., cytologic and nuclear abnormalities with long, enlarged, and hyperchromatic nuclei aligned to the basement membrane. Adenomatous polyps tend to develop through a top-down mechanism, with small lesions bearing adenomatous epithelium in their superficial portions.

Adenomatous polyps mostly have tubular architecture with atypical glands but may acquire filamentous/finger-like projections as they grow, thereby making them villous or tubulovillous (Conteduca et al., 2013). Villous adenomas contain >75% villous component, while tubulovillous adenomas comprise 25%–75% villous characteristics (Figure 2.2).

Adenomatous polyps that are larger than 1 cm in size, poorly differentiated, or showing villiform growth patterns are of clinical concern and require aggressive endoscopic surveillance.

2.2.2 Hyperplastic/Inflammatory

Hyperplastic polyps (HPs) are generally benign polyps commonly found in the distal colon as multiple sessile lesions that are >5 mm in diameter. They are smooth and lobulated and cannot be distinguished from adenoma endoscopically. Histologically, HPs have a simple tubular morphology with long and straight crypts. The upper region of the crypts shows budding luminal serration, while proliferation is confined to the narrow base of the crypts that is not serrated. While HPs do not usually develop into CRC and were previously considered non-malignant, larger polyps >1 cm have been shown to have neoplastic potential and thus require routine endoscopic surveillance.

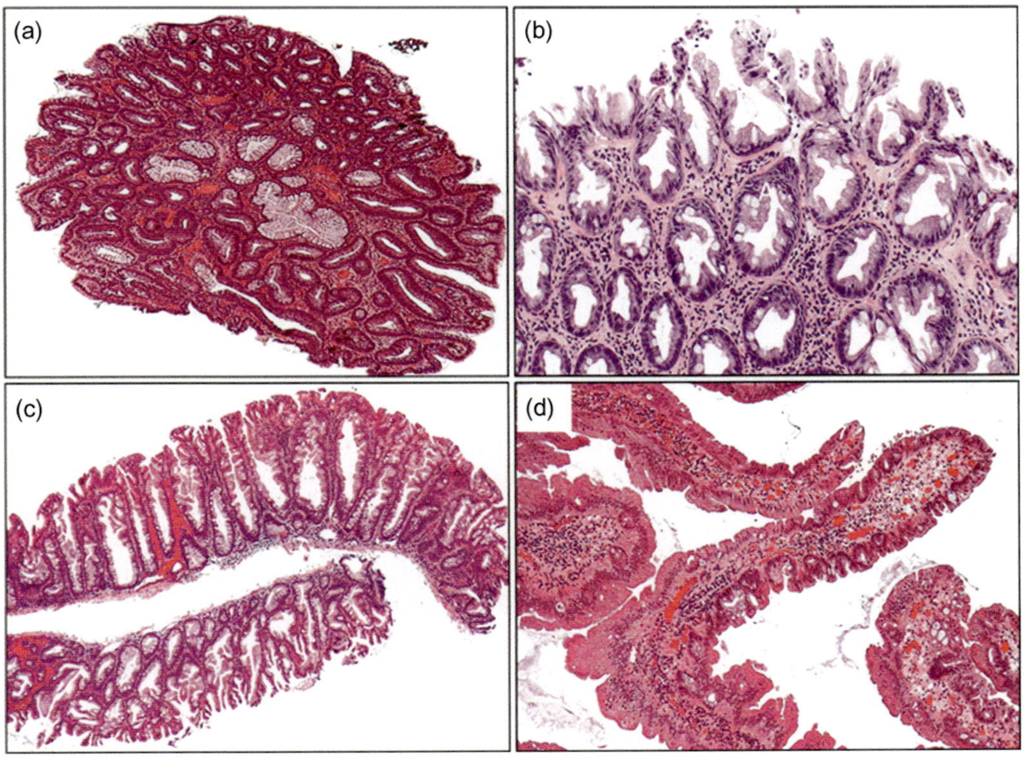

FIGURE 2.2 Microscopic images of the different types of polyps. (a) Adenomatous (b) Hyperplastic polyp (c) Sessile serrated polyp (d) Traditional serrated polyp

2.2.3 Sessile Serrated

Sessile/serrated polyps (SSP) are large polyps that mostly develop in the proximal colon and pose a risk of neoplastic progression via the serrated pathway. They are different from adenomas in having a flattened broad base instead of a thin stalk with/without a rigged or sawtooth appearance. Histologically, they are marked by excessive crypt serration that may cover the entire length of the crypt. Other distinct architectural features include a hypermucinous epithelium, a distorted crypt structure with dilations and branching, horizontal extension to the muscularis mucosa, and aberrant proliferation that is not restricted to the base (Bettington et al., 2013). However, dysplastic nuclear changes characterizing conventional adenomas are absent. SSP and HP may be histologically difficult to distinguish and until recently were used interchangeably.

2.2.4 Traditional Serrated Adenomas

Traditional serrated adenomas (TSAs) are precursor lesions that eventually develop into CRC, treading the alternate pathway (Bettington et al., 2015). They have been recognized as distinct morphological entities by the World Health Organization (WHO), frequently arising in the distal colon, with complex and villiform growth patterns and marked dysplasia. TSAs are typically characterized by ectopic crypts, i.e., crypts that are not oriented to the muscularis mucosae (Figure 2.2).

2.3 PATHOLOGY OF COLORECTAL CANCER

2.3.1 Histologic Features

2.3.1.1 Tumor Site

- Based on their relative position/site of development, CRC may be right-sided (RCRC) or left-sided (LCRC).
- Anatomically there are no defined boundaries differentiating the left- and right-sided CRC. In general, carcinomas arising from the rectum to the splenic flexure colon are categorized as LCRC, while those located within the splenic flexure colon to cecum are categorized as RCRC.
- The fundamental difference in the embryonic lineage of the right colon, originating from the midgut, and the left colon, arising from the hindgut, is reflected in disparities in gut micro bacteria, immunogenicity, and the expression profiles of over 200 genes, resulting in distinct clinicopathological features of LCRC and RCRC (Li et al., 2019).
- LCRCs are the most commonly diagnosed, with more than 70% frequency, while RCRCs present with an adverse prognosis.
- Epidemiologically, RCRCs are more frequent in females with older age at diagnosis and advanced stages of disease, as compared to LCRCs.
- Histologically, LCRCs are primarily tubular/villous adenocarcinoma and are relatively easy to detect in the early stages. RCRCs are essentially sessile serrated mucinous adenocarcinomas that are mostly flat, making them colonoscopically undetectable.
- Another notable difference is that RCRCs tend to metastasize to the peritoneum, while the liver and lung are the common sites for LCRC metastasis (Holch et al., 2017).
- Molecularly, RCRCs are associated with increased microsatellite instability (MSI), mismatch repair (MMR) deficiency, cytosine preceding guanine (CpG) island promotor phenotype, and *BRAF* mutations. LCRC, on the other hand, is attributed to chromosomal instability (CIN) with mutations in *TP53, KRAS, PI3CKA, SMAD2, SMAD4*, and *APC*.
- The prognostic utility of primary tumor position is maximum in metastatic settings, with LCRC showing superior OS in several studies independent of *KRAS* and *BRAF* mutational status.

2.3.1.2 Tumor Size

- Tumor size is defined by the maximum diameter of the biopsied tumor sample.
- Larger tumors are typically associated with poor prognosis, with patients having tumors >6 cm showing up to 46% increased mortality in comparison to patients with smaller tumors <2 cm (Saha et al., 2015).
- It is pertinent to note that a larger tumor is usually associated with several unfavorable features like higher grade, greater tumor penetration, and nodal involvement. It is however notable that maller tumors with T4b infiltration tend to present with a worse prognosis irrespective of their size (Kornprat et al., 2011).
- As such, the independent role of tumor size in CRC prognostication is debatable, unlike other solid tumors such as breast and lung.
- The depth of tumor penetration into the intestinal layers, rather than its actual dimension, is incorporated for T staging as per AJCC-UICC guidelines.
- However, considerable difference is observed in the prognosticative value of tumor size at different anatomical sites of the colon, with Kornprat et al. reporting a diminishing cut-off of optimal prognostication of tumor size from right to left.
- A tumor size cut-off of <4 cm/>=4 cm has been reported by different studies to be of prognostic significance.
- There is currently no consensus on the cut-off value of tumor size as per international guidelines.

2.3.1.3 Tumor Budding

- Tumor budding (TB) refers to the shedding of malignant cells from the invasive front of the tumors.
- It is recognized as an independent prognostic marker, with a higher budding score corresponding to reduced OS, disease relapse, and mortality.
- The prognostic value of TB is also evident by its ability to stratify stage II CRC patients compliant with adjuvant chemotherapy.
- Furthermore, TB has also been reported to identify patients at a higher risk of nodal metastasis, particularly in the early stages of CRC, i.e., stages I and II (Cangucu et al., 2020).
- The prognostic utility of TB, however, diminishes in higher stages of disease wherein other pathological features, such as nodal metastasis, become more aggressive.
- TB has been incorporated as a potential prognostic parameter in the latest Union for international Cancer Control (UICC) 8th edition of TNM classification of malignant tumors 2017.
- A three-tier system adopted by the International Tumor Budding Consensus Conference (ITBCC) 2016 is applied to describe TB as BD1-low (0–4 buds), BD2-intermediate (5–9 buds), and BD3-high (>10 buds).

2.3.1.4 Tumor-Infiltrating Lymphocytes

- Tumor-infiltrating lymphocytes (TIL) are indicators of an individual's immunogenicity and positively affect prognosis and survival in CRC.
- The protective effect of TIL in CRC is attributed to their role in mediating the recruitment and activation of immune cells, specifically T cells, NK cells, and macrophages, thereby impeding tumor progression.
- Several studies have strongly demonstrated the survival benefits of TIL, particularly CD3, CD8, and Fox3P subtypes (Zhao et al., 2019).
- Tumors with enriched TIL are usually associated with favorable prognostic features such as reduced vascular/lymphatic/perineural invasions (PNIs) and metastasis (nodal and distal).

- Populating the CD3+ and CD8+ T lymphocytes in the core and invasive margins of the tumor, known as the immune score, has demonstrated immense prognosticative potential in survival and guiding treatment strategies (Idos et al., 2020).
- Immunoclassification has been recently incorporated along with TNM staging in predicting the overall prognosis and relapse.
- Currently, standardized methods to evaluate TIL are lacking and are not routinely employed in clinical settings.

2.3.1.5 Lymph Node Yield

- Lymph node yield (LNY) is the number of lymph nodes removed during surgical resection, and it strongly affects the prognosis, particularly in non-metastatic CRC.
- A higher LNY positively affects the OS in CRC and is associated with reduced relapse, particularly for stage I and II cancers.
- Higher LNY possibly improves survival by clearing occult micrometastatic disease that can be missed during pathological examinations.
- Moreover, a larger number of lymph nodes indicates an active immune response to carcinogenic insult, making it a favorable prognostic feature.

2.3.1.6 Lymph Node Ratio

- It is the ratio of the number of lymph nodes positive for cancer to the total number of lymph nodes examined, with a higher lymph node ratio (LNR) indicative of advanced CRC stage.
- Prognostically, several studies have pointed to LNR being superior to N staging in stage III and IV CRC and being more significant in rectal than colon cancer (Poritz et al., 2011).
- There is no international guideline for LNR high threshold, with different studies adopting cut-off values between 0.125 and 0.3. Also, the minimum number of lymphocytes to be examined is not specified.

2.3.1.7 Apical Lymph Node

- Apical lymph nodes (ALNs) are the lymph nodes found in the major vessels that support tumors and occur in about 4%–19% of CRC.
- Although the presence of ALNs invariably suggests deep tumor infiltration, its ability to predict metastasis is not conclusive and is therefore not considered an independent prognostic factor.
- The AJCC/UICC counts ALNs within regional lymph nodes, and nodal staging is determined by the number of metastatic lymph nodes.

2.3.1.8 Lymphovascular Invasion

- Lymphovascular invasion (LVI) is histologically marked by the presence of small lymphatic and blood vessels in the tumors and is a key step in precipitating nodal metastasis.
- LVI is a strong indicator of poor prognosis in CRC at any given stage and is associated with significantly decreased OS and other unfavorable pathological features (Yuan et al., 2017).
- LVI is recommended as an additional prognostic parameter by AJCC/UICC TNM classification.

2.3.1.9 Perineural Invasion

- Perineural Invasion (PNI) is characterized by the neoplastic invasion of any of the three layers of the nerve cells by the tumor and is prevalent in 9%–30% of CRC patients.

- PNI is an aggressive pathological feature likely occurring in advanced stages of the disease and adversely affects prognosis and survival, independent of tumor grade and stage (Kerscher et al., 2013).
- Despite its high prognostic value, PNI tends to be underdiagnosed, and there is a lack of standard guidelines for its reporting.
- Pathologically, PNI is defined as tumor cells covering >33% of the neural boundaries.
- PNI is recognized by the AJCC/UICC as an additional prognostic factor, and documenting and reporting PNI status is routine in clinical practice.

2.3.1.10 Circumferential Resection Margin

- Circumferential resection margin (CRM) is the measure of the distance between the deepest point of tumor infiltration and the non-peritonealized surface of the surgically dissected specimen.
- Low rectal tumors are completely lined by a peritonealized margin, while upper rectal tumors do not have a peritonealized margin, with a peritonealized surface present only in the anterior. The circumferential margin of a colon tumor is formed by the mesenteric attachment and retroperitoneal segments.
- Tumors <1 mm from the tumor-free margin are considered CRM positive.
- CRM has stronger prognostic relevance in patients receiving radiotherapy before surgery than in patients undergoing surgery alone.
- CRM in rectal and colon cancers is not standardized, and CRM positivity reflects the quality of rectal cancer surgery. As per international guidelines, a negative margin >1 mm is desirable.
- CRM margins are the single most important prognostic factor predicting recurrence and survival in rectal carcinomas (Liu et al., 2018).
- As per the AJCC/UICC staging standards, positive CRM is indicative of T3 disease with R1 or R2 resection and is routinely documented.

2.3.2 Dysplasia

Dysplasia, also termed as intraepithelial neoplasia by WHO, may be defined as cytologic abnormalities, i.e., cytoplasmic and nuclear, associated with aberrations in the architecture of the crypt epithelium. Colorectal dysplasia is the neoplastic alteration of the intestinal epithelium confined to the basal membrane from which it arises and is considered as the earliest manifestation of CRC. Dysplastic lesions may be flat (endoscopically undetectable) or raised and may be histologically differentiated as negative, positive, or indefinite for dysplasia (Butt et al., 1983).

Positive dysplasia is characterized by nuclear atypism with increased nuclear/cytoplasmic ratio, loss of polarity, increased mitosis, nucleoli crowding, nuclear stratification within the epithelium, hyperchromasia, and nuclear pleomorphism, i.e., variation in shape and size of nuclei between the cells. Cytoplasmic abnormalities include mucinous degradation, distorted shape and distribution of goblet cells, and lack of surface maturation. Architectural aberrations manifest as abnormal growth patterns resulting in crypt budding, branching, and crowding; villiform changes in the epithelium; irregular crypt contours; and intraluminal bridges (Bressenot et al., 2014).

Negative for dysplasia indicates that the epithelium is regenerative, often seen during active inflammation in IBD, which can cause reactive changes that may resemble dysplasia. Nondysplastic epithelium usually shows mild to moderate atypia with preserved crypt structure; however, significant atypia may occur near ulcerated areas. A reactive crypt typically shows differential epithelial maturation at the base w.r.t. the surface, with mitotically active colonocytes dominating the base, while the surface harbors differentiated mature cells, which are distinctly absorptive/goblet cells, with normochromatic nuclei and absent mitoses.

2.3.3 TNM Staging

Staging of CRC is essential to describe the anatomical extent of the disease spread, i.e., whether the cancer is localized and confined within the region it developed or has invaded regional or distant tissues/organs. The pathologic stage, also known as the surgical stage, is established by examining tissues resected during surgeries. This is different from clinical staging, which is assessed based on the initial clinical examinations and biopsies. The pathologic stage gives a more precise estimate of the cancer spread and is a powerful tool for determining the prognosis and survival while also guiding treatments and predicting therapeutic response.

The TNM classification system, defined by the UICC and the American Joint Committee on Cancer (AJCC) (AJCC-UICC Tumor Node Metastasis system, 8th edition, 2017), is the most important pathological classification that is universally adopted to stage newly diagnosed CRC and guide adjuvant therapy (Amin et al., 2017). The TNM staging is based on three important anatomical features of tumors, i.e., the primary tumor (T), the presence of cancer in regional lymph nodes (N), and the distant spread of cancer (M).

CRCs are staged I to IV by giving each of the TNM components an individual score and integrating them together in a particular combination to generate the stage, known as stage grouping (Table 2.2). In general, a lower stage indicates a lesser disease load/spread and a better prognosis. Based on the TNM classification system, maximum prognostication is observed for the early (stage I) and late (stage IV) stages of the disease, while the intermediate stages of CRC show variability in prognosis and clinical outcome. The survival paradox is particularly pronounced for patients with stage IIb and IIc, demonstrating worse prognoses than patients with stage IIIa (Pei et al., 2021).

Occasionally, cancers need to be restaged either clinically or surgically to monitor disease progression/cure, assess the efficacy of neoadjuvant therapy/radiation, or in case of relapse. In either case, the original staging at the time of diagnosis remains the defining stage of the CRC and is not replaced. The new/upgraded stage is only appended to the original stage (Compton & Greene, 2004).

2.3.3.1 Primary Tumor (T)

The T stage is scored by analyzing the size and depth/penetration of the primary tumor into the intestinal wall, with the latter being the most important. The T stage may be a number or letter (Table 2.1) with the following annotations:

- T0 indicates no evidence of a primary tumor.
- Tis indicates a tumor confined within the inner layers from where it originated, i.e., in situ.
- T1–T4 is assigned based on the depth of tumor penetration through the different layers of the intestinal wall.
- TX means the T score could not be assessed due to insufficient information.

It is important to note that in CRC, carcinoma in situ refers both to intraepithelial and intramucosal tumors, in contrast to other gastrointestinal tumors, wherein evidence of mucosal invasion suffices for invasive diagnosis. This is because invasion confined to the lamina propria and muscularis mucosae does not pose a risk of metastasis in the colorectum. The tumor stage directly influences survival, with a higher stage corresponding to reduced OS and increased incidence of disease relapse. High-grade CRCs also exhibit an increased propensity for nodal and distant metastasis (Chen et al., 2021).

2.3.3.2 Regional Lymph Nodes (N)

Lymph nodes are small (> 5 mm), bean-shaped structures that are an important component of the immune system. Regional nodal invasion generally precedes cancer spread to distant parts of the body. Hence, lymph nodes near the primary tumors are checked to rule out the presence/absence

TABLE 2.1
TNM Variable Description for Colorectal Cancer

Tis	Carcinoma in situ, i.e., tumor confined within the mucosal lining and does not extend beyond the muscularis mucosa
T1	Tumor extending into the submucosa
T2	Tumor extending through submucosa into muscularis propria
T3	Tumor reaching the outermost layer of the colon, i.e., serosa but has not completely penetrated the walls of the colon
T4a	Tumor extending through the outermost layer of the colon including the visceral peritoneum but not into nearby tissue/organs
T4b	Tumors completely penetrating the walls of the colon and infiltrating the nearby tissues parietal peritoneum or organs
N1a	Cancer spread to 1 regional lymph node
N1b	Cancer spread up to 2–3 regional lymph nodes
N1c	Cancer spread into fatty tissues surrounding the lymph nodes without nodal involvement
N2a	Cancer spread to 4–6 regional lymph nodes
N2b	Cancer spread to >7 regional lymph nodes
M1a	Cancer spread to any one distant organ or multiple distant lymph nodes but not spread to distant part of the peritoneum
M1b	Cancer spread to more than one distant organ or multiple distant lymph nodes but not to the distant part of the peritoneum
M1c	Cancer spread to distant parts of the peritoneum with/without spreading to one or more distant organs or lymph nodes

T, primary tumor; N, regional nodes; M, metastasis.

TABLE 2.2
AJCC Colorectal Cancer Staging Based on Grouping of TNM Variables

Stage	TNM Stage Grouping		
Stage 0	Tis	N0	M0
Stage I	T1/2	N0	M0
Stage IIa	T3	N0	M0
Stage IIb	T4a	N0	M0
Stage IIc	T4b	N0	M0
Stage IIIa	T1/2	N1/N1c	M0
	T1	N2a	M0
Stage IIIb	T3	N1/N1c	M0
	T2/3	N2a	M0
	T1/2	N2b	M0
Stage IIIc	T4a	N2a	M0
	T3	N2b	M0
	T4b	N1/N2	M0
Stage 1Va	Any T	N0–2	M1a
Stage IVb	Any T	N0–2	M1b
Stage IVc	Any T	N0–2	M1c

T, primary tumor; N, regional nodes; M, metastasis.

of cancer invasion. It is noteworthy that cancer spread to anatomically distant nodes is discounted, and only the nodes near the primary site are assessed for N staging. The N score may be a letter or number (Table 2.1) with the following implication:

- N0 indicates nearby lymph nodes do not contain cancer.
- N1–N3 score is assigned based on the size, location, and number of lymph nodes involved (Table 2.1).
- NX indicates nodal involvement could not be assessed due to insufficient information.

Nodal invasion adversely affects the clinical outcome, with 5-year OS being reduced to 30%–60% in comparison to 70%–90% without nodal involvement. Tumor stage and histologic grade directly affect the adversity of lymph node involvement. The prognosis is negatively affected by the number of lymph nodes involved and low LNY. The AJCC/UICC recommends a minimum of 12 nodes for optimum staging. Paradoxically, the accuracy of staging diminishes with more than 12–17 lymph nodes (Compton, 2006).

2.3.3.3 Distant Metastasis (M)

The migration of cancer to a distant part or organ of the body, away from the original site of the primary tumor, is known as tumor metastasis. The metastasized tumor retains some of the characteristics of the primary and can be easily recognized and distinguished from the primary tumor of the metastasized organ. The presence of necrotic debris in glandular lumina is a marked feature of a primary CRC in a tumor of unknown origin. The liver and lungs are the most common sites of CRC metastasis, along with the bones and the peritoneum.

A designation of M0 is given in the absence of cancer spread, while M1 indicates the cancer has spread to distant organs and tissue, including distant lymph nodes (Table 2.1). CRC commonly metastasizes to the liver via the portal venous system. However, distal rectal tumors tend to metastasize to the lungs initially, as the inferior rectal veins directly drain into the inferior vena cava.

Distant metastasis at diagnosis (stage IV) presents the worst prognosis in CRC, with the OS being 10%. It is the strongest predictor of survival with therapies outlined mostly with palliative intent.

It should be noted the TNM classification system is not applied to residual tumors, i.e., tumors remaining after surgical resection. They are categorized by adopting an alternate system, i.e., R classification, annotated as below:

RX Residual tumor not assessed
R0 Absence of residual tumor
R1 Microscopic residual tumor
R2 Macroscopic residual tumor

At any given stage, R1 and R2 are associated with adverse outcomes and are pathologically evaluated based on the presence of tumors at the resection margins (Compton & Greene, 2004).

2.3.4 Tumor Grade

Tumor grade is a biological characteristic that is related to the degree of differentiation of the tumor cells. Histologic grading is based on the amount of glandular formation of the colonic epithelial cells and is a measure of their resemblance to the original tissue. Well-differentiated adenocarcinomas show >95% gland formation, while moderately differentiated tumors show 50%–95% gland formation. Poorly differentiated CRCs are almost solid, with <50% gland formation. Poorly differentiated cancers have the worst prognosis, with a greater tendency to grow and spread rapidly.

The College of American Pathologists (CAP) utilizes a four-tiered grading system for colorectal carcinoma based on the extent of gland formation:

Grade 1 is well-differentiated tumors with more than 95% gland formation.
Grade 2 is moderately differentiated tumors with 50%–95% gland formation.
Grade 3 is poorly differentiated tumors showing less than 50% gland formation.
Grade 4 is undifferentiated tumors with no gland or mucin formation.

The WHO classifies gastrointestinal tumors using a two-tiered system: low grade (≥50% gland formation) and high grade (poorly differentiated, <50% gland formation).

Tumor grade is considered an adverse prognostic variable in CRC independent of tumor stage, with a higher grade associated with poor clinical outcomes. An exception, however, is observed in histologic variants of CRC with MSI-H status, wherein poorly differentiated tumors exhibit low-grade phenotypes. Tumor grade strongly correlates with all three components of the TNM staging and the overall stage of the CRC (Derwinger et al., 2010).

A certain degree of subjectivity exists in grading tumors, which may be attributed to the clonal heterogeneity of tumors. Conventionally, the least differentiated cell observed in a resected sample defines the overall grade of the tumor. However, it is highly probable that sampled tissue may not include the clones of the highest grade and is therefore not an accurate representation of tumor grade (Compton, 2003).

Tumor grading is an important histopathological finding and is routinely reported in clinical practice. The AJCC/UICC TNM staging manual includes tumor grade as an additional prognostic factor.

2.4 MOLECULAR MARKERS OF COLORECTAL CANCER

2.4.1 Adenomatous Polyposis Coli

The adenomatous polyposis coli (*APC*) is a tumor suppressor gene that is most prominently implicated in CRC. The protein encoded by the *APC* gene negatively interacts with β-catenin, an intracellular molecule that stimulates cell proliferation via its transcriptional control of *c-myc*, *cyclin D1*, and *PPAR8*. In addition, β-catenin is also a component of the intracellular adhesion complex and plays a vital role in homotypic cell adhesion. *APC* protein phosphorylates β-catenin, resulting in its inactivation and subsequent degradation via the proteomic pathway.

APC is the most common mutation in CRC, with a prevalence of 80%–90%. It is considered a seeding genetic event, triggering the transformation of normal colonic epithelia into an adenoma (Powell et al., 1992). Inactivating mutations of *APC* result in nuclear translocation and accumulation of β-catenin, thereby driving the transcription of oncogenes (Narayan & Roy, 2003).

2.4.2 Kirsten Rat Sarcoma Viral Proto-Oncogene

Kirsten Rat Sarcoma Viral Proto-Oncogene (*KRAS*) is one of the most common oncogenic mutations in CRC, with a predominance of about 40%. Mutations in *KRAS* result in constitutively active *RAS* protein, GTP hydrolases, and a critical downstream component of the epidermal growth factor receptor (EGFR) pathway. Mutations in *KRAS* produce a hyperactive RAS-RAF-MEK-ERK signaling axis, ensuing uncontrolled cellular proliferation. Tumor progression is further driven by immune suppression and increased inflammation. Analysis of *KRAS* mutations is significant for consideration to anti-EGFR therapy, as they confer resistance, rendering them ineffective. The majority of *KRAS* mutations are point mutations occurring in codons 12 and 13 in exon 2 (118, 124, 125), with mutations in alternate loci such as codons 61 and 146 being rare.

2.4.3 V-Raf Murine Sarcoma Viral Oncogene Homolog B1

RAF encodes an important downstream molecule of the EGFR signaling pathway, a serine-threonine kinase, resulting in the sustained activation of the MAPK. Mutations in *BRAF* (V-Raf murine sarcoma viral oncogene homolog B1), specifically *BRAF* V600E, have been ascribed in almost 10% of CRC.

It is interesting to note that mutations in *KRAS* and *BRAF* in CRC are mutually exclusive. Moreover, *BRAF* mutations appear exclusively in sporadic MSI and not in Lynch syndrome. The prognosis of *BRAF* mutations is related to the MSI status of the tumor, with *BRAF* wild-type presenting a favorable prognosis in MSI-H tumors while *BRAF*-mutated microsatellite-stable (MSS) tumors demonstrating adverse clinical outcomes. Like KRAS, mutations in *BRAF* may show resistance to anti-EGFR therapy; however, there is a lack of sufficient evidence to underpin its role as a predictive molecular marker (Safaee Ardekani et al., 2012).

2.4.4 Tumor Protein 53

Tumor protein p53 (*TP53*), also known as the guardian of the genome gene, is one of the most important tumor suppressor genes. It encodes the protein p53 that regulates the mitotic entry of the cell, forming a cell cycle checkpoint that plays a crucial role in maintaining the integrity of the genome. In the event of any genomic insult, the p53 protein arrests cell division, inducing apoptosis in case of irreversible DNA damage.

Inactivating mutations of *TP53* result in sustained cycling of defective cells and are implicated in multiple cancers, including colon. Almost 50% of all invasive carcinomas show a mutant *TP53*. The most common type of mutations are inactivating mutations in one allele and loss of the other, due to chromosomal instability, and missense mutations in the DNA binding region, making it pro-oncogenic. Mutations in *TP53* are considered to be a later event in CRC pathogenesis, as evidenced by their predominance in invasive CRC and absence in precancerous lesions (Nakayama & Oshima, 2019).

2.5 COLORECTAL CANCER PATHOGENESIS

2.5.1 Adenoma-Carcinoma Sequence

Majority of CRCs arise from precancerous lesions or adenomatous polyps that accumulate genetic and epigenetic changes, progressively transforming into malignant adenocarcinoma. Approximately 10% of adenomas develop into invasive cancer, with the risk increasing with larger polyp size. Invasive cancers that originate from these polyps are adenocarcinomas and make up 96% of all CRCs.

The polyp-to-cancer progression is not abrupt and takes place in a multistage manner involving an initial polyp formation, which is benign, advancing into a histologically more complex neoplasm that subsequently transforms into invasive adenocarcinoma. Vogelstein et al. chartered a genetic map at different phases of this neoplastic transformation, primarily triggered by the loss of *APC* followed by activation of *KRAS* and subsequent loss of *TP53* and *DCC* (Fearon & Vogelstein, 1990). This is known as the adenoma-carcinoma sequence and is the most common mechanism of CRC pathogenesis (Figure 2.3).

Conventionally, adenomatous transformations are mediated by two molecular anomalies, i.e., inactivation of the *APC* tumor suppressor gene, which marks the key tumor-initiating event and is sustained by CSI (García-Solano et al., 2010), a type of genomic anomaly characterized by aneuploidy involving the gain or loss of large segments of chromosomes or entire chromosomes.

2.5.2 Serrated Pathway

CRCs evolve from hyperplastic/SSP with *BRAF* mutations as the primary underlying genetic modulation compounded with CpG methylator phenotype (CIMP). Interestingly, the progression from sessile adenoma to adenocarcinoma is associated with alteration in the gut microbiome, particularly an increased growth of *Fusobacterium nucleatum* (Park et al., 2016).

The alternative sequence pathway transforms normal mucosa into villous/partly serrated adenomas, is characterized by mutations in *KRAS*, *BRAF*, and *APC*, as well as CIMP, and is associated with a poor clinical prognosis.

At the molecular level, *KRAS* mutations and somatic copy number alterations (SCNAs) due to CIN enrich the adenoma-carcinoma sequence pathway. In contrast, the serrated polyp pathway is driven by *BRAF* mutations along with gene promoter hypermethylation via CIMP (De Palma, 2019

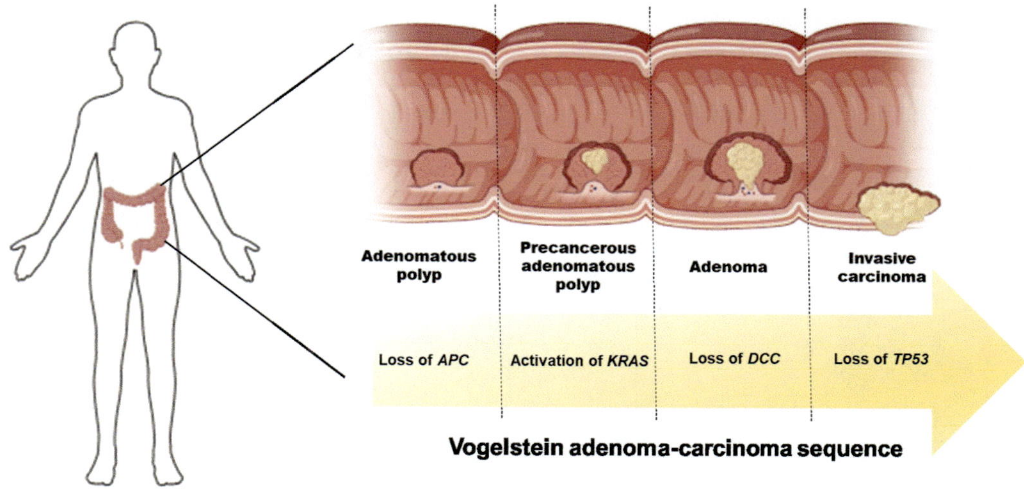

FIGURE 2.3 Morphological and molecular changes characterising Vogelstein adenoma-carcinoma transformation in colorectal cancer.

#38). CRCs that develop via the adenoma-carcinoma sequence tend to appear in the distal colon, while those following the serrated sequence are majorly proximal.

2.5.3 Inflammatory Pathway

Chronic inflammation in the luminal GI tract, associated with pathologic conditions such as IBD like ulcerative colitis and Crohn's disease, potentially drives and sustains neoplastic transformation in the colonic epithelia (Shah & Itzkowitz, 2022). Inflammation kindles oxidative stress-induced DNA defects, and a typical inflammation-dysplasia-carcinoma progression is observed with accumulating mutations in oncogenes and tumor suppressor genes, further aggravated by microenvironmental factors such as host immune response and gut microbiome.

IBD has been associated with a two- to threefold increased susceptibility to CRC, depending on the severity/duration of the condition and individual risk factors. About 1%–2% of CRCs are attributed to IBD. Dysplastic lesions characterizing IBD may be flat or elevated and demonstrate variable degrees of differentiation, i.e., indefinite, low, or high grade. IBD-related raised lesions are difficult to differentiate from adenomas endoscopically and can be clinically managed by polypectomy and surveillance. The majority of IBD-related CRCs are mucinous carcinoma, with a higher prevalence of signet cell carcinoma compared to the general population (Beaugerie & Itzkowitz, 2015). Owing to increased awareness and surveillance, IBD-related CRCs tend to be detected at relatively earlier stages, i.e., stages I and II. Moreover, CRCs induced by IBD and UC are typically of higher grade, tend to be multiple in number, and are uniformly distributed along the colon. Precedence of *TP53* mutation to *APC* is a notable molecular difference between IBD-associated CRC and sporadic CRC (Yaeger et al., 2016)

2.6 CLASSIFICATION OF COLORECTAL CANCER

2.6.1 Histologic Subtypes

Based on the distinct morphological features of the populating tumor cells, the WHO recognized the following histological subtypes of CRC (Figure 2.4) (Nagtegaal et al., 2020).

2.6.1.1 Adenomatous Carcinoma

Adenocarcinomas account for more than 90% of CRCs and are essentially the neoplasia of the glandular cells of the colorectal epithelia arising from adenomas. Unless otherwise specified, all CRCs are considered adenomatous.

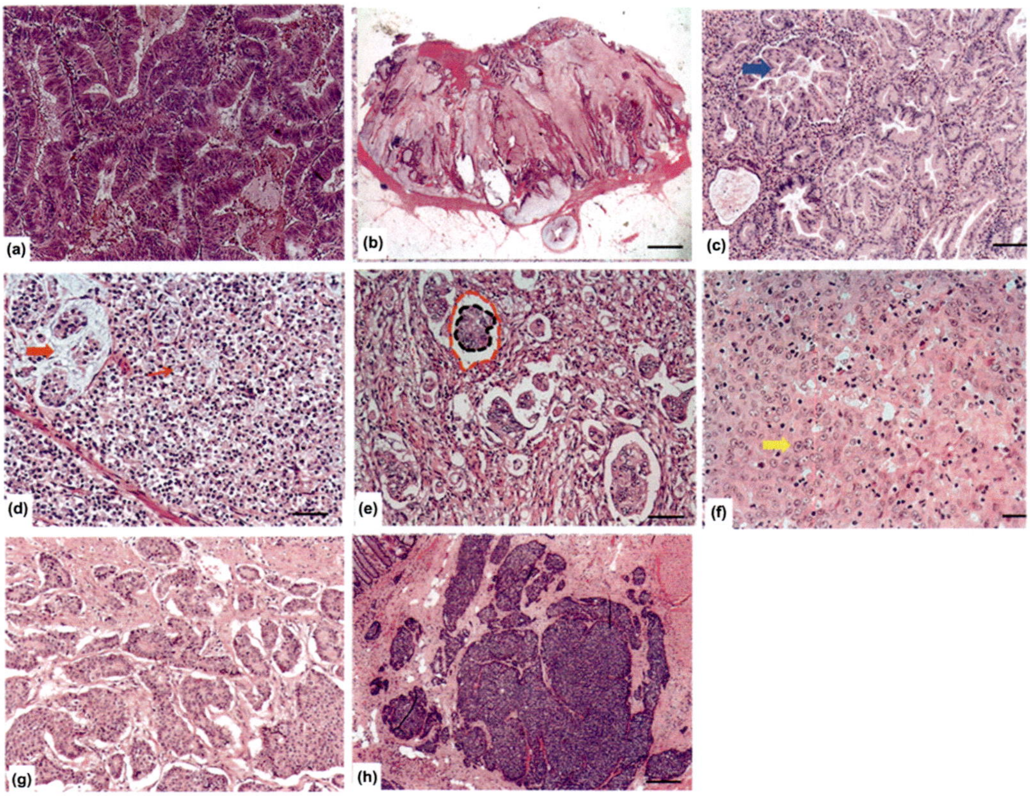

FIGURE 2.4 Histologic subtypes of colorectal cancer (a) Adenomatous CRC (b) Mucinous CRC (c) Serrated CRC with epithelial serrations marked by blue arrow (d) Signet cell CRC with red arrows indicating signet cell formation (e) Micropapillary CRC showing clusters of neoplastic cells circled in red (f) Medullary CRC showing syncytial appearing cells depicted by yellow arrow (g) Neuroendocrine CRC and (h) Undifferentiated CRC.

2.6.1.2 Mucinous Carcinoma

About 5%–10% of CRCs are of the mucinous subtype, characterized by the presence of tumor cells with an extracellular mucinous component, making up >50% of tumor volume. They exhibit enlarged glandular structures immersed in a matrix of extracellular mucin-containing discrete tumor cells. Mucinous adenoma-carcinoma is common in patients with Lynch syndrome, which demonstrates a type of genetic abnormality associated with microsatellite instability and typically behaves as low-grade tumors with a good prognosis (Nitsche et al., 2016). Microsatellite-stable tumors with mucinous adenocarcinoma, however, present a more aggressive phenotype.

2.6.1.3 Signet Cell Carcinoma

Signet cell CRCs are rarely encountered in CRC, with <2% prevalence. They are a type of mucinous adenocarcinoma, wherein >50% of the tumor is populated with signet cells, characterized by the presence of intracytoplasmic mucinous vacuoles squeezing the nucleus to the periphery. They are poorly differentiated tumors and present a clinically unfavorable prognosis (Nitsche et al., 2016). Signet ring cell carcinomas typically demonstrate MSI and are frequent with mutations in *BRAF* and CIMP. It is interesting to note that signet cell carcinoma with MSI-H may biologically behave as low-grade tumors.

2.6.1.4 Medullary Carcinoma

Medullary carcinoma is observed in less than 4% of CRCs. Medullary CRC is characterized by abnormal solid growths formed by sheets of neoplastic cells having vesicular nuclei, prominent nucleoli, and an excess of cytoplasm. They consist of a population of monotonous cells with a pushing border of invasion. These tumors are highly immunogenic with large immune cell trafficking and marked lymphoid response, prominent in the intratumoral regions (Scott et al., 2021). Despite the poorly differentiated or undifferentiated grade of the cells, these tumors have an overall favorable prognosis. They are typically MSI-H associated with *BRAF* mutations (Scott et al., 2021).

2.6.1.5 Serrated Carcinoma

About 8%–10% of carcinomas arise via the serrated pathway, recognized as serrated adenocarcinoma. They are characterized by cells having >5% epithelial serrations that may be mucinous, with a reduced nuclear-to-cytoplasmic ratio, an abundant cytoplasm, and the presence of cell balls and rods. At the molecular level, mutations in MMR genes and a CIMP phenotype typify serrated carcinoma and tend to be aggressive with poor prognosis (García-Solano et al., 2010).

2.6.1.6 Micropapillary Carcinoma

Tumors showing >5% micropapillary architecture, i.e., clusters of five or more tumor cells without fibrovascular cores and central lumen separated by cleared-out cleft spaces. Reversal of cell membrane polarity is an important feature of this phenotype, resulting in cell distortion and retraction from the stroma. Micropapillary carcinomas demonstrate a relatively aggressive phenotype and are predominantly metastatic (Lee et al., 2013).

2.6.1.7 Neuroendocrine Carcinoma

The WHO has classified Neuroendocrine Carcinoma (NEC) of the colon into two types: large cells and small cells. Small cell NECs are diffuse adenomas formed by atypical cells that are largely homogenous with reduced cytoplasm and hyperchromatic nuclei. Large cell NEC comprises solid aggregates of cells with little to no glandular lumen or mucin. They are characterized by regional necrosis and atypical nuclei with heightened abnormal mitotic activity (Smith et al., 2014).

2.6.1.8 Undifferentiated Carcinoma

Tumors cannot be classified in any of the distinct histologic subtypes and are a diagnosis of exclusion. These comprise <1% of CRCs and are primarily epithelial in origin, with solid sheets of tumor cells having normal nuclear features. Glands, intratumoral mucin, and microtubules are absent.

Other rare types of CRC include adenosquamous CRCs, which are a type of chimeric adenocarcinoma with dual glandular and squamous components, and cribriform comedo-type, which displays nest-like cribriform glands with marked comedo-type necrosis in the intervening spaces (Ono & Yilmaz, 2024)

2.6.2 Molecular Mechanism

CRC is mainly attributed to mutation in oncogenes, tumor suppressor genes, and DNA repair genes. Both genetic and epigenetic mechanisms have been identified to contribute to the development of CRC. Different genetic aberrations implicated in the pathogenesis of CRC have been categorized into three distinct molecular mechanisms: chromosomal instability, microsatellite instability (MSI), and CIMP. Although multiple genomic instability mechanisms may coexist, only a single type predominates in a specific CRC.

2.6.2.1 Chromosomal Instability

CIN characterizes 80%–85% of all CRC cases, making it the most common molecular modality of CRC pathogenesis. CIN is a molecular abnormality that is characterized by a shift in the number or structure of chromosomes resulting in aneuploidy (loss of chromosome 17p and 18q), loss of heterozygosity (LOH-tumor suppressor gene loci), deletions, duplications, and rearrangements along the length of the chromosome. This produces deficits in mechanisms that are critical in maintaining cellular integrity and function, such as chromosome segregation, telomere dysfunction, and DNA damage response (Walther et al., 2008). CIN tumors are distinguished by the presence of high-frequency SCNAs and mutations in *APC, KRAS, PIK3CA, TGF-β,* and *TP53*, which play an integral role in activating pathways favoring CRC initiation and progression. In CRC driven by CIN, inactivation of the *APC* gene is the first genetic event characterizing the hyperproliferative epithelium. This is followed by oncogenic mutations of *KRAS* and subsequent chromosomal alterations resulting in the inactivation of the *TP53* gene, driving the malignant transformation.

2.6.2.2 Microsatellite Instability

The MSI pathway is another common molecular pathway recognized in the pathogenesis of CRC, prevailing in about 15% of sporadic CRC (Thomas et al., 2007). Microsatellites (MS) are short tandem repeats of DNA sequences of 1–6 base pairs constituting 3% of the human genome. MS are highly susceptible to defects owing to their repetitive sequences. MSI is attributed to changes in the length of MS due to deletions or insertions in both coding and non-coding regions. MSI in CRC is thus characterized by the presence of a hypermutable phenotype essentially arising due to mutations in genes regulating DNA mismatch repair (dMMR).

The dMMR system is a highly conserved biological process that maintains the fidelity of the genome by eliminating any base-base mismatch that can take place during DNA replication and recombination. The cohort of dMMR proteins collectively scrutinizes erroneous base insertions, deletions, or substitutions and triggers their correction by activating and initiating DNA repair. Deficiency in dMMR genes thereby results in an increased frequency of genetic errors, particularly in the microsatellite regions of the genome that are small stretches of repetitive DNA sequences easily prone to mismatch, giving rise to a hypermutable phenotype.

Commonly implicated dMMR genes include MutL homolog *MLH1*, MutS homolog *MSH2*, *MSH6*, and mismatch repair endonucleases *PMS1* and *PMS2*. CRC with MSI usually has a diploid or near-diploid karyotype.

MSI is classified based on the size of the mono- or di-nucleotide repeat sequences altered, i.e., BAT25, BAT26, D2S123, D5S346, and D17S250, constituting the Bethesda panel (Boland et al., 1998). When two or more of the five microsatellite panel markers (>40%) are altered, the tumor is regarded as MSI-high (MSI-H,) while CRC with only one abnormal panel marker is considered MSI-low (MSI-L). When none of the markers are mutated, the tumor is deemed MSS. Frameshift mutations in genes containing nucleotide repeats in their exons, such as *TGFBR2*, *IGF23*, *E2F4*, *MSH6*, *MSH3*, and *CASP5*, are also frequently implied in MSI-H tumors. Mutations in *APC* and *BRAF* are common in MSI-H tumors, with *KRAS* mutations being rare. Germline mutations in dMMR genes are implicated in Lynch syndrome, while sporadic CRCs majorly arise due to the silencing of *MLH1* via promoter hypermethylation (O'Brien et al., 2006). MSI-H tumors are generally right-sided, mucinous, or medullary type with lymphocytic infiltration and present a better clinical prognosis with earlier stages at diagnosis.

2.6.2.3 CpG Island Methylator Phenotype

Epigenetic modification is a common feature of CRC, constituting the third molecular pathway, the CIMP pathway. CpG islands are short DNA sequences within the promoter region that are enriched with cytosine and guanine (Loughrey et al., 2021) residues and CpG dinucleotides (CpG) relative to the genome. The expression of more than 50% of human genes is regulated by these promoter sites. A CIMP tumor is characterized by extensive hypermethylation of CpG island sites, resulting in the inactivation of tumor suppressor genes, particularly dMMR.

There is a substantial overlap between the CIMP and MSI pathways, and CIMP tumors are frequently associated with MSI-H via hypermethylation of the *MLH1* promoter. However, they are rarely observed in Lynch syndrome, despite their high MSI. Mutations in *BRAF*, including *BRAF V600E,* are also prominent in CIMP-induced CRC (Koinuma et al., 2004). Tumors positive for CIMP tend to predominate the proximal colon and occur in higher frequency in females at an older age.

There is currently no consensus for an optimal panel of CpG sites to determine the CIMP status. CpG sites in *MLH1*, *CDKN2A* (p16), and *MINTs 1*, *2*, and *31* are classically analyzed to differentiate

CIMP-high and CIMP-low CRC. CIMP-high is associated with *MLH1* hypermethylation and *BRAF* mutations, while CIMP-low is characterized by mutations in *KRAS* in tumors that are essentially MSS (Weisenberger et al., 2006).

CIMP-negative tumors are MSS with mutations in *TP53*. An alternative panel including CACNA1G, IGF2, NEUROG1, RUNX3, and SOCS1 describes CIMP status with a strong correlation to *BRAF* mutation, specifically *BRAF V600E*, rather than *KRAS*.

2.6.3 Modality of Genetic Mutation

Genetic mutations implicated in CRC may be sporadic, hereditary, and familial. Majority of CRCs are sporadic in origin, accounting for more than 70% of CRC cases. Hereditary CRCs are rare, making up only 5% of the cases, and are associated with definite genetic mutations that are inheritable and increase the lifetime risk of CRC in individuals carrying them. About 20%–30% of CRCs are inherited with familial disposition, in the absence of identifiable germline mutation (Weledji, 2024). Understanding the etiology of inherited CRC is pivotal in identifying individuals at an increased risk of developing disease and improving outcomes by adopting stringent surveillance and prevention strategies.

2.6.3.1 Sporadic

Sporadic CRCs are colorectal tumors that develop incidentally due to somatic mutations without any genetic influence, i.e., germline mutations and family history, or pathological disposition, as in IBD. As the genetic abnormality is somatic in nature, it is restricted to the tumor cell and its descendants and is not inherited. Age (>50 years), environmental factors, diet, and lifestyle are the most important risk factors for its development, with age being the most prominent. Approximately 70% of sporadic CRCs develop in the distal colon. Sporadic CRCs are largely characterized by CIN with mutations in varied oncogenes and tumor suppressor genes. Sporadic CRCs generally comply with the adenoma-carcinoma blueprint, involving the inactivation of the *APC* gene followed by subsequent activating mutations in *BRAF* and eventual loss of function of *TP53* and *DCC*. Biallelic loss observed in the tumor suppressor gene is exemplified by *APC 5q21*, *DCC/SMAD2–4 18q*, and *TP53 17p* genes. In MSI-H sporadic CRC, hypermethylation of the *MLH1* gene promoter is the frequent somatic mutation. Somatic mutations in *RAS*, *SRC*, and *MYK* oncogenes are implicated, with RAS mutations being clinically most dominant (~50%) and relevant in sporadic CRC (Hermsen et al., 2002).

2.6.3.2 Hereditary Non-polyposis Colorectal Cancer/Lynch Syndrome

Lynch syndrome, also known as hereditary non-polyposis colorectal cancer, is the most common type of hereditary CRC, showing an autosomal dominant pattern of inheritance. It is clinically characterized by early onset and heightened propensity of colorectal and extra CRCs in multiple family members (Peltomaki et al., 2023).

Individuals with Lynch syndrome develop carcinomas mostly in the proximal colon at a comparatively young age, with the average age at diagnosis being 40 years. Colonic neoplasm is mucinous and poorly differentiated with peripheral lymphocytic infiltration and lymphoid reaction. Moreover, Lynch tumors develop without detectable adenomas, underlining an accelerated adenoma-carcinoma sequence in the syndrome. Synchronous and metachronous tumor formations with extracolonic malignancies, particularly those of the gastrointestinal tract and the female reproductive organs (i.e., endometrium and the ovaries), are common clinical features of Lynch syndrome.

At the molecular level, Lynch syndrome is characterized by a high level of MSI attributed to germline mutations in dMMR genes, i.e., *MLH1*, *MSH2*, *MSH6*, or *PMS2*.

Approximately 90% of patients with Lynch syndrome have mutations in either the *MLH1* or *MSH2* genes, while mutations in *MSH6* and *PMS2* are less frequently encountered. Diagnosis of Lynch syndrome typically follows the Amsterdam Criteria II, which requires a minimum of three relatives being affected with Lynch syndrome-related cancers (CRC, endometrial cancer, cancer of the ureters, bladder, and small bowel), with (i) at least one being a first-degree relative; (ii) at least one being diagnosed below the age of 5; and (iii) the syndrome appearing in two successive generations (Vasen et al., 1999). These criteria help identify families at risk and guide further genetic testing and management. Pathologically, tumors are verified by MSI testing for the final diagnosis, as per the revised Bethesda panel (Umar et al., 2004).

2.6.3.3 Familial Adenomatous Polyposis

Familial Adenomatous Polyposis (FAP) is characterized by the presence of exceedingly large numbers of adenomatous polyps, over hundreds to thousands, in the large bowel, i.e., the colorectum. FAP is relatively rare, affecting only about 1 in 7,000 individuals, mostly in their second or third decade, and follows an autosomal dominant inheritance (Thomas et al., 2007). In the absence of intervention, these polyps develop into CRC at a median age of 35–40 years, demonstrating 100% penetrance for the disease. Individuals with FAP are also susceptible to developing adenomatous polyps in other parts of the gastrointestinal tract, such as the small bowel and fundic glands of the stomach. Germline mutations in the *APC* and inactivation of *TP53* are strongly implicated in mediating the conversion from benign adenoma to malignant lesions in FAP.

Diagnosis for FAP is made based on the presence of >100 adenomatous polyps, *APC* germline mutation, or a history in the family. Young individuals with fewer adenomas, but with a family history, are also recognized as having FAP. In a certain subset of FAP patients with <100 colorectal adenomatous polyps, the lifetime risk of developing CRC is drastically reduced by 70%. Gardner syndrome, a variant of FAP, features extra-colonic manifestations such as epidermoid cysts, osteomas, dental anomalies, and desmoid tumors (Gardner, 1951). Turcot syndrome, also related to FAP, is typically associated with medulloblastoma.

2.6.3.4 MUTYH-Associated Polyposis Syndromes

MUTYH-associated polyposis (MAP) is an autosomal recessive syndrome that shows increased susceptibility to developing CRC. Biallelic mutations in the *MUTYH* gene (also known as the *MYH* gene) are essentially responsible for MAP (Morak et al., 2010). It is characterized by the presence of multiple adenomas with up to hundred or even thousand polyps. Although MAP may phenotypically present as attenuated FAP, it is characterized by certain atypical features such as the absence of *APC* gene mutation and the occurrence of single colorectal tumors. Since MAP is inherited in a recessive manner, it may not appear in all generations, making it difficult to detect, and tends to be diagnosed as sporadic CRC (Kastrinos & Syngal, 2011). Moreover, the presence of serrated polyps further confounds the diagnosis with serrated polyposis.

2.6.3.5 Familial CRC Type X

Familial non-syndromic CRC type X (FCCTX) represents a heterogeneous group of CRCs with undefined molecular pathogenesis that tends to appear and run in families without a specific germline mutation (Jasperson et al., 2010).

FCCTX may arise due to the inheritance of several relatively common low-penetrance genetic mutations in high frequency. Individually, the genetic mutation may not be highly carcinogenic, but the combination of several such mutations substantially increases susceptibility to CRC. The risk for FCCTX is thus subjective and increases with the number of first-degree relatives diagnosed with advanced CRC at a young age.

FCCTX may be considered sporadic CRCs that cluster in families and mostly predominate the rectum, frequently occurring in males of older age. FCCTX, like Lynch syndrome, is diagnosed as tumors that fulfill the Amsterdam criteria but do not present with identifiable mutations in the dMMR genes. FCCTX are characterized by MSS tumors without multiple/extracolonic manifestations and an overall unfavorable prognosis.

2.6.4 Consensus Molecular Classification

Molecular subtyping of CRC not only enables identification of the key molecular markers but also helps in elucidating the signaling pathways underpinning the tumor development, which is integral to predicting response to a particular therapy, facilitating better therapeutic decisions.

The first comprehensive multi-omics classification was attempted by The Cancer Genome Atlas Project (TCGA) based on whole genome sequencing, mRNA, and microRNA expression profiles, which classified CRC into two broad subtypes:

Non-hypermutated cancer: They constitute about 85% of CRC, with CIN being the prominent molecular mechanism. This subtype is associated with mutations in genes, such as *APC, TP53, KRAS, PIK3CA*, and *SMAD4*. They exhibit a large number of SCNAs and are MSS. This definition, however, integrated colon and rectal cancers due to similar patterns of genomic alterations.

Hypermutated cancers: They comprise about 15% of CRCs with characteristic MSI and defective dMMR, primarily arising due to *MLH1* promoter silencing, with only a small fraction showing somatic mutation in dMMR genes. They are further stratified into ultra-mutated cancers, with mutations in DNA polymerase epsilon, which render a defective proofreading function.

The Colorectal Cancer Subtyping Consortium compiled existing molecular classifications and databases together with gene-transcriptomic profiles and unified Consensus Molecular Subtypes (CMS 1–4) in 2015 (Table 2.3). The CMS subtypes better complied with the histomorphic and

TABLE 2.3
Biological Characteristics of CMS1–4 with Their Overall Prognosis (Phipps et al., 2015)

CMS Subtype	Prevalence	Type	Molecular Characteristic	Pathological Characteristic	Prognosis
CMS-1	15% Proximal colon	Immune type	MSI-H CIMP positive hypermethylation *BRAF* mutations	Immune infiltrated Serrated pathway Medullary growth	Immunotherapy
CMS-2	40%	Canonical type	CIN-Somatic CNV MSS Activated WNT and Myc EGFR HFN4A	Non mucinous/ glandular adenoma Adenoma-carcinoma sequence	
CMS-3	15%	Metabolic type	*KRAS* Deregulated metabolic pathways (glutaminolysis and lipidogenesis) MSI CIMP		
CMS-4	30% Distal colon	Mesenchymal type "Inflamed"	CIMP Somatic SCNV *TGF-β* activation MSS Activation of VEGF pathway	Invasive EMT Angiogenesis Stromal invasion Stemness	Poor response to standard therapy Distant relapse Death

CMS, Consensus Molecular Classification; CIMP, CpG methylator phenotype; MSS, microsatellite stable; EGFR, epidermal growth factor receptor.

clinical differences and effectively captured the genetic and pathological heterogeneity of CRC (Kasi et al., 2020; Phipps et al., 2015). A fifth subtype showing undefined or mixed phenotype is also included in the classification.

There is, however, inconsistency in clinical compliance, especially for CMS-2 and CMS-3, with CMS-1 and CMS-2 showing maximum prognostic significance (Noack & Langer, 2023). As such, CMS subtyping is not routinely performed.

2.7 Immunohistochemistry Markers in Colorectal Cancer Diagnosis

The ease with which IHC markers can be integrated into diagnostic settings in a cost- and time-effective manner makes them ideal for routine molecular testing. Its utility is, however, subjective, as pathogenic mutations that do not lead to loss of protein expression are missed by this method.

MMR deficiency characterizing MSI is preferably assessed by IHC analysis for the four important MMR proteins, i.e., MLH1, MSH2, PMS2, and MSH6. Tumors expressing all four proteins in the nucleus are considered MSS, while deficiency of any one or two proteins indicates MSI. MLH and MSH2 form stable complexes with PMS2 and MSH6, respectively. Thus, the loss of any one corresponds to the loss of its functional counterpart as well.

A panel comprising five IHC markers (FRMD6, ZEB1, HTR2B, CDX2, and cytokeratin), in addition to standard MSI/MMR testing, closely reflected CMS classification, showing an 87% concordance (Loughrey et al., 2021). CMS-1 was defined based on MSI/dMMR status, while the remaining three CMS subtypes could be grouped into epithelial (CMS-2/3) or mesenchymal (CMS-4) categories. However, this system could not mark a distinction between CMS-2 and CMS-3.

The relative distribution of cytokeratins across different epithelia, particularly K20 and CK7, is utilized to determine tumor origin in case of metastatic presentation (Hrudka et al., 2021). Although the $CK20^{+}/CK7^{-}$ expression profile is a common feature of primary CRC, many CRCs show the $CK20^{-}$ or $CK7^{+}$ immunotype. Several studies showed CK7 positivity related to a higher grade and more advanced stage of disease with lymph node involvement (Hrudka et al., 2021). CDX2, a transcription factor that regulates differentiation of the enterocytic cell, is another marker that is positively expressed in CRC. However, several losses of CDX2 have been reported in the advanced stages of the disease (Ilieva et al., 2020).

2.8 SUMMARY AND CONCLUSION

CRC is a major health concern worldwide, contributing to about 8% of the total annual deaths. Although early screening and advancements in treatment strategies have considerably decreased morbidity and mortality in developed countries, the prognosis is dismal for the majority of patients due to poor therapeutic compliance, resulting in disease reoccurrence and death.

Substantial variability in the survival rates among patients having similar clinical presentations and therapies underscores the complex manifestation of the disease. The current chapter details the histological and molecular irregularities identified in CRC, with a key focus on their prognostic significance. It also describes the parameters that are employed in CRC staging, pivotal for clinical diagnosis and curating therapeutic strategies. The various mechanisms of CRC pathogenesis have been discussed, and an attempt has been made to navigate through CRC types based on different modalities, which form the core of CRC classification.

In conclusion, the classification of CRC, particularly CMS, brings about compelling patient stratification and drug response prediction. However, addressing the inter- and intra-tumor heterogeneity and managing patients with unique clinicopathological features remain daunting clinical challenges in minimizing the global burden of the disease. An accurate and efficient diagnosis, based on concrete classification methods that can effectively capture the histological, genetic, and clinical diversity of CRC, is primary for attaining the objectives of personalized medicine.

REFERENCES

Amin, M. B., Greene, F. L., Edge, S. B., Compton, C. C., Gershenwald, J. E., Brookland, R. K., … Winchester, D. P. (2017). The Eighth Edition AJCC Cancer Staging Manual: Continuing to build a bridge from a population-based to a more "personalized" approach to cancer staging. *CA Cancer J Clin, 67*(2), 93–99. doi:10.3322/caac.21388

Beaugerie, L., & Itzkowitz, S. H. (2015). Cancers complicating inflammatory bowel disease. *New Engl J Med, 372*(15), 1441–1452. doi:10.1056/NEJMra1403718

Bettington, M., Walker, N., Clouston, A., Brown, I., Leggett, B., & Whitehall, V. (2013). The serrated pathway to colorectal carcinoma: current concepts and challenges. *Histopathology, 62*(3), 367–386. doi:10.1111/his.12055

Bettington, M. L., Walker, N. I., Rosty, C., Brown, I. S., Clouston, A. D., McKeone, D. M., … Whitehall, V. L. (2015). A clinicopathological and molecular analysis of 200 traditional serrated adenomas. *Mod Pathol, 28*(3), 414–427. doi:10.1038/modpathol.2014.122

Boland, C. R., Thibodeau, S. N., Hamilton, S. R., Sidransky, D., Eshleman, J. R., Burt, R. W., … Srivastava, S. (1998). A National Cancer Institute Workshop on Microsatellite Instability for cancer detection and familial predisposition: development of international criteria for the determination of microsatellite instability in colorectal cancer. *Cancer Res, 58*(22), 5248–5257. Retrieved from https://www.ncbi.nlm.nih.gov/pubmed/9823339

Bressenot, A., Cahn, V., Danese, S., & Peyrin-Biroulet, L. (2014). Microscopic features of colorectal neoplasia in inflammatory bowel diseases. *World J Gastroenterol, 20*(12), 3164–3172. doi:10.3748/wjg.v20.i12.3164

Butt, J. H., Konishi, F., Morson, B. C., Lennard-Jones, J. E., & Ritchie, J. K. (1983). Macroscopic lesions in dysplasia and carcinoma complicating ulcerative colitis. *Dig Dis Sci, 28*(1), 18–26. doi:10.1007/BF01393356

Cangucu, A. L., Valerio, E., Peixoto, R. B. P., Felismino, T. C., de Mello, C. A. L., Neotti, T., … Riechelmann, R. (2020). The prognostic influence of tumour budding in Western patients with stage II colorectal cancer. *Ecancermedicalscience, 14*, 1130. doi:10.3332/ecancer.2020.1130

Cervantes, A., Adam, R., Rosello, S., Arnold, D., Normanno, N., Taieb, J., … ESMO Guidelines Committee. (2023). Metastatic colorectal cancer: ESMO Clinical Practice Guideline for diagnosis, treatment and follow-up. *Ann Oncol, 34*(1), 10–32. doi:10.1016/j.annonc.2022.10.003

Chen, K., Collins, G., Wang, H., & Toh, J. W. T. (2021). Pathological features and prognostication in colorectal cancer. *Curr Oncol, 28*(6), 5356–5383. doi:10.3390/curroncol28060447

Compton, C. C. (2003). Colorectal carcinoma: diagnostic, prognostic, and molecular features. *Mod Pathol, 16*(4), 376–388. doi:10.1097/01.MP.0000062859.46942.93

Compton, C. C. (2006). Key issues in reporting common cancer specimens: problems in pathologic staging of colon cancer. *Arch Pathol Lab Med, 130*(3), 318–324. doi:10.5858/2006-130-318-KIIRCC

Compton, C. C., & Greene, F. L. (2004). The staging of colorectal cancer: 2004 and beyond. *CA Cancer J Clin, 54*(6), 295–308. doi:10.3322/canjclin.54.6.295

Conlin, A., Smith, G., Carey, F. A., Wolf, C. R., & Steele, R. J. (2005). The prognostic significance of K-ras, p53, and APC mutations in colorectal carcinoma. *Gut, 54*(9), 1283–1286. doi:10.1136/gut.2005.066514

Conteduca, V., Sansonno, D., Russi, S., & Dammacco, F. (2013). Precancerous colorectal lesions (Review). *Int J Oncol, 43*(4), 973–984. doi:10.3892/ijo.2013.2041

Derwinger, K., Kodeda, K., Bexe-Lindskog, E., & Taflin, H. (2010). Tumour differentiation grade is associated with TNM staging and the risk of node metastasis in colorectal cancer. *Acta Oncol, 49*(1), 57–62. doi:10.3109/02841860903334411

Fearon, E. R., & Vogelstein, B. (1990). A genetic model for colorectal tumorigenesis. *Cell, 61*(5), 759–767. doi:10.1016/0092-8674(90)90186-i

García-Solano, J., Pérez-Guillermo, M., Conesa-Zamora, P., Acosta-Ortega, J., Trujillo-Santos, J., Cerezuela-Fuentes, P., & Mäkinen, M. J. (2010). Clinicopathologic study of 85 colorectal serrated adenocarcinomas: further insights into the full recognition of a new subset of colorectal carcinoma. *Hum Pathol, 41*(10), 1359–1368. doi:10.1016/j.humpath.2010.04.002

Gardner, E. J. (1951). A genetic and clinical study of intestinal polyposis, a predisposing factor for carcinoma of the colon and rectum. *Am J Hum Genet, 3*(2), 167–176. Retrieved from https://www.ncbi.nlm.nih.gov/pubmed/14902760

Hermsen, M., Postma, C., Baak, J., Weiss, M., Rapallo, A., Sciutto, A., … Meijer, G. (2002). Colorectal adenoma to carcinoma progression follows multiple pathways of chromosomal instability. *Gastroenterology, 123*(4), 1109–1119. doi:10.1053/gast.2002.36051

Holch, J. W., Ricard, I., Stintzing, S., Modest, D. P., & Heinemann, V. (2017). The relevance of primary tumour location in patients with metastatic colorectal cancer: a meta-analysis of first-line clinical trials. *Eur J Cancer, 70*, 87–98. doi:10.1016/j.ejca.2016.10.007

Hrudka, J., Fiserova, H., Jelinkova, K., Matej, R., & Waldauf, P. (2021). Cytokeratin 7 expression as a predictor of an unfavorable prognosis in colorectal carcinoma. *Sci Rep, 11*(1), 17863. doi:10.1038/s41598-021-97480-4

Idos, G. E., Kwok, J., Bonthala, N., Kysh, L., Gruber, S. B., & Qu, C. (2020). The prognostic implications of tumor infiltrating lymphocytes in colorectal cancer: a systematic review and meta-analysis. *Sci Rep, 10*(1), 3360. doi:10.1038/s41598-020-60255-4

Ilieva, N. I., Galvanetto, N., Allegra, M., Brucale, M., & Laio, A. (2020). Automatic classification of single-molecule force spectroscopy traces from heterogeneous samples. *Bioinformatics, 36*(20), 5014–5020. doi:10.1093/bioinformatics/btaa626

Jasperson, K. W., Tuohy, T. M., Neklason, D. W., & Burt, R. W. (2010). Hereditary and familial colon cancer. *Gastroenterology, 138*(6), 2044–2058. doi:10.1053/j.gastro.2010.01.054

Kasi, A., Handa, S., Bhatti, S., Umar, S., Bansal, A., & Sun, W. (2020). Molecular pathogenesis and classification of colorectal carcinoma. *Curr Colorectal Cancer Rep, 16*(5), 97–106. doi:10.1007/s11888-020-00458-z

Kastrinos, F., & Syngal, S. (2011). Inherited colorectal cancer syndromes. *Cancer J, 17*(6), 405–415. doi:10.1097/PPO.0b013e318237e408

Kerscher, A. G., Chua, T. C., Gasser, M., Maeder, U., Kunzmann, V., Isbert, C., … Pelz, J. O. (2013). Impact of peritoneal carcinomatosis in the disease history of colorectal cancer management: a longitudinal experience of 2406 patients over two decades. *Br J Cancer, 108*(7), 1432–1439. doi:10.1038/bjc.2013.82

Koinuma, K., Shitoh, K., Miyakura, Y., Furukawa, T., Yamashita, Y., Ota, J., … Mano, H. (2004). Mutations of BRAF are associated with extensive hMLH1 promoter methylation in sporadic colorectal carcinomas. *Int J Cancer, 108*(2), 237–242. doi:10.1002/ijc.11523

Kornprat, P., Pollheimer, M. J., Lindtner, R. A., Schlemmer, A., Rehak, P., & Langner, C. (2011). Value of tumor size as a prognostic variable in colorectal cancer: a critical reappraisal. *Am J Clin Oncol, 34*(1), 43–49. doi:10.1097/COC.0b013e3181cae8dd

Lee, H. J., Eom, D. W., Kang, G. H., Han, S. H., Cheon, G. J., Oh, H. S., … Han, M. S. (2013). Colorectal micropapillary carcinomas are associated with poor prognosis and enriched in markers of stem cells. *Mod Pathol, 26*(8), 1123–1131. doi:10.1038/modpathol.2012.163

Li, Y., Feng, Y., Dai, W., Li, Q., Cai, S., & Peng, J. (2019). Prognostic effect of tumor sidedness in colorectal cancer: a SEER-based analysis. *Clin Colorectal Cancer, 18*(1), e104–e116. doi:10.1016/j.clcc.2018.10.005

Liu, Q., Luo, D., Cai, S., Li, Q., & Li, X. (2018). Circumferential resection margin as a prognostic factor after rectal cancer surgery: a large population-based retrospective study. *Cancer Med, 7*(8), 3673–3681. doi:10.1002/cam4.1662

Loughrey, M. B., McGrath, J., Coleman, H. G., Bankhead, P., Maxwell, P., McGready, C., … James, J. A. (2021). Identifying mismatch repair-deficient colon cancer: near-perfect concordance between immunohistochemistry and microsatellite instability testing in a large, population-based series. *Histopathology, 78*(3), 401–413. doi:10.1111/his.14233

Morak, M., Laner, A., Bacher, U., Keiling, C., & Holinski-Feder, E. (2010). MUTYH-associated polyposis - variability of the clinical phenotype in patients with biallelic and monoallelic MUTYH mutations and report on novel mutations. *Clin Genet, 78*(4), 353–363. doi:10.1111/j.1399-0004.2010.01478.x

Nagtegaal, I. D., Odze, R. D., Klimstra, D., Paradis, V., Rugge, M., Schirmacher, P., … WHO Classification of Tumours Editorial Board (2020). The 2019 WHO classification of tumours of the digestive system. *Histopathology, 76*(2), 182–188. doi:10.1111/his.13975

Nakayama, M., & Oshima, M. (2019). Mutant p53 in colon cancer. *J Mol Cell Biol, 11*(4), 267–276. doi:10.1093/jmcb/mjy075

Narayan, S., & Roy, D. (2003). Role of APC and DNA mismatch repair genes in the development of colorectal cancers. *Mol Cancer, 2*, 41. doi:10.1186/1476-4598-2-41

Nitsche, U., Friess, H., Agha, A., Angele, M., Eckel, R., Heitland, W., … Engel, J. (2016). Prognosis of mucinous and signet-ring cell colorectal cancer in a population-based cohort. *J Cancer Res Clin Oncol, 142*(11), 2357–2366. doi:10.1007/s00432-016-2224-2

Noack, P., & Langer, R. (2023). Molecular pathology of colorectal cancer. *memo - Magazine of European Medical Oncology, 16*, 116–121. doi:10.1007/s12254-023-00893-2

O'Brien, M. J., Yang, S., Mack, C., Xu, H., Huang, C. S., Mulcahy, E., … Farraye, F. A. (2006). Comparison of microsatellite instability, CpG island methylation phenotype, BRAF and KRAS status in serrated polyps and traditional adenomas indicates separate pathways to distinct colorectal carcinoma end points. *Am J Surg Pathol, 30*(12), 1491–1501. doi:10.1097/01.pas.0000213313.36306.85

Ono, Y., & Yilmaz, O. (2024). Emerging and under-recognised patterns of colorectal carcinoma morphologies: a comprehensive review. *J Clin Pathol, 77*(7), 439–451. doi:10.1136/jcp-2023–208816

Park, C. H., Han, D. S., Oh, Y. H., Lee, A. R., Lee, Y. R., & Eun, C. S. (2016). Role of Fusobacteria in the serrated pathway of colorectal carcinogenesis. *Sci Rep, 6*, 25271. doi:10.1038/srep25271

Pei, J. P., Zhang, C. D., Fu, X., Ba, Y., Yue, S., Zhao, Z. M., & Dai, D. Q. (2021). A novel TNM classification for colorectal cancers based on the metro-ticket paradigm. *J Cancer, 12*(11), 3299–3306. doi:10.7150/jca.55097

Peltomaki, P., Nystrom, M., Mecklin, J. P., & Seppala, T. T. (2023). Lynch syndrome genetics and clinical implications. *Gastroenterology, 164*(5), 783–799. doi:10.1053/j.gastro.2022.08.058

Phipps, A. I., Limburg, P. J., Baron, J. A., Burnett-Hartman, A. N., Weisenberger, D. J., Laird, P. W., … Newcomb, P. A. (2015). Association between molecular subtypes of colorectal cancer and patient survival. *Gastroenterology, 148*(1), 77–87 e72. doi:10.1053/j.gastro.2014.09.038

Poritz, L. S., Sehgal, R., Hartnett, K., Berg, A., & Koltun, W. A. (2011). Tumor volume and percent positive lymph nodes as a predictor of 5-year survival in colorectal cancer. *Surgery, 150*(4), 649–655. doi:10.1016/j.surg.2011.07.049

Powell, S. M., Zilz, N., Beazer-Barclay, Y., Bryan, T. M., Hamilton, S. R., Thibodeau, S. N., … Kinzler, K. W. (1992). APC mutations occur early during colorectal tumorigenesis. *Nature, 359*(6392), 235–237. doi:10.1038/359235a0

Rawla, P., Sunkara, T., & Barsouk, A. (2019). Epidemiology of colorectal cancer: incidence, mortality, survival, and risk factors. *Prz Gastroenterol, 14*(2), 89–103. doi:10.5114/pg.2018.81072

Safaee Ardekani, G., Jafarnejad, S. M., Tan, L., Saeedi, A., & Li, G. (2012). The prognostic value of BRAF mutation in colorectal cancer and melanoma: a systematic review and meta-analysis. *PLoS One, 7*(10), e47054. doi:10.1371/journal.pone.0047054

Saha, S., Shaik, M., Johnston, G., Saha, S. K., Berbiglia, L., Hicks, M., … Wiese, D. (2015). Tumor size predicts long-term survival in colon cancer: an analysis of the National Cancer Data Base. *Am J Surg, 209*(3), 570–574. doi:10.1016/j.amjsurg.2014.12.008

Scott, N., West, N. P., Cairns, A., & Rotimi, O. (2021). Is medullary carcinoma of the colon underdiagnosed? An audit of poorly differentiated colorectal carcinomas in a large national health service teaching hospital. *Histopathology, 78*(7), 963–969. doi:10.1111/his.14310

Shah, S. C., & Itzkowitz, S. H. (2022). Colorectal cancer in inflammatory bowel disease: mechanisms and management. *Gastroenterology, 162*(3), 715–730 e713. doi:10.1053/j.gastro.2021.10.035

Smith, J. D., Reidy, D. L., Goodman, K. A., Shia, J., & Nash, G. M. (2014). A retrospective review of 126 high-grade neuroendocrine carcinomas of the colon and rectum. *Ann Surg Oncol, 21*(9), 2956–2962. doi:10.1245/s10434-014-3725-3

Thomas, T., Abrams, K. A., Robinson, R. J., & Mayberry, J. F. (2007). Meta-analysis: cancer risk of low-grade dysplasia in chronic ulcerative colitis. *Aliment Pharmacol Ther, 25*(6), 657–668. doi:10.1111/j.1365–2036.2007.03241.x

Umar, A., Boland, C. R., Terdiman, J. P., Syngal, S., de la Chapelle, A., Ruschoff, J., … Srivastava, S. (2004). Revised Bethesda Guidelines for hereditary nonpolyposis colorectal cancer (Lynch syndrome) and microsatellite instability. *J Natl Cancer Inst, 96*(4), 261–268. doi:10.1093/jnci/djh034

Vasen, H. F., Watson, P., Mecklin, J. P., & Lynch, H. T. (1999). New clinical criteria for hereditary nonpolyposis colorectal cancer (HNPCC, Lynch syndrome) proposed by the International Collaborative group on HNPCC. *Gastroenterology, 116*(6), 1453–1456. doi:10.1016/s0016-5085(99)70510-x

Walther, A., Houlston, R., & Tomlinson, I. (2008). Association between chromosomal instability and prognosis in colorectal cancer: a meta-analysis. *Gut, 57*(7), 941–950. doi:10.1136/gut.2007.135004

Weisenberger, D. J., Siegmund, K. D., Campan, M., Young, J., Long, T. I., Faasse, M. A., … Laird, P. W. (2006). CpG island methylator phenotype underlies sporadic microsatellite instability and is tightly associated with BRAF mutation in colorectal cancer. *Nat Genet, 38*(7), 787–793. doi:10.1038/ng1834

Weledji, E. P. (2024). The etiology and pathogenesis of colorectal cancer. *Clin Oncol, 9*, 2046.

Yaeger, R., Shah, M. A., Miller, V. A., Kelsen, J. R., Wang, K., Heins, Z. J., … Kelsen, D. (2016). Genomic alterations observed in colitis-associated cancers are distinct from those found in sporadic colorectal cancers and vary by type of inflammatory bowel disease. *Gastroenterology, 151*(2), 278–287 e276. doi:10.1053/j.gastro.2016.04.001

Yuan, H., Dong, Q., Zheng, B., Hu, X., Xu, J. B., & Tu, S. (2017). Lymphovascular invasion is a high risk factor for stage I/II colorectal cancer: a systematic review and meta-analysis. *Oncotarget, 8*(28), 46565–46579. doi:10.18632/oncotarget.15425

Zhao, Y., Ge, X., He, J., Cheng, Y., Wang, Z., Wang, J., & Sun, L. (2019). The prognostic value of tumor-infiltrating lymphocytes in colorectal cancer differs by anatomical subsite: a systematic review and meta-analysis. *World J Surg Oncol, 17*(1), 85. doi:10.1186/s12957-019-1621-9

3 Cellular and Molecular Pathology, Signaling Pathways, and Genetic and Epigenetic Roles Related to Colorectal Cancer

Alaa A. A. Aljabali[1], *Almuthanna Alkaraki*[2], *Mohammad Obeid*[1], *Omar Gammoh*[3], *and Mohamed El-Tanani*[4]

[1]Faculty of Pharmacy, Department of Pharmaceutics & Pharmaceutical Technology, Yarmouk University, Irbid, Jordan

[2]Department of Biological Sciences, Faculty of Science, Yarmouk University, Irbid, Jordan

[3]Department of Clinical Pharmacy and Pharmacy Practice, Faculty of Pharmacy, Yarmouk University

[4]College of Pharmacy, Ras Al Khaimah Medical and Health Sciences University, Ras Al Khaimah, UAE

3.1 INTRODUCTION

3.1.1 CRC Is a Major Global Health Concern Worldwide

Colorectal cancer (CRC) ranks third among all cancers in terms of the frequency of diagnosis. It represents the second most prevalent cause of cancer-related mortality worldwide (Jayarathna et al. 2022). Several studies have shown that the occurrence and mortality rates of CRC are increasing. The incidence, which is on the rise, brings about a need for proper allocation of health service resources. However, it should be mentioned that, based on screening programs as well as overall management and care, the mortality trends for CRC improve over time in most countries (Sung et al. 2021).

CRC can be viewed as a huge public health crisis, with over one million new cases estimated in 2008. The global incidence of CRC is projected to reach approximately 1.9 million cases by 2020, resulting in a staggering 900,000 fatalities each year. CRC represents the third most prevalent form of cancer and the tenth leading cause of mortality and morbidity worldwide, posing a significant threat to public health (Harras and Gaweesh 2019; Sung et al. 2021). CRC is widely believed to be the second leading cause of cancer-related deaths globally, according to various sources.

This phenomenon represents a significant public health concern, as evidenced by approximately 1.2 million new cases and 600,000 deaths that occur annually on average (Barresi 2021). According to the American Cancer Society (ACS), it is projected that in 2024, approximately 152,810 new cases of CRC will be diagnosed in the United States, with approximately 81,540 cases occurring in men and 71,270 in women. Of these new diagnoses, approximately 106,590 are for colon cancer and approximately 46,220 are for rectal cancer, respectively (Siegel et al. 2024).

 DOI: 10.1201/9781003503842-3

Therefore, it is essential to acquire a comprehensive understanding of the global incidence of CRC and to develop novel therapeutic approaches (Lu et al. 2022). The current chapter focuses on the development of innovative treatment techniques, monoclonal antibody therapy, natural products in the gut microbiota, and genetic modifications for the prediction of response in CRC (Hwang et al. 2021; Proietto et al. 2023). Subsequent studies further confirmed the complexity of CRC and revealed the existence of abnormal splicing patterns and molecular markers, namely, tensin-like, Twist1, and Ki-67 (Martins et al. 2015).

CRC has a constant global health effect, with increasing incidence and fatality rates. As a result, the time spent on research, prevention, and efficacy of treatment options should be enhanced.

3.1.2 The Importance of Understanding the Cellular and Molecular Pathology of CRC

Consequently, elucidating the cellular and molecular pathogenesis of CRC is of paramount importance for enhancing diagnostic approaches, therapeutic interventions, and patient outcomes. Knowledge of the molecular mechanisms of CRC is an area of intense research, especially where it provides insights into how cancer cells acquire invasive and metastatic potential (Vu and Datta 2017). Therefore, this knowledge is important and instrumental for the formulation of appropriate CRC treatment strategies. Greater knowledge of the origin and course of CRC would assist in the early detection of molecular markers and risk stratification approaches, which would eventually lead to advances in the clinical care of patients with CRC (Malki et al. 2020).

Molecular pathologies have advanced to the extent that they allow the classification of CRC into distinct molecular subtypes, thus enabling the adoption of personalized therapies and improved patient management (Inamura 2018). There have been significant improvements in the care and outcomes of patients with CRC using targeted and conventional therapies based on molecular testing (Sepulveda et al. 2017). CRC is a genetic and molecular disease. This understanding opens a new path for clinical pathways. They aim to prevent the disease, make better predictions, and improve the prognosis (Hagland et al. 2013).

This is particularly important for understanding tumorigenesis, development, and prognosis, which forms the basis of a clinical approach that needs to be tailored for treatment (Mármol et al. 2017). Developments in molecular pathology will soon provide non-invasive CRC screening assays with several potentially effective early-detection molecular markers (Lao and Grady 2011). This link connects the molecular and clinicopathological characteristics of CRC. These findings may enrich our understanding. This can help us to understand the causes of CRC. This helps block and cure it simultaneously (Pancione et al. 2012).

Consequently, it is imperative to elucidate the molecular and cellular mechanisms of CRC. This understanding is essential for advancing the diagnosis and treatment of diseases. Research on the molecular mechanisms of CRC by investigators and molecular pathologists has created the potential for personalized therapies, early detection, and better prognostic outcomes for this highly incident cancer (Heinimann et al. 2000). CRC is a complex disease. Multiple genetic alterations drive its progression. These modifications involve mutations in oncogenes and tumor suppressor genes. It can also cause this disease. Some of the key genes were COX2, p16, MGMT, hMLH1, TIMP3, and CDH13 (Figure 3.1). They play a key role in colorectal carcinoma development (Glasgow and Hardiman 2022).

Additionally, other studies have indicated that genes encoding drug metabolism enzymes such as MDR1 participate in the progression of CRC (Jonges et al. 2001). The development of CRC is dependent on critical genes, such as APC, STX2, PODXL, KLK6, GRB10, EHBP1, CREB5, ANOS1, and CTHRC1 (Gayet et al. 2001; Gu et al. 2014; Pätzold et al. 2017), as shown in Figure 3.2. Finally, the roles of these key genes should be well understood for targeted therapies and individual treatment strategies against CRC (Hibi et al. 2002).

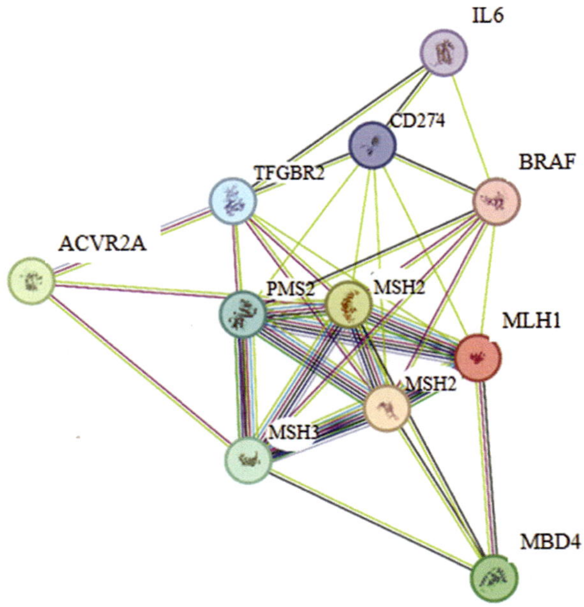

FIGURE 3.1 The microsatellite instability (MSI) pathway in colorectal cancer elucidates the interactions among mismatch repair proteins, including MLH1, MSH2, MSH6, PMS2, and MSH3. Impaired DNA mismatch repair mechanisms result in microsatellite instability, which plays a crucial role in the tumorigenesis of a subset of colorectal cancers. Notably, these cancers often exhibit a more favorable prognosis because of their enhanced responsiveness to immunotherapy.

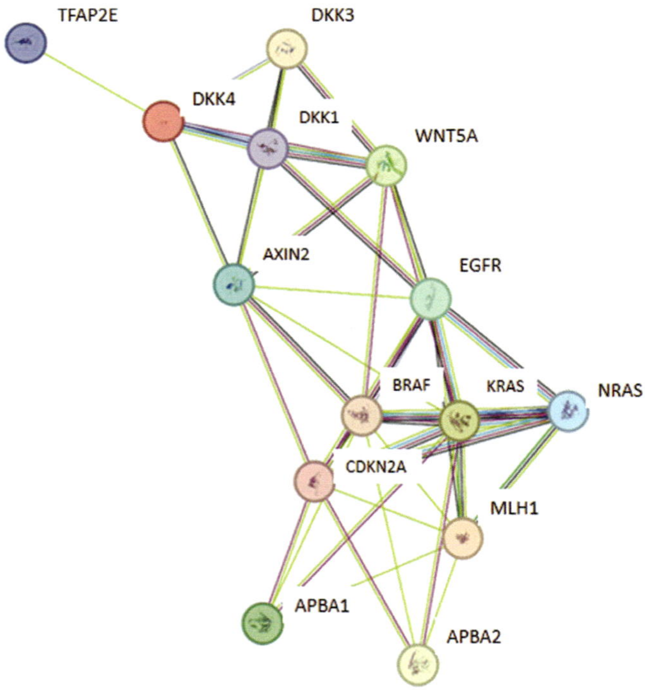

FIGURE 3.2 Schematic representation of the CpG island methylator phenotype pathway. This diagram illustrates the methylation-mediated silencing of tumor suppressor genes and their subsequent impact on colorectal cancer progression.

In addition, DNA repair genes, such as MUTYH and KLF6, affect CRC risk. The same was true for single-letter changes in the genes related to one-carbon metabolism. This includes genes such as MTHFR and MTRR. They perform this process by DNA repair and altering nucleotide synthesis. Some of these genes, such as TYMS and BCL2L1, have expression and regulation networks related to prognosis in patients (Tsuchiya et al. 1997; Sareeboot et al. 2011; Saplacan et al. 2015). Further insight into the molecular processes of CRC progression came from a regulatory network for highly expressed genes in CRC cells created with CRC-specific enhancers (Young and Jass 2006).

Many genetic alterations involve dominant oncogenes and tumor suppressor genes. They interact with one another in a complex manner. This interaction sheds light on the underlying causes of CRC. Understanding how these key genes contribute to carcinogenesis may help develop effective targeted therapies and personalized treatment approaches in patients with CRC.

3.1.3 Purpose and Scope

This chapter presents a comprehensive examination of cellular and molecular CRC pathology and its implications for patient diagnosis, prognosis, and treatment planning. This chapter offers insights into the molecular causes of CRC. It focuses on those factors that cause invasion and spread. In addition, further discussion on the importance of knowledge in early-stage CRC and the development of the disease will be discussed for its identification as an early-detection molecular marker and risk stratification strategy. Therefore, the main aim of this study was to consolidate current research on molecular pathology. The objective of this chapter was to categorize CRC into distinct molecular subtypes. These subtypes are essential for personalized therapeutic approaches and effective patient management. This chapter emphasizes the significance of molecular testing in patient care and proposes several potential clinical applications for CRC prevention, prediction, and prognosis.

3.1.4 Cellular Pathology of CRC

CRC is a multifaceted disease that encompasses numerous cellular pathways and genetic factors. Several studies have strongly suggested this. They stated that signaling pathways drive CRC (Mirza-Aghazadeh-Attari et al. 2020). The etiology of this type of cancer has been linked to genetic instability, particularly in APC genes (Fodde et al. 2001). An optimal approach to investigate these networks is to examine those that contribute to the development of CRC. CRC poses a complex challenge, and key signaling pathways control the process of proliferation, survival, and metastasis. Knowledge of these pathways is crucial because they are representative of the key targets that need to be targeted therapeutically. The Wnt/β-catenin pathway is also involved in CRC pathogenesis. Mutations in the adenomatous polyposis coli gene initiate β-catenin accumulation, which causes the activation of genes that promote cell growth (Liu et al. 2022). Similarly, mutant KRAS and BRAF genes cause deviations in the RAS/RAF/MAPK pathway, resulting in constantly active signaling cascades that drive unrestrained cellular proliferation (Figure 3.3). This would indicate that modifications in the PI3K/AKT/mTOR pathway are further amplified to result in enhanced cellular proliferation, survival, and angiogenesis (Huang et al. 2021).

The TGF-β pathway, which acts as a tumor suppressor during the early phases of cancer, is paradoxically converted into a promoter of invasion and metastasis during the later phases of CRC development. These interconnected pathways underlie the complex landscape of CRC, with potential routes for targeted therapies aimed at reducing disease progression and enhancing patient outcomes. Further dissection of the molecular mechanisms underlying the deregulation of such pathways holds promise for the advancement of precision medicine approaches in the battle against this formidable disease (Liu et al. 2021).

Chromosome segregation 1 is a critical regulator of chromosome segregation. It modulates the apoptosis, migration, and invasion of colorectal cells. It promotes malignancy (Alnabulsi et al. 2012). This chapter also suggests that studying the key genes involved in slow-transit constipation may reveal the underlying molecular mechanisms. These mechanisms play a role in the progression of CRC (Figure 3.4).

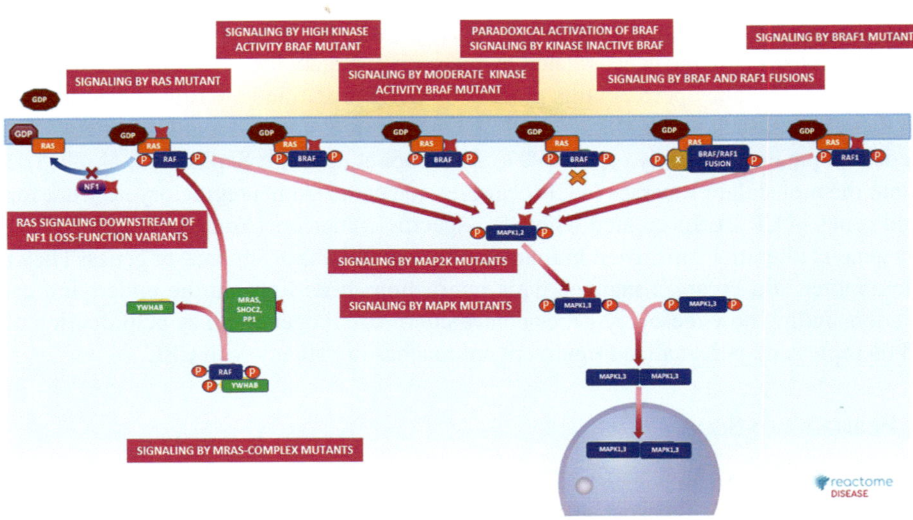

FIGURE 3.3 Oncogenic MAPK signaling pathway in colorectal cancer. The figure shows how mutations in KRAS and BRAF lead to continuous cell proliferation and tumor growth.

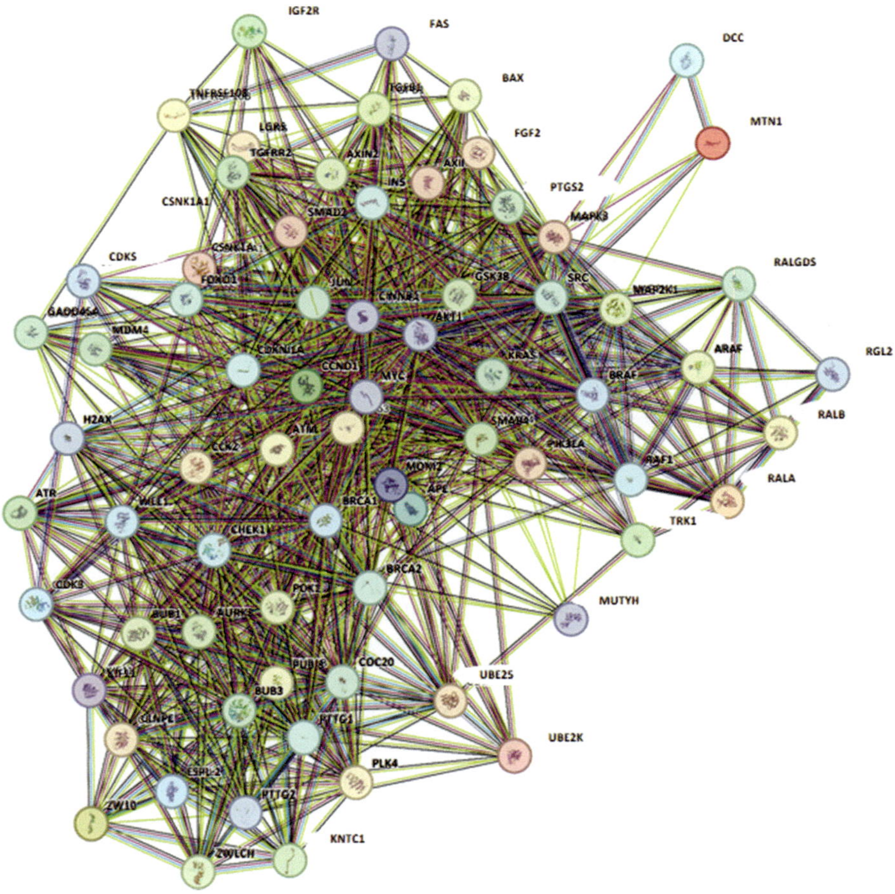

FIGURE 3.4 Protein-protein interaction network elucidating the role of chromosomal instability in colorectal cancer development. This network delineates the key protein interactions involved in chromosomal alterations, contributing to tumor heterogeneity and disease progression.

TABLE 3.1
Cellular and Molecular Pathology of CRC

Feature	Description	Clinical Significance
Adenoma-carcinoma sequence	Progression from benign polyp to cancer	Model for CRC development, importance of early detection
Microsatellite instability (MSI)	Defective DNA mismatch repair	Better prognosis, potential response to immunotherapy
Chromosomal instability (CIN)	Numerical and structural chromosomal abnormalities	Tumor heterogeneity, progression, and poorer prognosis
Tumor microenvironment	Stromal cells, immune cells, vasculature	Influences tumor growth, invasion, and metastasis

CRC, colorectal cancer.

The molecular changes associated with CRC are related to its spread. Patients with a poor prognosis have these symptoms. This is not surprising. Later studies found that most molecular changes in the process cause cancer. These studies have focused on cancer processes (Yu et al. 2022). This introduces additional molecular complexity to CRC pathogenesis. Signet ring cell carcinoma represents an uncommon subtype of CRC. This malignancy exhibits distinctive characteristics that distinguish it from other forms.

Artificial intelligence and deep learning methodologies have shown promising results in the accurate diagnosis of CRC from histopathological images and may enhance diagnostic accuracy and outcomes for patients (Gupta et al. 2021; Wang et al. 2021b). Recent studies have elucidated the role of RNA in the inhibition of motility and invasion of CRC cells. The pathology of CRC is complex. This involves a combination of genetics, molecules, and cells. Understanding this complex interplay is, therefore, crucial. It involves signaling pathways, genetic alterations, and shifts in gene expression. To improve CRC care, we must understand this complex relationship. It affects diagnosis, treatment, and management (Table 3.1).

3.1.4.1 Histological Characteristics of Cancer Cells

The histological features of the cells in colorectal carcinoma vary and are important for understanding the disease. CRC is a histological subtype. These encompass mucinous adenocarcinomas, signet ring cell carcinomas, medullary carcinomas, and undifferentiated carcinomas (Martínez et al. 2009; Hyngstrom et al. 2012; Fatima et al. 2021). Subtypes have unique characteristics that may alter prognosis and treatment. For example, these two types of CRC have poor prognoses. They are mucinous adenocarcinomas and signet ring cell carcinomas (Sung et al. 2008). Histological analysis may help to identify therapeutic targets and biomarkers. It has also been reported to upregulate its expression. This occurred in the cancerous features of the colorectal tissue. This upregulation affects the prognosis and treatment of the disease (Horst et al. 2008). Histological analysis may help to identify therapeutic targets and biomarkers.

Researchers have found that its expression is elevated in malignant CRC tissues. This elevation is associated with disease prognosis and treatment outcomes (Wang et al. 2015). In addition, hollow spheroids have been observed beyond the invasive margin in the colorectum, indicating malignant potential and emphasizing the value of histological aspects in estimating the aggressiveness of a disease (Tamura et al. 2011). Other studies have advanced histological criteria regarding the differentiation of colonic and rectal carcinomas. The studies find that these differences may be due to different carcinogenic mechanisms; thus, they indicate that the two cancers may be histopathologically heterogeneous (Konishi et al. 1999). In summary, the histological features of CRC cells help in prognostication and guide therapeutic decisions, as well as enlighten the mechanisms underlying the disease. These histological features provide information regarding the behavioral traits of CRC in terms of personalized therapeutic approaches.

3.1.4.2 Abnormal Cellular Proliferation and Differentiation Mechanisms in CRC

Several mechanisms drive CRC progression. This results in abnormal cell growth and subsequent cell changes. These abnormalities have been studied, and several molecular mechanisms have been identified. For example, most colorectal tumors exhibit c-Myc overexpression, indicating that c-Myc may promote abnormal cell proliferation (Hartman et al. 2009). Dysregulation of the PRL-3/AURKA/FZR1 circuitry axis is also associated with low cellular proliferation during CRC progression (Zhang et al. 2019). These findings suggest that CRC exhibits a complex network of disrupted signaling pathways. Therefore, genetic and molecular modifications may serve as molecular markers for CRC development (Worthley et al. 2007). The cell regulator SETDB1 has also been shown to be an important inhibitor of cellular differentiation in CRC cells, emphasizing the role of oncogenic mutations and microenvironmental changes in blocking normal cell differentiation (Lee et al. 2020). Additionally, miR-107 and miR-366 regulate the growth and movement of CRC cells. Hence, these small non-coding RNAs are important for CRC pathogenesis (Li et al. 2018; Fu et al. 2019). For example, the overexpression of DHX32 leads to the initiation and development of cancer owing to the mis-proliferation, differentiation, and apoptosis of cells (Wei et al. 2021). Diet primarily shapes the development of CRC in a major way. It also affects the cell growth and division. For example, studies have shown that consumption of black raspberries can be beneficial. Compounds such as Alisol A affect these processes (Han et al. 2022). Another related area of complexity is cholesterol metabolism, vitamin D receptor activity, and cholesterol metabolites in lipid metabolism, which have been implicated in cell differentiation and proliferation (Winandar et al. 2021). Overall, the interplay between these different molecular factors, signaling pathways, and genetic alterations underlines the complicated mechanism of abnormal cell proliferation and differentiation in CRC.

3.1.5 The Role of Inflammatory Processes in CRC Pathogenesis

Inflammation plays a crucial role in the etiology and progression of CRC. Chronic inflammation is a key risk in CRC. This has many implications for the development of this disease (Jiang et al. 2018). Researchers have identified certain novel inflammatory genes as clinical prognostic biomarkers for CRC (Ieda et al. 2019). In addition, inflammation and EMT are involved in the CRC microenvironment. Inflammation and metabolic dysfunction, especially in the interplay with tumor-associated macrophages, contribute to the etiology of early-onset CRC (Scheurlen et al. 2020). Most molecular species associated with inflammation, such as HMGB1, have versatile functions in the pathogenesis of CRC and provide strict evidence for their participation in tumorigenesis (Cheng et al. 2020).

Furthermore, NF-κB signaling is associated with chronic inflammation, which is one of the most significant risk factors for CRC, and serves as a therapeutic target for the suppression of inflammatory pathways in CRC (Morrison et al. 2023). Toll-like receptors are upregulated in CRC tissues, highlighting the need to enhance the inflammatory response to drive CRC prognosis (Bhardwaj et al. 2024).

Previous studies have examined the relationship between dietary factors and inflammation in CRC patients. The Dietary Inflammatory Index has provided evidence that links diet to the inflammatory process and the development of CRC (Malcomson et al. 2020). However, it should be noted that the association between these diseases becomes stronger if it concerns inflammatory bowel diseases (IBD) because these diseases have very strong similarities in terms of long-standing chronic inflammation, genetic alterations, and environmental factors of the two diseases (Ghobashi et al. 2023). This shows some shared pathways between IBD and CRC, in which common inflammatory mechanisms result (Derakhshani et al. 2022). Therefore, the comprehension and targeting of inflammation-related pathways play a pivotal role in the prevention, diagnosis, and treatment of CRC, as this inflammatory process plays a complex role in the pathogenesis of CRC. The tumor microenvironment promotes CRC progression.

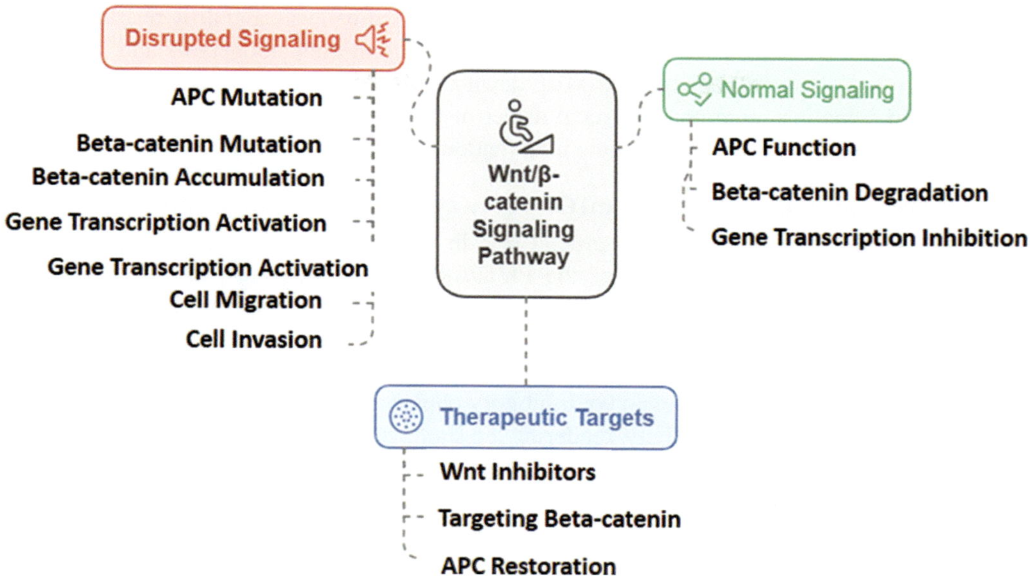

FIGURE 3.5 The Wnt/β-catenin signaling pathway in colorectal cancer (CRC) development. This figure delineates the normal functioning of the Wnt/β-catenin pathway and elucidates how mutations in APC or β-catenin result in β-catenin accumulation within the nucleus, consequently promoting cellular proliferation, migration, and invasion. Furthermore, this figure identifies potential therapeutic targets for the inhibition of Wnt signaling.

3.1.6 Molecular Pathways in Colorectal Cancer

3.1.6.1 Dysregulated Wnt/β-Catenin Signaling Pathway and Its Implications in CRC

Wnt/β-catenin signaling is a critical molecular pathway for CRC pathogenesis because dysregulation of this pathway directly affects tumor initiation, promotion, and metastasis. In the canonical Wnt signaling pathway, as illustrated in Figure 3.5, β-catenin is stabilized within the nucleus, where it functions as a co-transcriptional activator for multiple Wnt target genes, thereby conferring properties associated with cellular proliferation, survival, and invasion (Cheriyamundath and Ben-Ze'ev 2020; Garrett et al. 2024). However, significant attention has been given to the Wnt/β-catenin signaling pathway as a crucial pathway in CRC. Further summary indicated that this pathway is associated with resistance to chemotherapeutic agents (Cheriyamundath and Ben-Ze'ev 2020), features of cancer stem cells (CSCs) (Vieira and Conte-Junior 2024), and radioresistance (Xu et al. 2023b; Mollanoori et al. 2024).

The Wnt/β-catenin pathway exhibits aberrant activation, is associated with CRC cell proliferation, migration, and invasion, and promotes carcinogenesis through epithelial-mesenchymal transition (Xu et al. 2023b). Furthermore, dysregulation of its pathway has been demonstrated to be associated with the progression of CRC growth and metastasis. Inhibition of Wnt/β-catenin signaling represents a promising mechanism for the potential treatment of CRC (He and Gan 2023). Consequently, the principal molecules in this pathway are UXT-V1 and JMJD1A, which regulate the Wnt/β-catenin pathway to inhibit CRC progression (He and Gan 2023).

Furthermore, investigations have been conducted to ascertain the potential involvement of small molecule drugs and novel inhibitors of the Wnt/β-catenin pathway in CRC treatment. These investigations have been carried out in multiple studies (Mook et al. 2019). Moreover, activation of the canonical Wnt/β-catenin signaling pathway has been demonstrated to induce epithelial-mesenchymal transition and enhance cellular motility and invasiveness in CRC (Pan et al. 2019).

The observed relationship between the Wnt/β-catenin signaling pathway and drug resistance, stemness, and metastasis significantly underscores its importance in the pathogenesis and progression of CRC (Correa-Arzate et al. 2023; Gajos-Michniewicz and Czyz 2023). Collectively, dysregulated Wnt/β-catenin signaling represents a critical driver of colorectal carcinogenesis; consequently, it constitutes a potential target for therapeutic interventions in CRC (Table 3.2).

3.1.6.2 Alterations in the PI3K/Akt/mTOR Pathway and Its Association with CRC Cancer

The PI3K/Akt/mTOR pathway plays a crucial role in CRC and regulates cellular proliferation, survival, and invasion. Dysregulation of the PI3K/Akt/mTOR signaling pathway is associated with the occurrence of CRC and the development of resistance to therapeutic interventions. The PIK3CA gene contains mutations that can predict the response to its inhibitors in advanced cancers, including CRC (Byun et al. 2023). Alterations in the PI3K/Akt/mTOR pathway, as illustrated in Figure 3.6, such as decreased expression of inhibitory phosphatases and gain-of-function mutations in PIK3CA, enhance signaling activity, rendering it a significant candidate for CRC carcinogenesis (El-Tanani et al. 2023).

TABLE 3.2
Key Signaling Pathways in CRC

Pathway	Key Components	Role in CRC	Therapeutic Targets
Wnt/β-catenin	APC, β-catenin, TCF/LEF	Cell proliferation, stemness	Wnt inhibitors, porcupine inhibitors
RAS/RAF/MAPK	KRAS, BRAF, MEK, ERK	Cell growth, survival, proliferation	KRAS inhibitors (G12C specific), MEK inhibitors, BRAF inhibitors
PI3K/AKT/mTOR	PI3K, PTEN, AKT, mTOR	Cell growth, survival, angiogenesis	PI3K inhibitors, AKT inhibitors, mTOR inhibitors
TGF-β	TGF-β receptors, SMADs	Tumor suppression (early), metastasis (late)	TGF-β receptor inhibitors, anti-TGF-β antibodies

CRC, colorectal cancer.

FIGURE 3.6 Alterations in the PI3K/Akt/mTOR pathway in colorectal cancer. The schematic representation illustrates the amplification of the PI3K/Akt/mTOR pathway through mutations in key components, such as PIK3CA, resulting in enhanced cell proliferation, survival, and angiogenesis. Potential therapeutic interventions, including PI3K and mTOR inhibitors, have been identified as prospective targets for treatment.

The PI3K/Akt/mTOR pathway has also been reported to be associated with invasiveness in urothelial carcinoma of the urinary bladder and is also linked to poor overall survival; therefore, it is believed to be a major modulator of cancer behavior (de Kort et al. 2023; Pezzicoli et al. 2024).

This pathway is associated with the pathogenesis of diseases, including carcinogenesis, invasion, and metastasis in CRC (Bahrami et al. 2017). The potential of this pathway for CRC treatment is very high because the study showed it downregulates important molecules in the pathway to hinder further CRC growth and progression (Arqués et al. 2016). The PI3K/Akt/mTOR signaling pathway is involved in multiple cellular processes in CRC, including cell proliferation, adhesion, migration, and survival (Xu et al. 2023a).

In the evidentiary basis of CRC cells, inhibition of this pathway results in cellular apoptosis and autophagy, thus guiding researchers on possible therapeutic interventions (Dharmaraj et al. 2024). Furthermore, this pathway has been demonstrated to be involved in cellular proliferation, differentiation, apoptosis, metabolism, and autophagy in CRC and other malignancies (Su et al. 2020). Consequently, alterations associated with the initiation and progression of CRC are stringently regulated by the PI3K/Akt/mTOR signaling axis. This emphasizes the therapeutic potential of targeting this pathway for treatment.

3.1.6.3 Role of TP53 Tumor Suppressor Pathway in CRC Development and Progression

TP53 functions as a primary downstream effector of the tumor suppressor pathway during CRC development and progression. TP53 is known to regulate major cell processes such as cell cycle progression, DNA repair, and apoptosis, which are involved in tumor repression. While the TP53 pathway is often compromised in CRC, it is also associated with several tumorigenesis events (Tomičić et al. 2021). Consequently, this investigation established a pivotal role of TP53 mutations in the pathogenesis of CRC. Mutations in TP53 predominantly occur at an early stage in the progression from inflamed colonic epithelium to dysplasia and ultimately to CRC, thus suggesting a function in tumor initiation. In CRC, TP53 mutations were found to have a worse prognosis, and the gene exhibited remarkable effects on disease outcomes (Do et al. 2022). The TP53 pathway plays a vital role in the regulation of important biological processes involved in tumor suppression. Due to the loss of TP53, the respective functionalities of both stromal and non-stromal cells in the tumor microenvironment are compromised, thereby exerting a significant influence on tumor behavior and progression. The implications of TP53 mutations in CRC vary among studies. Missense mutations in TP53 exhibit diverse contributions at various stages of cancer; consequently, they have differential effects on patient management (Hirsch et al. 2021).

TP53 mutations also drive CRC development, including drug resistance, in both diseases. Mutated TP53 induces the activation of mechanisms that increase the severity of disease outcomes and reduce the effectiveness of applied therapies. Mutations in the TP53 gene have been consistently associated with overall survival in stage II/III CRC patients, further underscoring their significance in prognostic determination. Consequently, a comprehensive understanding of TP53 mutations is essential for the development of more effective targeted treatments and improved clinical outcomes in CRC (Hirsch et al. 2021).

3.1.6.4 Crosstalk between Signaling Pathways and Their Impact on CRC Biology

Multiple recent investigations have elucidated the interconnectivity between diverse signaling pathways in CRC. A notable illustration of this phenomenon, particularly in relation to cellular proliferation, is an interaction between the Wnt/β-catenin and ERK pathways (Gajos-Michniewicz and Czyz 2023). A further instance is the interaction between the Wnt/β-catenin and PI3K/Akt pathways, which, according to several studies, influences CRC tumorigenesis and progression (Cheriyamundath and Ben-Ze'ev 2020; He and Gan 2023; Gajos-Michniewicz and Czyz 2023). Furthermore, the interaction between the PI3K-AKT-mTOR and Wnt/β-catenin signaling pathways, specifically mediated by GSK3β and the TSC1:TSC2 complex, influences the malignant potential of cells derived from the prostate (de Kort et al. 2023; El-Tanani et al. 2023; Dharmaraj et al. 2024).

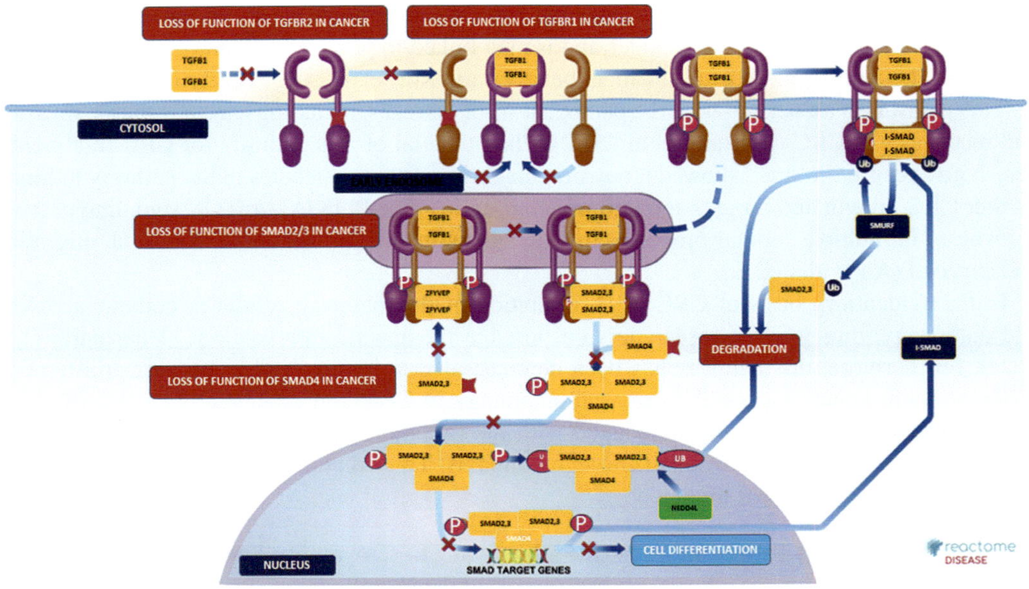

FIGURE 3.7 The TGF-beta signaling pathway in cancer. This diagram demonstrates the transition of TGF-beta from a tumor suppressor in early cancer stages to a promoter of metastasis in advanced colorectal cancer.

In CRC, crosstalk between PI3K/Akt/mTOR and Wnt/β-catenin is upregulated, independent of the mutational status (Fleming-de-Moraes et al. 2022. In summary, the interplay between Wnt/β-catenin, PI3K/Akt, ERK, and TGF-β1 signaling pathways is intricate and significantly influences a broad spectrum of CRC biology, as illustrated in Figure 3.7, particularly those aspects associated with cellular processes such as proliferation, invasion, stemness, and therapeutic response (Fleming-de-Moraes et al. 2022). Thus, a good understanding of the interplay between these elements is of great importance in guiding therapeutic strategies for enhancing outcomes in patients with CRC.

3.1.7 Genetic Factors in Colorectal Cancer

3.1.7.1 Overview of Common Genetic Mutations Implicated in CRC (e.g., APC, KRAS, and BRAF)

Genetic alterations constitute pivotal events in CRC pathogenesis. Mutations in APC, KRAS, and BRAF are considered critical markers of disease progression and treatment response (Corcoran et al. 2012). APC mutations are among the earliest events that occur during tumor development in the colonic epithelium. These mutations initiate adenomas, which subsequently lead to the development of CRC (Malki et al. 2020). Disruption of Wnt pathway function by mutations in APC leads to abnormal growth and tumorigenesis. Frequent observations of KRAS gene mutations in CRC are related to adverse outcomes such as decreased patient survival (Phipps et al. 2013). It influences cell proliferation, differentiation, and survival in a manner that leads to the development of CRC. BRAF mutations predominate in CRCs, particularly in the V600E subset, and are associated with a worse prognosis (Bond and Whitehall 2018). They occur in approximately 10%–15% of cases of CRC and are mostly found in MSI tumors with other associated factors. These genetic changes can occur singly or in combination and have potent effects on the clinical behavior of CRC, patient survival, and responsiveness to therapeutics. Clinicians must understand the genetic information on CRC, including APC, KRAS, and BRAF mutations, to provide personalized care and treatment planning that results in improved treatment outcomes (Huang et al. 2021; Sinkala 2023).

3.1.7.2 Hereditary Colorectal Cancer Syndromes (e.g., Lynch Syndrome, Familial Adenomatous Polyposis)

Familial adenomatous polyposis (FAP) is a major challenge in estimating and managing cancer risk (Fearon 2011). Lynch syndrome, previously referred to as hereditary non-polyposis CRC, is attributed to the dysfunction of MMR genes. This autosomal dominant condition predisposes individuals to CRC and other malignancies. The screening process for Lynch syndrome is primarily based on age of onset and familial history. Lynch syndrome represents the most prevalent hereditary CRC; female patients exhibit a 40%–60% lifetime risk of developing endometrial cancer (Munteanu and Mastalier 2014; Chen et al. 2018; Marx et al. 2023).

FAP is characterized by the development of multiple adenomatous polyps in the colon. These polyps serve as precursors to adenocarcinomas and are inherited in an autosomal dominant manner in the lower gastrointestinal tract. FAP is further classified into classic and attenuated FAP, based on the number of polyps present and mutations or deletions in the APC tumor suppressor gene (Half et al. 2009; Aihara et al. 2014). Identification of high-risk individuals and inappropriate surveillance and prevention will require an understanding of the genetic basis of Lynch syndrome and FAP. This establishes the importance of individualized approaches for the screening and management of hereditary CRC (Cohen and Leininger 2014).

3.1.8 Genetic Polymorphisms and Their Contribution to Susceptibility

These factors significantly contribute to an individual's susceptibility to pathology. Numerous genetic polymorphisms have been identified and investigated for their significance in CRC susceptibility (Huang et al. 2007). This gene demonstrated low penetrance and susceptibility to CRC in the MTHFR C677T and A1298C polymorphisms. These polymorphisms may exert a protective effect and decrease the risk of colorectal adenomas in patients with CRC (Huang et al. 2007).

For instance, single nucleotide polymorphisms (SNPs) in DNA repair genes contribute genetically to individual DNA repair capacity and, consequently, to susceptibility to certain types of cancers such as CRC (Pardini et al. 2008). This ALDH2 polymorphism suggests a potentially broader role of ALDH2 in elevating the risk of CRC and other human malignancies (Guo et al. 2013). The GSTM1 polymorphism has been associated with susceptibility to CRC in the Asian population, suggesting its potential role in CRC risk (Guo et al. 2013). This suggests that GSTM1 polymorphism is involved in CRC susceptibility in Asian populations. These data suggest that a locus on 15q13.3 is associated with susceptibility to colorectal adenomas as well as CRC (Tu et al. 2015). Other studies have defined a relationship between CRC susceptibility and PTPRJ gene polymorphisms, emphasizing the involvement of genetic variation at the PTPRJ locus in PTC susceptibility (Adel Fahmideh et al. 2014; Chiarella et al. 2023). The subsequent subtopics in this chapter elaborate on these genetic polymorphisms and their consequences for CRC susceptibility to help identify individuals who are at risk and aid in the development of appropriate prevention and screening strategies.

3.1.9 Advances in Genetic Screening and Personalized Medicine Approaches for CRC Management

Technological advancements have modernized genetic screening and personalized medicine; the latter should revolutionize the management of CRC by providing strategic approaches for its prevention, detection, and treatment. According to Underhill et al. (2016), the contribution of genetic research to hereditary GI cancers could have an impact on the implementation of precision medicine for the effective prevention of CRC and pancreatic cancer. This is actually changing the understanding of CRC, as reviewed by Hardiman (2018), and is bound to change the therapeutic and surveillance strategies. Trials such as those conducted by Imperiale et al. (2014) have demonstrated that multitarget stool DNA is efficacious in CRC screening. It has been validated to be very sensitive and non-invasive and can detect cancer, among other advanced precancerous lesions. Therefore, it may offer early detection and intervention, eliciting favorable clinical outcomes. Building on the recommendation made by Ramdzan et al. (2019), personalized CRC screening through genetic testing should primarily target individuals at moderate-to-high risk. Such a strategy will ensure that surveillance of high-risk populations is appropriately performed, with interventions taken accordingly, but at the same time reduce unnecessary surveillance among low-risk individuals. In addition, Jenkins et al. (2016) suggested that new estimates of personal cancer risk through the use of markers, specifically SNPs, could be enacted in a more informed and personalized fashion for recommendations related to cancer screening, thus increasing the effectiveness of colorectal screening. Such a critical role for genetic screening and personalized medicine in the management of CRC has been outlined above, with the prospect of offering an improved evaluation of the risk, early cancer detection, and interventional designs for individuals who are at a high risk of developing CRC.

3.1.10 Epigenetic Regulation of Colorectal Cancer

3.1.10.1 DNA Methylation Patterns in CRC and Their Roles in Gene Silencing

Among the molecular mechanisms responsible for regulating CRC, epigenetic phenomena, especially the DNA methylation patterns involved in the mechanism, participate in gene silencing and thus change the expression patterns of such genes. Other important epigenetic mechanisms in CRC include methylation of cytosine bases in CpG islands that inactivate genes and their related gene expression patterns. This may contribute to different aspects of CRC biology, particularly early carcinogenesis (Lao and Grady 2011). Another important epigenetic factor critically involved in colorectal carcinogenesis and progression is histone modification of gene activities and expression in CRC by changing chromatin structure or gene control (Qin et al. 2020). Moreover, there is increasing evidence that epigenetic silencing by promoter hypermethylation of microRNAs such as miR-137 also occurs in the early stages of colorectal carcinogenesis. Aberrant DNA methylation of the promoters of microRNA genes in CRC contributes to the dysregulated expression of target genes, which subsequently affects cancer development and progression (Mukohyama et al. 2017). MicroRNAs have been found to play a major role in CRC development and progression, mostly by regulating the signaling of oncogenes related to stem cells, epithelial-mesenchymal transition, and metastasis. MicroRNAs are epigenetic regulators that serve as key mediators in nearly all critical aspects of CRC biology (Guo et al. 2017). The Wnt/β-catenin pathway is the main signaling pathway in CRC and is a target of epigenetic regulation. Hence, the role of histone modifications and DNA methylation in the pathogenesis and progression of CRC is to affect the Wnt/β-catenin pathway. The epigenetic regulation of CRC pathogenesis occurs through DNA methylation patterns and histone modifications (Jiang et al. 2008). Consequently, elucidating these mechanisms is essential for developing targeted therapies that yield improved outcomes in CRC patients (Table 3.3).

TABLE 3.3
Genetic and Epigenetic Alterations in CRC

Gene/Epigenetic Mark	Alteration	Function	Clinical Significance
APC	Mutation (loss of function)	Tumor suppressor; Wnt pathway regulation	Initiation of adenoma formation
KRAS	Mutation (gain-of*function)	Oncogene; RAS/MAPK pathway activation	Resistance to EGFR inhibitors
BRAF	Mutation (gain-of-function)	Oncogene; RAS/MAPK pathway activation	Poor prognosis, potential for targeted therapy
TP53	Mutation (loss of function)	Tumor suppressor; DNA repair, apoptosis	Aggressive tumor behavior
MMR genes	Mutation (loss of function)	DNA mismatch repair	Microsatellite instability (MSI), better prognosis
CpG island methylation	Hypermethylation	Silencing of tumor suppressor genes	Tumor initiation and progression

3.1.10.2 Histone Modification and Chromatin Remodeling During CRC Development

Chromatin remodeling and histone modification are critical processes in gene regulation and chromatin structure. Several studies have reported on the implications of these two processes in CRC development. Fang et al. (2014) could demonstrate in their study how the oncoprotein BRAF mediates the CpG island methylator phenotype in CRC through an MAFG-dependent transcriptional mechanism. This study supports the role of chromatin remodeling in gene regulation. One study demonstrated that histone modifications play a crucial role in the regulation of CRC development by modulating chromatin structure and gene expression. It has been reported that histone modifications potentially influence CRC development through the modulation of chromatin structure and gene expression.

Additionally, Amirkhah et al. (2019) presented the histone marks and the chromatin remodeling in CRC and the complex interplay of epigenetic events, including alternative splicing and non-coding RNAs. Füllgrabe et al. (2011) proposed a comprehensive explanation of aberrant covalent histone modifications in cancer, particularly CRC, and the associated changes in histone-modifying enzymes, which lead to the pathogenesis of a large number of human cancers. A recent study by Jancewicz et al. (2019) detailed the involvement of the BRM ATPase subunit within the SWI/SNF chromatin-remodeling complex in the development of cancer.

This underlines the important role of this subunit in chromatin remodeling, DNA accessibility, and CRC expression. Taken together, these studies strongly support the importance of histone modifications and chromatin remodeling signaling in mediating gene regulation, chromatin structure, and epigenetic changes during CRC development.

3.1.10.3 Non-Coding RNAs (e.g., microRNAs and Long Non-Coding RNAs) Are Epigenetic Regulators of CRC

They also function as non-coding RNAs, including microRNAs and long non-coding RNAs (lncRNAs), which are key players in the epigenetic regulation of normal and diseased conditions. Xie et al. (2016) have compiled the importance of lncRNAs in CRC development and highlighted their critical roles in the process of CRC progression. Another study concluded that lncRNAs regulate the biological traits of residual CRC cells during chemoradiotherapy and may hence be used as biomarkers and regulatory molecules in this tumor type (Chen et al. 2018). Cheng et al. demonstrated that miRNAs are important regulators of CRC and potential biomarkers of colorectal carcinoma (Cheng et al. 2011). One of the drivers of the EMT process was shown by Zheng et al. (2023) to be Wnt5a, further confirming the role of M2-polarized TAMs in CRC metastasis. Together, these studies highlight the importance of ncRNAs, either miRNAs or lncRNAs, in the epigenetic control of CRC, thereby modifying numerous mechanisms of tumorigenesis and tumor development.

3.1.11 Therapeutic Targeting of Epigenetic Alterations in CRC Treatment Strategies

Therapeutic targeting of epigenetic alterations shows promise in CRC treatment. Zeuner et al. (2014) stressed the significance of comprehending the plasticity of colorectal CSCs to enhance preclinical models and develop improved cancer therapies for CRC. Cervena et al. (2020) drew attention to the potential use of methylation-based therapies in CRC, with particular emphasis on the reversibility of epigenetic changes for therapeutic purposes. For example, the treatment of CRC through epigenetic modifications could provide insights into the potential of drugs targeting epigenetic mechanisms and the development of anti-tumor strategies (Patnaik 2019). Rezapour et al. (2019) discussed epigenetics-based therapy for CRC and underscored the importance of targeting epigenetic modifications as a potential treatment approach. Han et al. (2022) identified the lncRNA/Wnt axis as a diagnostic, prognostic, and therapeutic target for flavonoids in CRC, highlighting the effectiveness of this approach for treating CRC. Bardhan and Liu (2013) discussed the current usage of epigenetic inhibitors decitabine and vorinostat as sensitizers in CRC and suggested their potential use in combination therapy to overcome drug resistance. Taken together, these studies demonstrate the significance of considering epigenetic alterations in treatment strategies for CRC, as well as the potential of epigenetic-based therapies and novel drugs for the effective management of cancer.

3.1.12 Interplay between Genetic, Epigenetic, and Environmental Factors

3.1.12.1 Impact of Lifestyle Factors (e.g., Diet, Smoking, Physical Activity) on CRC Risk

Genetic, epigenetic, and environmental factors contribute significantly to CRC susceptibility. Prominent determinants of gene-environment interactions encompass the risks associated with diet, smoking, and physical activity, and these factors demonstrate a strong correlation with CRC exposure (Hull 2021). Thus, this gene-environment interaction is a major determinant of an individual's risk of developing CRC, although its development is largely influenced by environmental factors (Rezapour et al. 2019). Whole precision medicine involves the genetic, socio-environmental, and clinical spheres for a full comprehension of health and its impact on the individual level of treatment and public health interventions (Velmovitsky et al. 2021). By contrast, precision public health refers to interventions tailored at the population level by harnessing Big Data and is one of the most important mechanisms for fighting health inequalities (Dolley 2018). For example, the observation of an effective intervention is difficult to translate into an effective intervention because of the limited mechanistic understanding. Public health interventions in most CRC screening programs target risk factor reduction and are underpinned by patient involvement (Golding et al. 2021). Precision public health involves the combination of precision medicine with public health and calls for the collection of unbiased data on both genetics and the environment to enable early-stage interventions. This demands an integrated approach to precision medicine and public health strategies to understand the interplay between genetic, epigenetic, and environmental factors in CRC, which could allow the development of targeted interventions to improve outcomes and reduce the burden of CRC (Figure 3.8).

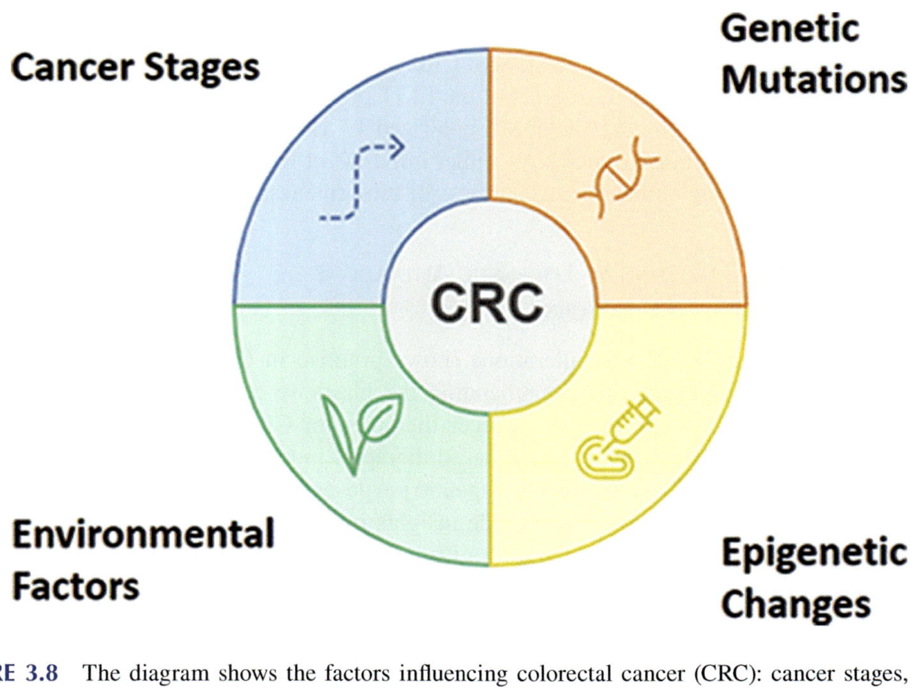

FIGURE 3.8 The diagram shows the factors influencing colorectal cancer (CRC): cancer stages, genetic mutations, epigenetic changes, and environmental factors.

3.1.13 Gene-Environment Interactions in CRC Susceptibility and Progression

Gene-environment interactions are key in determining the susceptibility and progression of CRC. Jia et al. (2013) and Ying et al. (2020) showed specific SNPs in KRAS and APC genes that might predispose individuals to CRC. These genetic variations interact with the environment owing to factors such as diet, lifestyle, and smoking, thereby increasing the susceptibility of an individual to CRC (Song et al. 2017). Notably, miRNA gene-environment interactions, including the use of alcohol and tobacco, are also important for CRC development (Wang et al. 2021a). Recently, an additional role of the gut microbiome in CRC development and treatment response has been confirmed. This has been emphasized as a complicated interaction between genetic, epigenetic, and environmental risk factors (Sánchez-Alcoholado et al. 2020). Extensive research has been conducted on the modulation of gut microbiota and the impact of their metabolites on the immune response and disease development, including CRC (Zhang et al. 2018b). The modulation and role of the gut microbiota in the immune response and development of diseases, such as CRC, have been widely investigated. The major epigenetic mechanisms involved in CRC development include DNA methylation and histone modifications. Understanding these interactions is imperative for the development of a precise medical approach to CRC treatment. This would enable the generation of personalized interventions based on genetic constitution and individual environmental exposures, culminating in targeted treatment or prevention strategies (Jeon et al. 2015). The addition of genetic information, lifestyle, and environmental exposure data provides a CRC risk assessment that is much more effective and would allow the same in terms of screening (Weinberg et al. 2014). All these genetic, epigenetic, and environmental inputs in CRC result in complex interactions that are critical in personalized medicine and public health interventions against this disease.

3.1.14 Epigenetic Mechanisms Mediating Environmental Influences on CRC Development

Epigenetic mechanisms are among the most significant factors through which environmental influences contribute to the development of CRC. For example, environmental factors have increasingly indicated that environmental factors, including diet and microbiota, can lead to CRC through various epigenetic pathways. They affect the patterns of gene expression, metabolic regulation, and immune responses, with consequences on cancer development. Epigenetic changes and histone modifications are important processes in CRC development (Song and Chan 2019). A complex interaction exists between genetic and environmental factors in CRC development. Although genetic mutations are significant, the pathogenesis of this disease is predominantly influenced by environmental factors. Research has indicated that genetic susceptibility and environmental exposure may determine an individual's predisposition to CRC (An et al. 2023). Genetic variations may interact with other risk factors, such as obesity, dietary habits, and infections, resulting in CRC risk and clinical results (Hwang et al. 2020). Epigenetic changes appear to be involved in the transition from normal cells to cancer progenitor cells, as observed in most cancer types. DNA methylation, histone modifications, and non-coding RNAs are critical factors in the initiation and progression of cancer due to their capacity to induce dysregulation of gene expression and cellular functions, ultimately resulting in CRC initiation and progression. In summary, CRC development is a complex process involving genetic predisposition, which is modulated by environmental influences through epigenetic mechanisms. Accordingly, the characterization of these interactions and their contribution to the pathogenesis of CRC will be relevant for the development of CRC prevention and treatment approaches (Sarkar et al. 2013; Sun et al. 2019).

3.1.15 Implications for Precision Medicine and Public Health Interventions

Understanding the interactions between genetic and epigenetic changes in CRC will be important not only to advance precision medicine and public health interventions but also to develop new prevention strategies. Generally known is the fact that epigenetic modifications encompass

DNA methylation and histone modifications, while non-coding RNAs play a crucial role in CRC development. Consequently, these factors can serve as biomarkers for diagnosis, prognosis, and treatment, potentially facilitating the development of personalized therapeutic strategies (Lao and Grady 2011). Therefore, knowledge of epigenetic mechanisms can be used to design public health interventions that target lifestyle factors associated with CRC risk, such as exercise, diet, and smoking. Understanding these risk factors and their epigenetic implications has reduced the incidence of CRC. Moreover, it offers a new target site for novel therapies and another opportunity for CRC treatment because epigenetic modifications are reversible (Ghosh et al. 2022). The detection of epigenetic outliers in the normal mucosa of CRC patients will improve screening, early detection, and prevention strategies. Screening is beneficial for identifying individuals at a higher risk of CRC by detecting unique patterns of epigenetic alterations. For instance, it can also reveal epigenetic dysregulation in CRC, leading to targeted therapy that directly addresses the molecular mechanisms underlying the disease (Harriss et al. 2009). The integration of these insights from epigenetic studies into precision medicine and public health interventions for CRC will improve patient care, preventive strategies, and innovation in the discovery of new therapeutics. In fact, the complexity of CRC at both the individual and population levels is exposed, allowing for personal treatment and prevention (Hong 2018).

3.1.16 Emerging Technologies and Future Directions

3.1.16.1 Application of Omics Technologies (e.g., Genomics, Transcriptomics, Epigenomics) in CRC Research

These include omics technologies—genomics, transcriptomics, and epigenomics—applied in CRC research to advance knowledge about the disease and open new areas of possible development in its diagnosis, treatment, and prevention. Technologies that have been discussed in relation to the molecular landscape of CRC include the genetics, transcriptomics, and epigenomics of the genome, as well as the mechanisms of its development and progression (Wu et al. 2022; Zhao et al. 2022). Genomic studies have elucidated key genetic mutations and alterations that play significant roles in CRC, thereby illuminating the molecular mechanisms underlying tumor initiation and progression (Zhao et al. 2022). The identification of specific gene expression profiles for different CRC subtypes from transcriptome data allows these differences to be considered as biomarkers for prognosis and prediction (Berg et al. 2017). These genomic studies, along with those based on genomic DNA methylation and histone marks, highlight epigenetic changes in developing cancers and provide insights into the development of epigenetic therapies for CRC (Wu et al. 2022). As mentioned above, there is huge potential for putting omics technologies into practice at the genomic, transcriptomic, and epigenomic levels to understand CRC heterogeneity and develop personalized strategies (Zhao et al. 2022). This includes genetic mutations, gene expression profiles, and epigenetic changes when classifying CRC patients into molecular subtypes that exhibit characteristic clinical responses to therapy (Liang et al. 2022). The use of omics methods to identify novel biomarkers and therapeutic targets for CRC treatment is advancing. For example, proteogenomics, in which genomic alterations are related to the proteome of CRC tumors, is a new frontier in targeted therapies (Vasaikar et al. 2019). Moreover, genome-wide studies of transcriptomes have shown that a significant share of non-coding RNAs regulates the development of CRC, most likely offering a wide spectrum of novel candidates in the fight against this disease (Yu et al. 2022). Overall, the integration of genomics, transcriptomics, and epigenomics has revolutionized our understanding of CRC at the molecular level, providing a detailed molecular signature of the disease. By utilizing omics technologies to comprehend the complex pathogenesis of CRC, identify biomarkers for early detection, and personalize treatment approaches, patient outcomes can be significantly improved.

3.1.17 Development of Novel Biomarkers for Early Detection and Prognosis of CRC

In this regard, extensive research has been conducted to identify novel strategies for detecting CRC biomarkers for early diagnosis and prognosis. These include circulating miRNAs, DNA methylation markers, long non-coding RNAs, and protein biomarkers. Within this framework, miR-24, miR-320a, and miR-423-5p are among the most studied and most promising miRNAs for the detection of CRC, especially in its early stages. In fact, they exhibited an astonishingly high sensitivity of 92% for stage I CRC, indicating the role of DNA methylation markers in the early detection of CRC (Oh et al. 2013). Additional studies have reported that ncRNA is also associated with the progression of CRC and has been correlated with poor prognosis in CRC cases, indicating its potential as a prognostic biomarker for CRC. Other novel protein biomarkers demonstrating the significant potential for the early detection and prognosis of CRC include aldolase A and TMP2 (Ma et al. 2016; Yamamoto et al. 2016). Recent advancements in proteomics and targeted proteomics have allowed researchers to identify new biomarkers, including TPM2, which can assist in the development of diagnostic procedures (Ma et al. 2016). Advanced proteomics and targeted proteomics of the present time have made it possible for new biomarkers, such as TPM2, to be useful for diagnostic procedure development (Xiao et al. 2020; Liu et al. 2023). Recently, the non-coding RNAs circFADS2 and LINC01594 were found to be associated with CRC and may serve as new prognostic biomarkers for the disease, thereby suggesting their pathogenesis, progression, and metastasis. For example, omics technologies supported by several researchers have facilitated research on biomarker signatures in CRC. Zhang et al. (2018a) found that global and targeted profiling of circulating miRNAs would discriminate between CRC patients and adenoma patients, hence indicating potential for this approach as a novel biomarker for early-stage CRC detection. In this regard, multi-omic studies should provide a full view of the molecular alterations in CRC and would lead to the identification of a robust set of biomarkers to enhance diagnostic specificity and improve patient management. Thus, the discovery of new biomarkers composed of microRNAs, DNA methylation markers, lncRNAs, and proteins, along with modern omics technologies, may hold promise for the development of novel biomarkers for the early detection and prognosis of CRC. The use of such biomarkers would altogether change the scenario of CRC diagnosis, prognosis, and treatment; it would provide non-invasive detection methods, personalized therapeutic strategies, and patient management with enhanced results (Table 3.4).

3.1.18 Targeted Therapies and Immunotherapeutic Approaches for CRC Treatment

Immunotherapy has now been on the frontline for most solid tumors other than a certain subgroup of CRC, owing to the success of the treatment in melanoma (Ganesh et al. 2019). Targeted therapy has improved the OS of CRC patients (Xie et al. 2020). Combined targeted therapies, such

TABLE 3.4
Emerging Biomarkers and Therapeutic Targets

Biomarker/Target	Role in CRC	Potential Clinical Application
Circulating tumor DNA (ctDNA)	Tumor-derived DNA in blood	Early detection, monitoring response to therapy
MicroRNAs	Small non-coding RNAs with gene regulatory function	Prognosis prediction, potential therapeutic targets
Immune checkpoints (PD-L1)	Regulate anti-tumor immune response	Predictive biomarker for immunotherapy response
Cancer stem cells	Tumor-initiating cells with self-renewal capacity	Targeting may prevent recurrence and metastasis

as anti-angiogenic agents and inhibitors of epidermal growth factor receptor or vascular endothelial growth factor receptor, have demonstrated efficacy for certain subtypes of CRC (Lichtenstern et al. 2020; Xie et al. 2020). Current strategies include enhancing the immunogenicity of the tumors, which elicits more pro-inflammatory cytokines, and the recruitment of tumor-infiltrating lymphocytes (TILs) that could be directed against the tumor by various therapeutic agents (Manz et al. 2021). Novel potential immunotherapeutic approaches for inflamed, transforming growth factor-β (TGF-β)-mediated mesenchymal tumors may encompass the targeting of myeloid-derived suppressor cell (MDSC) depletion and recently identified immunosuppressive chemokine signaling loops and pathways observed in various immunosuppressive tumor microenvironments (Gang et al. 2018). Autologous adoptive cell therapy with TILs is a powerful strategy for immunotherapy in CRC patients (Kitsou et al. 2020). The KEYNOTE 028 clinical trial has made a strong contribution to identifying important immunotherapeutic agents specifically for advanced chemotherapy-resistant colorectal malignancies, including those with mismatch repair deficiencies (Koustas et al. 2022). Immunotherapy has shown some clinical efficacy in the treatment of CRC; however, many patients are not satisfied with the limited benefits in terms of overall survival (Zheng et al. 2022). Thus, these two pathways represent immunotherapeutic targets, even in liver metastatic cancers, most commonly in the absence of mismatch repair deficiency (Zhou et al. 2018).

3.1.19 Integration of Multi-Omics Data for a Comprehensive Understanding of CRC Pathogenesis

The integration of multi-omics data and its relationship to CRC pathogenesis likely contributes to improved outcomes. This approach facilitates the examination of the extensive and complex molecular landscape of CRC and enables the identification of significant molecular alterations that drive its pathology (Hasin et al. 2017; Subramanian et al. 2020). Currently, multi-omics data integration promises a wider view of the biological information underlying disease and permits in-depth research into disease biology. This will enable a deeper understanding of tumor heterogeneity, regulatory pathways, and molecular interactions at a detailed level (Hasin et al. 2017). As such, this would pave the way for better accuracy in CRC classification and patient stratification (Zhong et al. 2022; Wang et al. 2023). Furthermore, it has been demonstrated that better predictions of patient clinical outcomes can be realized by combining data from many omics technologies compared with single omics data types (Wang et al. 2021c). It is possible to implement multi-omic approaches that give rise to more coherent biological signatures that can predict the behavior of the disease and gene-protein-metabolite responses to biotic and/or abiotic stimuli. Furthermore, multi-omics data can be embedded in protein-protein interaction networks that delineate the molecular traits of the disease and contribute to understanding the molecular mechanisms of CRC pathogenesis (Al-Harazi et al. 2021). This allows for an integrative approach toward the identification of cancer subnetwork markers and boosting the understanding of the disease at the molecular level (Al-Harazi et al. 2021). This integration of multi-omics data provides a powerful way of disentangling the complexity of CRC pathogenesis. This opens a way for the integration of disparate datasets with novel molecular insights to identify putative biomarkers and pathways that open avenues for personalized and precise approaches in the diagnosis, prognosis, and treatment of CRC.

3.1.20 Conclusion

Genetic, environmental, and epigenetic factors all influence the development of CRC. CRC development is based on a mixture of internal genetic predispositions and external agents in the body, such as diet, gut microbiota, lifestyle, and potential epigenetic mechanisms. CRC initiation and progression involve complex chain reactions involving DNA methylation, histone modifications, and other epigenetic changes. The complexity involved in the development of CRC allows one to appreciate the crossroads between genetic and environmental factors through epigenetic mechanisms. Since then, there has been a growing interest in using omics technologies, especially genomics, transcriptomics, and epigenomics, in CRC studies to show the molecular landscape of the disease.

The integration of multi-omics data will facilitate insights into the heterogeneity, regulatory pathways, and molecular interactions in CRC and further improve the accuracy of the classification and stratification of CRC patients. Integrating diverse omics datasets can identify novel biomarkers that may enhance the prediction of disease progression and inform the selection of appropriate targeted therapies to improve patient outcomes. Targeted therapies and immunotherapeutic interventions represent promising modalities for CRC treatment.

For example, the use of vascular endothelial growth factor inhibitors has been found to contribute to the overall survival among CRC patients with CRC, especially in those with certain subtypes of the disease. Immunotherapy, which includes checkpoint inhibitors and adoptive cell therapy, provides new potential for personalized and precise treatment of CRC. Integration of multi-omics data with protein-protein interaction networks aids in identifying potential targets for immunotherapy and in understanding the development of CRC at the molecular level. The integration of genetic, environmental, and epigenetic factors with omics technology-based targeted therapies and immunotherapeutic approaches promises to expand our understanding of this type of cancer and to facilitate effective treatment.

Analysis of multi-omics data also holds the possibility of developing comprehensive biomedical information through integrated analysis, fostering multidisciplinary research for new insights into CRC development, innovative treatment strategies, and improved patient outcomes in the battle against CRC disease.

Authors' Contribution: All authors contributed equally to the data collection, writing, and reviewing of the chapter.

REFERENCES

Adel Fahmideh, M., J. Schwartzbaum, P. Frumento and M. Feychting (2014). "Association between DNA repair gene polymorphisms and risk of glioma: a systematic review and meta-analysis." *Neuro-Oncology* **16**(6): 807–814.

Aihara, H., N. Kumar and C. C. Thompson (2014). "Diagnosis, surveillance, and treatment strategies for familial adenomatous polyposis: rationale and update." *European Journal of Gastroenterology & Hepatology* **26**(3): 255–262.

Al-Harazi, O., I. H. Kaya, A. El Allali and D. Colak (2021). "A network-based methodology to identify subnetwork markers for diagnosis and prognosis of colorectal cancer." *Frontiers in Genetics* **12**: 721949.

Alnabulsi, A., A. Agouni, S. Mitra, I. Garcia-Murillas, B. Carpenter, S. Bird and G. I. Murray (2012). "Cellular apoptosis susceptibility (chromosome segregation 1-like, CSE1L) gene is a key regulator of apoptosis, migration and invasion in colorectal cancer." *The Journal of Pathology* **228**(4): 471–481.

Amirkhah, R., H. Naderi-Meshkin, J. S. Shah, P. D. Dunne and U. Schmitz (2019). "The intricate interplay between epigenetic events, alternative splicing and noncoding RNA deregulation in colorectal cancer." *Cells* **8**(8): 929.

An, X., X. Lan, Z. Feng, X. Li and Q. Su (2023). "Histone modification: biomarkers and potential therapies in colorectal cancer." *Annals of Human Genetics* **87**(6): 274–284.

Arqués, O., I. Chicote, I. Puig, S. P. Tenbaum, G. Argilés, R. Dienstmann, N. C. Fernández, G. Caratù, J. Matito, D. Silberschmidt, J. Rodón, S. Landolfi, A. Prat, E. Espín, R. Charco, P. Nuciforo, A. Vivancos, W. Shao, J. Tabernero and H. G. Pálmer (2016). "Tankyrase inhibition blocks WNT/B-catenin pathway and reverts resistance to PI3K and AKT inhibitors in the treatment of colorectal cancer." *Clinical Cancer Research* **22**(3): 644–656.

Bahrami, A., F. Amerizadeh, S. Shahidsales, H. R. Sadeghnia, M. Maftouh and S. M. Hassanian (2017). "Therapeutic potential of targeting PI3K/AKT pathway in treatment of colorectal cancer: rational and progress." *Journal of Cellular Biochemistry* **119**(3): 2460–2469.

Bardhan, K. and K. Liu (2013). "Epigenetics and colorectal cancer pathogenesis." *Cancers* **5**(2): 676–713.

Barresi, V. (2021). "Colorectal cancer: from pathophysiology to novel therapeutic approaches." *Biomedicines* **9**(12): 1858.

Berg, K. C. G., P. W. Eide, I. A. Eilertsen, B. Johannessen, J. Bruun, S. A. Danielsen, M. Bjørnslett, L. A. Meza-Zepeda, M. Eknæs and G. E. Lind (2017). "Multi-omics of 34 colorectal cancer cell lines-a resource for biomedical studies." *Molecular Cancer* **16**: 1–16.

Bhardwaj, A., D. Prasad and S. Mukherjee (2024). "Role of toll-like receptor in the pathogenesis of oral cancer." *Cell Biochemistry and Biophysics* **82**(1): 91–105.

Bond, C. E. and V. L. J. Whitehall (2018). "How the BRAF V600E mutation defines a distinct subgroup of colorectal cancer: molecular and clinical implications." *Gastroenterology Research and Practice* **2018**: 9250757.

Byun, J., N. J.-y. Park, G. Yoon, M. K. Kang, H. J. Kim, S. Y. Park, J. S. Park, G. S. Choi, J. H. Baek and J. G. Kim (2023). "PIK3CA mutations as a prognostic factor in patients with residual rectal cancer after neoadjuvant chemoradiotherapy." *Anticancer Research* **43**(4): 1513–1520.

Cervena, K., A. Siskova, T. Buchler, P. Vodicka and V. Vymetalkova (2020). "Methylation-based therapies for colorectal cancer." *Cells* **9**(6): 1540.

Chen, E., X. Xu and T. Liu (2018). "Hereditary nonpolyposis colorectal cancer and cancer syndromes: recent basic and clinical discoveries." *Journal of Oncology* **2018**: 3979135.

Chen, Z., X. Cai, L. Chang, Y. Xia, L. Wang, Y. Hou, L. Li, D. Pan, F. Li and S. Liu (2018). "LINC00152 is a potential biomarker involved in the modulation of biological characteristics of residual colorectal cancer cells following chemoradiotherapy." *Oncology Letters* **15**(4): 4177–4184.

Cheng, H., L. Zhang, D. E. Cogdell, H. Zheng, A. J. Schetter, M. Nykter, C. C. Harris, K. Chen, S. R. Hamilton and W. Zhang (2011). "Circulating plasma MiR-141 is a novel biomarker for metastatic colon cancer and predicts poor prognosis." *PloS One* **6**(3): e17745.

Cheng, K. J., M. A. Alshawsh, E. H. Mejia Mohamed, S. Thavagnanam, A. Sinniah and Z. A. Ibrahim (2020). "HMGB1: an overview of its versatile roles in the pathogenesis of colorectal cancer." *Cellular Oncology* **43**: 177–193.

Cheriyamundath, S. and A. Ben-Ze'ev (2020). "WNT/B-catenin target genes in colon cancer metastasis: the special case of L1CAM." *Cancers* **12**(11): 3444.

Chiarella, P., P. Capone and R. Sisto (2023). "Contribution of genetic polymorphisms in human health." *International Journal of Environmental Research and Public Health* **20**(2): 912.

Cohen, S. A. and A. Leininger (2014). "The genetic basis of Lynch syndrome and its implications for clinical practice and risk management." *The Application of Clinical Genetics* **7**: 147–158.

Corcoran, R. B., H. Ebi, A. B. Turke, E. M. Coffee, M. Nishino, A. P. Cogdill, R. D. Brown, P. Della Pelle, D. Dias-Santagata and K. E. Hung (2012). "EGFR-mediated reactivation of MAPK signaling contributes to insensitivity of BRAF-mutant colorectal cancers to RAF inhibition with vemurafenib." *Cancer Discovery* **2**(3): 227–235.

Correa-Arzate, L., J. Portilla-Robertson, J. O. Ramírez-Jarquín, L. F. Jacinto-Alemán, C. P. Mejía-Velázquez, F. G. Villanueva-Sánchez and M. Rodríguez-Vázquez (2023). "LRP5, SLC6A3, and SOX10 expression in conventional ameloblastoma." *Genes* **14**(8): 1524.

de Kort, W. W. B., E. J. de Ruiter, W. E. Haakma, E. Driehuis, L. A. Devriese, R. J. J. van Es and S. M. Willems (2023). "p-mTOR, p-ERK and PTEN expression in tumor biopsies and organoids as predictive biomarkers for patients with HPV negative head and neck cancer." *Head and Neck Pathology* **17**(3): 697–707.

Derakhshani, A., D. Javadrashid, N. Hemmat, A. Dufour, A. G. Solimando, M. A. Shadbad, P. H. G. Duijf, O. Brunetti, N. Silvestris and B. Baradaran (2022). "Identification of common and distinct pathways in inflammatory bowel disease and colorectal cancer: a hypothesis based on weighted gene co-expression network analysis." *Frontiers in Genetics* **13**: 848646.

Dharmaraj, S., A. K. Swaroop, M. Esakkimuthukumar, P. Negi and S. Jubie (2024). "In-silico design and development of novel hydroxyurea lipid drug conjugates for breast cancer therapy targeting PI3K/AKT/mTOR pathway." *Drug Research* **74**(01): 32–41.

Do, M. D., T. H. Nguyen, K. Le, L.-T. Le, B. H. Nguyen, K. T. Le, T. P. T. Doan, C. Ho, H.-N. Nguyen, T. Tran and H. A. Vũ (2022). "Molecular characteristics of young-onset colorectal cancer in Vietnamese patients." *Asia-Pacific Journal of Clinical Oncology* **18**(6): 678–685.

Dolley, S. (2018). "Big data's role in precision public health." *Frontiers in Public Health* **6**: 297813.

Eilertsen, I. A., J. Langerud, B. Johannessen, A. Nesbakken, R. A. Lothe and A. Sveen (2023). "Tumor Heterogeneity and Prognostic Impact of the Aberrant Splicing Burden of Colorectal Cancers."

El-Tanani, M., H. Nsairat, A. A. Aljabali, A. Serrano-Aroca, V. Mishra, Y. Mishra, G. A. Naikoo, W. Alshaer and M. M. Tambuwala (2023). "Role of mammalian target of rapamycin (mTOR) signalling in oncogenesis." *Life Sciences* **323**: 121662.

Fang, M., J. Ou, L. Hutchinson and M. R. Green (2014). "The BRAF oncoprotein functions through the transcriptional repressor MAFG to mediate the CpG Island Methylator phenotype." *Molecular Cell* **55**(6): 904–915.

Fatima, Z., P. Sharma, B. Youssef and K. Krishnan (2021). "Medullary carcinoma of the colon: a histopathologic challenge." *Cureus* **13**(6): e15831.

Fearon, E. R. (2011). "Molecular genetics of colorectal cancer." *Annual Review of Pathology: Mechanisms of Disease* **6**: 479–507.

Fleming-de-Moraes, C. D., M. R. Rocha, J. W. Tessmann, W. M. de Araujo and J. A. Morgado-Diaz (2022). "Crosstalk between PI3K/Akt and Wnt/β-catenin pathways promote colorectal cancer progression regardless of mutational status." *Cancer Biology & Therapy* **23**(1): 1–13.

Fodde, R., R. Smits and H. Clevers (2001). "APC, signal transduction and genetic instability in colorectal cancer." *Nature Reviews Cancer* **1**(1): 55–67.

Fu, Y., L. Lin and L. Xia (2019). "MiR-107 function as a tumor suppressor gene in colorectal cancer by targeting transferrin receptor 1." *Cellular & Molecular Biology Letters* **24**: 1–11.

Füllgrabe, J., E. Kavanagh and B. Joseph (2011). "Histone onco-modifications." *Oncogene* **30**(31): 3391–3403.

Gajos-Michniewicz, A. and M. Czyz (2023). "WNT/β-catenin signaling in hepatocellular carcinoma: the aberrant activation, pathogenic roles, and therapeutic opportunities." *Genes & Diseases* **11**(2): 727–746.

Ganesh, K., Z. K. Stadler, A. Cercek, R. B. Mendelsohn, J. Shia, N. H. Segal and L. A. Diaz Jr (2019). "Immunotherapy in colorectal cancer: rationale, challenges and potential." *Nature Reviews Gastroenterology & Hepatology* **16**(6): 361–375.

Gang, W., J. J. Wang, R. Guan, S. Yan, F. Shi, J. Y. Zhang, Z. M. Li, J. Gao and X. L. Fu (2018). "Strategy to targeting the immune resistance and novel therapy in colorectal cancer." *Cancer Medicine* **7**(5): 1578–1603.

Garrett, A. A., S. Bai, S. Cascio, N. Gupta, D. Yang and R. J. Buckanovich (2024). "EGFL6 promotes endometrial cancer cell migration and proliferation." *Gynecologic Oncology* **185**: 75–82.

Gayet, J., X.-P. Zhou, A. Duval, S. Rolland, J.-M. Hoang, P. Cottu and R. Hamelin (2001). "Extensive characterization of genetic alterations in a series of human colorectal cancer cell lines." *Oncogene* **20**(36): 5025–5032.

Ghobashi, A. H., R. Lanzloth, C. A. Ladaika and H. M. O'Hagan (2023). "Single-cell profiling reveals the impact of genetic alterations on the differentiation of inflammation-induced colon tumors." *Cancers* 16(11): 2040. bioRxiv: 2023-2011.

Ghosh, J., B. M. Schultz, J. Chan, C. Wultsch, R. Singh, I. Shureiqi, S. Chow, A. Doymaz, S. Varriano and M. Driscoll (2022). "Epigenome-wide study identifies epigenetic outliers in normal mucosa of patients with colorectal cancer." *Cancer Prevention Research* **15**(11): 755–766.

Glasgow, S. C. and K. M. Hardiman (2022). "Sporadic and inherited colorectal cancer: how epidemiology and molecular biology guide screening and treatment." *The ASCRS Textbook of Colon and Rectal Surgery* **2022**: 397–412.

Golding, H., C. E. Webber and P. A. Groome (2021). "Factors contributing to time to diagnosis in symptomatic colorectal cancer: a scoping review." *European Journal of Cancer Care* **30**(3): e13397.

Gu, J., Y. Chen, H. Huang, L. Yin, Z. Xie and M. Q. Zhang (2014). "Gene module based regulator inference identifying miR-139 as a tumor suppressor in colorectal cancer." *Molecular BioSystems* **10**(12): 3249–3254.

Guo, X.-F., J. Wang, S.-J. Yu, J. Song, M.-Y. Ji, J.-X. Zhang, Z. Cao, J. Wang and W.-G. Dong (2013). "Meta-analysis of the ADH1B and ALDH2 polymorphisms and the risk of colorectal cancer in East Asians." *Internal Medicine* **52**(24): 2693–2699.

Guo, Y., Y. Bao and W. Yang (2017). "Regulatory miRNAs in colorectal carcinogenesis and metastasis." *International Journal of Molecular Sciences* **18**(4): 890.

Gupta, P., Y. Huang, P. K. Sahoo, J.-F. You, S.-F. Chiang, D. D. Onthoni, Y.-J. Chern, K.-Y. Chao, J.-M. Chiang and C.-Y. Yeh (2021). "Colon tissues classification and localization in whole slide images using deep learning." *Diagnostics* **11**(8): 1398.

Hagland, H. R., M. Berg, I. W. Jolma, A. Carlsen and K. Søreide (2013). "Molecular pathways and cellular metabolism in colorectal cancer." *Digestive Surgery*.

Half, E., D. Bercovich and P. Rozen (2009). "Familial adenomatous polyposis." *Orphanet Journal of Rare Diseases* **4**: 22.

Han, S., Y. Cao, T. Guo, Q. Lin and F. Luo (2022). "Targeting lncRNA/Wnt axis by flavonoids: a promising therapeutic approach for colorectal cancer." *Phytotherapy Research* **36**(11): 4024–4040.

Han, W., W. Xing, K. Wang, B. Wang and K. Bai (2022). "Alisol A attenuates malignant phenotypes of colorectal cancer cells by inactivating PI3K/Akt signaling." *Oncology Letters* **24**(2): 1–10.

Hardiman, K. M. (2018). "Update on sporadic colorectal cancer genetics." *Clinics in Colon and Rectal Surgery* **31**(03): 147–152.

Harras, M. E. A. H. F. and O. K. R. K. Gaweesh (2019). "Immunohistochemical study of KPNA2 expression in correlation to PCNA expression in colorectal carcinoma." *The Medical Journal of Cairo University* **87**(September): 2993–3001.

Harriss, D. J., G. Atkinson, A. Batterham, K. George, N. Tim Cable, T. Reilly, N. Haboubi, A. G. Renehan, L. E. Colorectal Cancer and G. Research (2009). "Lifestyle factors and colorectal cancer risk (2): a systematic review and meta-analysis of associations with leisure-time physical activity." *Colorectal Disease* **11**(7): 689–701.

Hartman, J., K. Edvardsson, K. Lindberg, C. Zhao, C. Williams, A. Ström and J.-A. Gustafsson (2009). "Tumor repressive functions of estrogen receptor β in SW480 colon cancer cells." *Cancer Research* **69**(15): 6100–6106.

Hasin, Y., M. Seldin and A. Lusis (2017). "Multi-omics approaches to disease." *Genome Biology* **18**: 1–15.

He, K., and Gan, W.-J. (2023). Wnt/β-catenin signaling pathway in the development and progression of colorectal cancer. *Cancer Management and Research* **15**: 435–448. https://doi.org/10.2147/cmar.s411168

Heinimann, K., H. Müller and Z. Dobbie (2000). "Microsatellite instability in colorectal cancer." *The New England Journal of Medicine* **342**(21): 1607–1608.

Hibi, K., H. Nakayama, M. Koike, Y. Kasai, K. Ito, S. Akiyama and A. Nakao (2002). "Colorectal cancers with both p16 and p14 methylation show invasive characteristics." *Japanese Journal of Cancer Research* **93**(8): 883–887.

Hirsch, D., J. Hardt, C. Sauer, K. Heselmeyer-Hadded, S. H. Witt, P. Kienle, T. Ried and T. Gaiser (2021). "Molecular characterization of ulcerative colitis-associated colorectal carcinomas." *Modern Pathology* **34**(6): 1153–1166.

Hong, S. N. (2018). "Genetic and epigenetic alterations of colorectal cancer." *Intestinal Research* **16**(3): 327.

Horst, D., L. Kriegl, J. Engel, T. Kirchner and A. Jung (2008). "CD133 expression is an independent prognostic marker for low survival in colorectal cancer." *British Journal of Cancer* **99**(8): 1285–1289.

Huang, L., Z. Guo, F. Wang and L. Fu (2021). "KRAS mutation: from undruggable to druggable in cancer." *Signal Transduction and Targeted Therapy* **6**(1): 386.

Huang, Y., S. Han, Y. Li, Y. Mao and Y. Xie (2007). "Different roles of MTHFR C677T and A1298C polymorphisms in colorectal adenoma and colorectal cancer: a meta-analysis." *Journal of Human Genetics* **52**(1): 73–85.

Hull, M. A. (2021). "Nutritional prevention of colorectal cancer." *Proceedings of the Nutrition Society* **80**(1): 59–64.

Hwang, K., J. h. Yoon, J. H. Lee and S. Lee (2021). "Recent advances in monoclonal antibody therapy for colorectal cancers." *Biomedicines* **9**(1): 39.

Hwang, S., C. G. Lee, M. Jo, C. O. Park, S.-Y. Gwon, S. Hwang, H. C. Yi, S.-Y. Lee, Y.-B. Eom and B. Karim (2020). "Enterotoxigenic Bacteroides fragilis infection exacerbates tumorigenesis in AOM/DSS mouse model." *International Journal of Medical Sciences* **17**(2): 145.

Hyngstrom, J. R., C.-Y. Hu, Y. Xing, Y. N. You, B. W. Feig, J. M. Skibber, M. A. Rodriguez-Bigas, J. N. Cormier and G. J. Chang (2012). "Clinicopathology and outcomes for mucinous and signet ring colorectal adenocarcinoma: analysis from the National Cancer Data Base." *Annals of Surgical Oncology* **19**: 2814–2821.

Ieda, T., H. Tazawa, H. Okabayashi, S. Yano, K. Shigeyasu, S. Kuroda, T. Ohara, K. Noma, H. Kishimoto and M. Nishizaki (2019). "Visualization of epithelial-mesenchymal transition in an inflammatory microenvironment–colorectal cancer network." *Scientific Reports* **9**(1): 16378.

Imperiale, T. F., D. F. Ransohoff, S. H. Itzkowitz, T. R. Levin, P. Lavin, G. P. Lidgard, D. A. Ahlquist and B. M. Berger (2014). "Multitarget stool DNA testing for colorectal-cancer screening." *New England Journal of Medicine* **370**(14): 1287–1297.

Inamura, K. (2018). "Colorectal cancers: an update on their molecular pathology." *Cancers* **10**(1): 26.

Jancewicz, I., J. A. Siedlecki, T. J. Sarnowski and E. Sarnowska (2019). "BRM: the core ATPase subunit of SWI/SNF chromatin-remodelling complex—a tumour suppressor or tumour-promoting factor?" *Epigenetics & Chromatin* **12**(1): 68.

Jayarathna, D., M. E. Rentería, P. F. Kho, J. Batra and N. S. Gandhi (2022). "Dehydroepiandrosterone sulfate and colorectal cancer risk: a mendelian randomization analysis." *Twin Research and Human Genetics* **25**(4–5): 180–186.

Jenkins, M. A., E. Makalic, J. G. Dowty, D. F. Schmidt, G. S. Dite, R. J. MacInnis, D. Ait Ouakrim, M. Clendenning, L. B. Flander and O. K. Stanesby (2016). "Quantifying the utility of single nucleotide polymorphisms to guide colorectal cancer screening." *Future Oncology* **12**(4): 503–513.

Jeon, Y. J., J. W. Kim, H. M. Park, J. O. Kim, H. G. Jang, J. Oh, S. G. Hwang, S. W. Kwon, D. Oh and N. K. Kim (2015). "Genetic variants in 3′-UTRs of methylenetetrahydrofolate reductase (MTHFR) predict colorectal cancer susceptibility in Koreans." *Scientific Reports* **5**(1): 11006.

Jia, W.-H., B. Zhang, K. Matsuo, A. Shin, Y.-B. Xiang, S. H. Jee, D.-H. Kim, Z. Ren, Q. Cai and J. Long (2013). "Genome-wide association analyses in East Asians identify new susceptibility loci for colorectal cancer." *Nature Genetics* **45**(2): 191–196.

Jiang, H., L. Dong, F. Gong, Y. Gu, H. Zhang, D. Fan and Z. Sun (2018). "Inflammatory genes are novel prognostic biomarkers for colorectal cancer." *International Journal of Molecular Medicine* **42**(1): 368–380.

Jiang, X., J. Tan, J. Li, S. Kivimäe, X. Yang, L. Zhuang, P. L. Lee, M. T. W. Chan, L. W. Stanton and E. T. Liu (2008). "DACT3 is an epigenetic regulator of Wnt/β-catenin signaling in colorectal cancer and is a therapeutic target of histone modifications." *Cancer Cell* **13**(6): 529–541.

Jonges, L. E., J. F. Nagelkerke, N. G. Ensink, E. A. van der Velde, R. A. E. M. Tollenaar, G. J. Fleuren, C. J. H. van de Velde, H. Morreau and P. J. K. Kuppen (2001). "Caspase-3 activity as a prognostic factor in colorectal carcinoma." *Laboratory Investigation* **81**(5): 681–688.

Kitsou, M., G. D. Ayiomamitis and A. Zaravinos (2020). "High expression of immune checkpoints is associated with the TIL load, mutation rate and patient survival in colorectal cancer." *International Journal of Oncology* **57**(1): 237–248.

Konishi, K., T. Fujii, N. Boku, S. Kato, I. Koba, A. Ohtsu, H. Tajiri, A. Ochiai and S. Yoshida (1999). "Clinicopathological differences between colonic and rectal carcinomas: are they based on the same mechanism of carcinogenesis?" *Gut* **45**(6): 818–821.

Koustas, E., E.-M. Trifylli, P. Sarantis, N. Papadopoulos, G. Aloizos, A. Tsagarakis, C. Damaskos, N. Garmpis, A. Garmpi and A. G. Papavassiliou (2022). "Implication of gut microbiome in immunotherapy for colorectal cancer." *World Journal of Gastrointestinal Oncology* **14**(9): 1665.

Lao, V. V. and W. M. Grady (2011). "Epigenetics and colorectal cancer." *Nature Reviews Gastroenterology & Hepatology* **8**(12): 686–700.

Lee, S., C. Lee, C. Y. Hwang, D. Kim, Y. Han, S. N. Hong, S.-H. Kim and K.-H. Cho (2020). "Network inference analysis identifies SETDB1 as a key regulator for reverting colorectal cancer cells into differentiated normal-like cells." *Molecular Cancer Research* **18**(1): 118–129.

Li, C., H.-S. Luo, H. Hu, Y. Tian and Y. Zhu (2018). "The influence of miR-366 on colorectal cancer cell proliferation and migration by targeting CTHRC1." *Biomedical Research (0970–938X)*. https://www.alliedacademies.org/articles/the-influence-of-mir366-on-colorectal-cancer-cell-proliferation-and-migration-by-targeting-cthrc1-7888.html

Liang, Y., H.-X. Sun, B. Ma and Q.-K. Meng (2022). "Identification of a genomic instability-related long noncoding RNA prognostic model in colorectal cancer based on bioinformatic analysis." *Disease Markers* **2022**(1): 4556585.

Lichtenstern, C. R., R. K. Ngu, S. Shalapour and M. Karin (2020). "Immunotherapy, inflammation and colorectal cancer." *Cells* **9**(3): 618.

Liu, B., A. Song, P. Gui, J. Wang, Y. Pan, C. Li, S. Li, Y. Zhang, T. Jiang and Y. Xu (2023). "Long noncoding RNA LINC01594 inhibits the CELF6-mediated splicing of oncogenic CD44 variants to promote colorectal cancer metastasis." *Cell Death & Disease* **14**(7): 427.

Liu, J., Q. Xiao, J. Xiao, C. Niu, Y. Li, X. Zhang, Z. Zhou, G. Shu and G. Yin (2022). "Wnt/β-catenin signalling: function, biological mechanisms, and therapeutic opportunities." *Signal Transduction and Targeted Therapy* **7**(1): 3.

Liu, S., J. Ren and P. ten Dijke (2021). "Targeting TGFβ signal transduction for cancer therapy." *Signal Transduction and Targeted Therapy* **6**(1): 8.

Lu, L., J. Dong, Y. Liu, Y. Qian, G. Zhang, W. Zhou, A. Zhao, G. Ji and H. Xu (2022). "New insights into natural products that target the gut microbiota: effects on the prevention and treatment of colorectal cancer." *Frontiers in Pharmacology* **13**: 964793.

Ma, Y., T. Xiao, Q. Xu, X. Shao and H. Wang (2016). "iTRAQ-based quantitative analysis of cancer-derived secretory proteome reveals TPM2 as a potential diagnostic biomarker of colorectal cancer." *Frontiers of Medicine* **10**: 278–285.

Malcomson, F. C., N. D. Willis, I. McCallum, N. Shivappa, M. D. Wirth, J. R. Hébert, B. Kocaadam, A. Özturan, S. B. Kelly, D. M. Bradburn, N. J. Belshaw, I. T. Johnson and J. C. Mathers (2020). "Relationships between dietary inflammatory index score and markers of inflammation and WNT signalling in the healthy colorectal mucosa." *Proceedings of the Nutrition Society* **79**(OCE1): E16.

Malki, A., R. A. ElRuz, I. Gupta, A. Allouch, S. Vranic and A. E. Al Moustafa (2020). "Molecular mechanisms of colon cancer progression and metastasis: recent insights and advancements." *International Journal of Molecular Sciences* **22**(1): 130.

Manz, S. M., M. Losa, R. Fritsch and M. Scharl (2021). "Efficacy and side effects of immune checkpoint inhibitors in the treatment of colorectal cancer." *Therapeutic Advances in Gastroenterology* **14**: 17562848211002018.

Mármol, I., C. Sánchez-de-Diego, A. Pradilla-Dieste, E. Cerrada and M. J. R. Yoldi (2017). "Colorectal carcinoma: a general overview and future perspectives in colorectal cancer." *International Journal of Molecular Sciences* **18**(1): 197.

Martínez, M. E., J. A. Baron, D. A. Lieberman, A. Schatzkin, E. Lanza, S. J. Winawer, A. G. Zauber, R. Jiang, D. J. Ahnen and J. H. Bond (2009). "A pooled analysis of advanced colorectal neoplasia diagnoses after colonoscopic polypectomy." *Gastroenterology* **136**(3): 832–841.

Martins, S., R. Amorim, S. C. Mota, L. Costa, F. Pardal, M. Rodrigues and A. Longatto-Filho (2015). "Ki-67 expression in CRC lymph node metastasis does not predict survival." *BioMed Research International* **2015**(1): 131685.

Marx, O., M. Mankarious and G. Yochum (2023). "Molecular genetics of early-onset colorectal cancer." *World Journal of Biological Chemistry* **14**(2): 13.

Mirza-Aghazadeh-Attari, M., A. Mohammadzadeh, S. Mostavafi, A. Mihanfar, S. Ghazizadeh, S. Sadighparvar, S. Gholamzadeh, M. Majidinia and B. Yousefi (2020). "Melatonin: an important anticancer agent in colorectal cancer." *Journal of Cellular Physiology* **235**(2): 804–817.

Mollanoori, H., Y. Ghelmani, B. Hassani and M. Dehghani (2024). "Integrated whole transcriptome profiling revealed a convoluted circular RNA-based competing endogenous RNAs regulatory network in colorectal cancer." *Scientific Reports* **14**(1): 91.

Mook, R. A., J. Wang, X.-R. Ren, H. Piao, H. K. Lyerly and W. Chen (2019). "Identification of novel triazole inhibitors of WNT/Β-catenin signaling based on the niclosamide chemotype." *Bioorganic & Medicinal Chemistry Letters* **29**(2): 317–321.

Morrison, H. A., B. Trusiano, A. J. Rowe and I. C. Allen (2023). "Negative regulatory NLRs mitigate inflammation via NF-κB pathway signaling in inflammatory bowel disease." *Biomedical Journal* **46**(5): 100616.

Mukohyama, J., Y. Shimono, H. Minami, Y. Kakeji and A. Suzuki (2017). "Roles of microRNAs and RNA-binding proteins in the regulation of colorectal cancer stem cells." *Cancers* **9**(10): 143.

Munteanu, I. and B. Mastalier (2014). "Genetics of colorectal cancer." *Journal of Medicine and Life* **7**(4): 507.

Oh, T., N. Kim, Y. Moon, M. S. Kim, B. D. Hoehn, C. H. Park, T. S. Kim, N. K. Kim, H. C. Chung and S. An (2013). "Genome-wide identification and validation of a novel methylation biomarker, SDC2, for blood-based detection of colorectal cancer." *The Journal of Molecular Diagnostics* **15**(4): 498–507.

Pan, B., T. Zhang, W. Yang, Y. Liu, Y. Chen, Z. Zhou, Y. Tang, J. Zeng, Y. Liu, C. Zhao and Y. Guo (2019). "SNX3 suppresses the migration and invasion of colorectal cancer cells by reversing epithelial-to-mesenchymal transition via the Β-catenin pathway." *Oncology Letters* **18**(5): 5332–5340.

Pancione, M., A. Remo and V. Colantuoni (2012). "Genetic and epigenetic events generate multiple pathways in colorectal cancer progression." *Pathology Research International* **2012**(1): 509348.

Pardini, B., A. Naccarati, J. Novotny, Z. Smerhovsky, L. Vodickova, V. Polakova, M. Hanova, J. Slyskova, E. Tulupova and R. Kumar (2008). "DNA repair genetic polymorphisms and risk of colorectal cancer in the Czech Republic." *Mutation Research/Fundamental and Molecular Mechanisms of Mutagenesis* **638**(1–2): 146–153.

Patnaik, S. (2019). "Drugs targeting epigenetic modifications and plausible therapeutic strategies against colorectal cancer." *Frontiers in Pharmacology* **10**: 448742.

Pätzold, L. A., D. Bērziņa, Z. Daneberga, J. Gardovskis and E. Miklaševičs (2017). "Detection of allelic variants of the and genes in colorectal cancer patients." *Balkan Journal of Medical Genetics* **20**(2): 83–87.

Pezzicoli, G., F. Ciciriello, V. Musci, S. Minei, A. Biasi, A. Ragno, P. Cafforio and M. Rizzo (2024). "Genomic profiling and molecular characterisation of metastatic urothelial carcinoma." *Medicina* **60**(4): 585.

Phipps, A. I., D. D. Buchanan, K. W. Makar, A. K. Win, J. A. Baron, N. M. Lindor, J. D. Potter and P. A. Newcomb (2013). "KRAS-mutation status in relation to colorectal cancer survival: the joint impact of correlated tumour markers." *British Journal of Cancer* **108**(8): 1757–1764.

Proietto, M., Crippa, M., Damiani, C., Pasquale, V., Sacco, E., Vanoni, M., & Gilardi, M. (2023). Tumor heterogeneity: Preclinical models, emerging technologies, and future applications. *Frontiers in Oncology*, 13. https://doi.org/10.3389/fonc.2023.1164535

Qin, J., B. Wen, Y. Liang, W. Yu and H. Li (2020). "Histone modifications and their role in colorectal cancer." *Pathology & Oncology Research* **26**: 2023–2033.

Ramdzan, A. R., M. A. Abd Rahim, A. M. Zaki, Z. Zaidun and A. M. Nawi (2019). "Diagnostic accuracy of FOBT and colorectal cancer genetic testing: a systematic review & meta-analysis." *Annals of Global Health* **85**(1): 70.

Rezapour, S., E. Hosseinzadeh, F. Marofi and A. Hassanzadeh (2019). "Epigenetic-based therapy for colorectal cancer: prospect and involved mechanisms." *Journal of Cellular Physiology* **234**(11): 19366–19383.

Sánchez-Alcoholado, L., B. Ramos-Molina, A. Otero, A. Laborda-Illanes, R. Ordóñez, J. A. Medina, J. Gómez-Millán and M. I. Queipo-Ortuño (2020). "The role of the gut microbiome in colorectal cancer development and therapy response." *Cancers* **12**(6): 1406.

Saplacan, R. M. M., P. A. Mircea, L. Balacescu and O. Balacescu (2015). "MicroRNAs as non-invasive screening biomarkers of colorectal cancer." *Clujul Medical* **88**(4): 453.

Sareeboot, T., P. Punyarit and S. Petmitr (2011). "DNA amplification on chromosome 13q31. 1 correlated with poor prognosis in colorectal cancer." *Clinical and Experimental Medicine* **11**: 97–103.

Sarkar, S., G. Horn, K. Moulton, A. Oza, S. Byler, S. Kokolus and M. Longacre (2013). "Cancer development, progression, and therapy: an epigenetic overview." *International Journal of Molecular Sciences* **14**(10): 21087–21113.

Scheurlen, K. M., A. T. Billeter, S. J. O'Brien and S. Galandiuk (2020). "Metabolic dysfunction and early-onset colorectal cancer–how macrophages build the bridge." *Cancer Medicine* **9**(18): 6679–6693.

Sepulveda, A. R., S. R. Hamilton, C. J. Allegra, W. W. Grody, A. M. Cushman-Vokoun, W. K. Funkhouser, S. Kopetz, C. H. Lieu, N. M. Lindor, B. D. Minsky, F. A. Monzon, D. J. Sargent, V. Singh, J. E. Willis, J. J. Clark, C. Colasacco, R. B. Rumble, R. Temple-Smolkin, C. B. Ventura and J. A. Nowak (2017). "Molecular biomarkers for the evaluation of colorectal cancer: guideline from the american society for clinical pathology, college of american pathologists, association for molecular pathology, and the american society of clinical oncology." *Journal of Clinical Oncology* **147**(3): 221–260.

Siegel, R. L., A. N. Giaquinto and A. Jemal (2024). "Cancer statistics, 2024." *CA: A Cancer Journal for Clinicians* **74**(1): 12–49.

Sinkala, M. (2023). "Mutational landscape of cancer-driver genes across human cancers." *Scientific Reports* **13**(1): 12742.

Song, M. and A. T. Chan (2019). "Environmental factors, gut microbiota, and colorectal cancer prevention." *Clinical Gastroenterology and Hepatology* **17**(2): 275–289.

Song, N., A. Shin, H. S. Jung, J. H. Oh and J. Kim (2017). "Effects of interactions between common genetic variants and smoking on colorectal cancer." *BMC Cancer* **17**: 1–9.

Su, T., L. Huang, N. Zhang, S. Peng, X. Li, G. Wei, E. Zhai, Z. Zeng and L. Xu (2020). "FGF14 functions as a tumor suppressor through inhibiting PI3K/AKT/mTOR pathway in colorectal cancer." *Journal of Cancer* **11**(4): 819.

Subramanian, I., S. Verma, S. Kumar, A. Jere and K. Anamika (2020). "Multi-omics data integration, interpretation, and its application." *Bioinformatics and Biology Insights* **14**: 1177932219899051.

Sun, D., Y. Chen and J.-Y. Fang (2019). "Influence of the microbiota on epigenetics in colorectal cancer." *National Science Review* **6**(6): 1138–1148.

Sung, C. O., J. W. Seo, K.-M. Kim, I.-G. Do, S. W. Kim and C.-K. Park (2008). "Clinical significance of signet-ring cells in colorectal mucinous adenocarcinoma." *Modern Pathology* **21**(12): 1533–1541.

Sung, H., J. Ferlay, R. L. Siegel, M. Laversanne, I. Soerjomataram, A. Jemal and F. Bray (2021). "Global cancer statistics 2020: GLOBOCAN estimates of incidence and mortality worldwide for 36 cancers in 185 countries." *Ca: A Cancer Journal for Clinicians* **71**(3): 209–249.

Tamura, K., S. Yokoyama, J. Ieda, K. Takifuji, T. Hotta, K. Matsuda, Y. Oku, T. Watanabe, T. Nasu and S. Kiriyama (2011). "Hollow spheroids beyond the invasive margin indicate the malignant potential of colorectal cancer." *BMJ Open* **1**(1): e000179.

Tomičić, M., M. Dawood and T. Efferth (2021). "Epigenetic alterations upstream and downstream of P53 signaling in colorectal carcinoma." *Cancers* **13**(16): 4072.

Tsuchiya, A., T. Nomizu, M. Onda, S. Ohki, H. Sato and R. Abe (1997). "Molecular genetic alteration and DNA ploidy in hereditary nonpolyposis colorectal cancer." *International Journal of Clinical Oncology* **2**: 224–229.

Tu, L., B. Yan and Z. Peng (2015). "Common genetic variants (rs4779584 and rs10318) at 15q13. 3 contributes to colorectal adenoma and colorectal cancer susceptibility: evidence based on 22 studies." *Molecular Genetics and Genomics* **290**: 901–912.

Underhill, M. L., K. A. Germansky and M. B. Yurgelun (2016). "Advances in hereditary colorectal and pancreatic cancers." *Clinical Therapeutics* **38**(7): 1600–1621.

Vasaikar, S., C. Huang, X. Wang, V. A. Petyuk, S. R. Savage, B. Wen, Y. Dou, Y. Zhang, Z. Shi and O. A. Arshad (2019). "Proteogenomic analysis of human colon cancer reveals new therapeutic opportunities." *Cell* **177**(4): 1035–1049.

Velmovitsky, P. E., T. Bevilacqua, P. Alencar, D. Cowan and P. P. Morita (2021). "Convergence of precision medicine and public health into precision public health: toward a big data perspective." *Frontiers in Public Health* **9**: 561873.

Vieira, I. R. S. and C. A. Conte-Junior (2024). "Nano-delivery systems for food bioactive compounds in cancer: prevention, therapy, and clinical applications." *Critical Reviews in Food Science and Nutrition* **64**(2): 381–406.

Vu, T. and P. K. Datta (2017). "Regulation of EMT in colorectal cancer: a culprit in metastasis." *Cancers* **9**(12): 171.

Wang, C., H. Zhang, Y. Liu, Y. Wang, H. Hu and G. Wang (2023). "Molecular subtyping in colorectal cancer: a bridge to personalized therapy." *Oncology Letters* **25**(6): 1–13.

Wang, D., Q. Liu, Y. Ren, Y. Zhang, X. Wang and B. Liu (2021a). "Association analysis of miRNA-related genetic polymorphisms in miR-143/145 and KRAS with colorectal cancer susceptibility and survival." *Bioscience Reports* **41**(4): BSR20204136.

Wang, J., F. Cui, X. Wang, Y. Xue, J. Chen, Y. Yu, H. Lu, M. Zhang, H. Tang and Z. Peng (2015). "Elevated kinesin family member 26B is a prognostic biomarker and a potential therapeutic target for colorectal cancer." *Journal of Experimental & Clinical Cancer Research* **34**: 1–10.

Wang, K.-S., G. Yu, C. Xu, X.-H. Meng, J. Zhou, C. Zheng, Z. Deng, L. Shang, R. Liu and S. Su (2021b). "Accurate diagnosis of colorectal cancer based on histopathology images using artificial intelligence." *BMC Medicine* **19**: 1–12.

Wang, T., W. Shao, Z. Huang, H. Tang, J. Zhang, Z. Ding and K. Huang (2021c). "MOGONET integrates multi-omics data using graph convolutional networks allowing patient classification and biomarker identification." *Nature Communications* **12**(1): 3445.

Wei, Q., J. Geng, Y. Chen, H. Lin, J. Wang, Z. Fang, F. Wang and Z. Zhang (2021). "Structure and function of DEAH-box helicase 32 and its role in cancer." *Oncology Letters* **21**(5): 1–8.

Weinberg, D. S., R. E. Myers, E. Keenan, K. Ruth, R. Sifri, B. Ziring, E. Ross and S. L. Manne (2014). "Genetic and environmental risk assessment and colorectal cancer screening in an average-risk population: a randomized trial." *Annals of Internal Medicine* **161**(8): 537–545.

Winandar, T., A. Raharjo, H. Wujoso, B. R. A. Sidharta and B. I. Bagus (2021). "Relationship between 25 (OH) D and colorectal cancer prevalence at Dr. Moewardi Hospital Surakarta-Indonesia: a cross-sectional study." *Open Access Macedonian Journal of Medical Sciences* **9**(B): 1591–1595.

Worthley, D. L., V. L. Whitehall, K. J. Spring and B. A. Leggett (2007). "Colorectal carcinogenesis: road maps to cancer." *World Journal of Gastroenterology: WJG* **13**(28): 3784.

Wu, Y.-L., T. Jiang, W. Huang, X.-Y. Wu, P.-J. Zhang and Y.-P. Tian (2022). "Genome-wide methylation profiling of early colorectal cancer using an illumina infinium methylation EPIC BeadChip." *World Journal of Gastrointestinal Oncology* **14**(4): 935.

Xiao, Y.-S., H.-Z. Tong, X.-H. Yuan, C.-H. Xiong, X.-Y. Xu and Y.-F. Zeng (2020). "CircFADS2: a potential prognostic biomarker of colorectal cancer." *Experimental Biology and Medicine* **245**(14): 1233–1241.

Xie, X., B. Tang, Y.-F. Xiao, R. Xie, B.-S. Li, H. Dong, J.-Y. Zhou and S.-M. Yang (2016). "Long non-coding RNAs in colorectal cancer." *Oncotarget* **7**(5): 5226.

Xie, Y.-H., Y.-X. Chen and J.-Y. Fang (2020). "Comprehensive review of targeted therapy for colorectal cancer." *Signal Transduction and Targeted Therapy* **5**(1): 22.

Xu, J., S. Hao, K. Han, W. Yang and H. Deng (2023a). "How is the AKT/mTOR pathway involved in cell migration and invasion?" *Biocell* **47**(4): 773–788.

Xu, Y., D. Cheng, L. Hu, X. Dong, L. Lv, C. Zhang and J. Zhou (2023b). "Single-cell sequencing analysis reveals the molecular mechanism of promotion of SCAP proliferation upon AZD2858 treatment." *Biocell* **47**(4): 825-836.

Yamamoto, T., M. Kudo, W.-X. Peng, H. Takata, H. Takakura, K. Teduka, T. Fujii, K. Mitamura, A. Taga and E. Uchida (2016). "Identification of aldolase A as a potential diagnostic biomarker for colorectal cancer based on proteomic analysis using formalin-fixed paraffin-embedded tissue." *Tumor Biology* **37**: 13595–13606.

Ying, R., Z. Wei, Y. Mei, S. Chen and L. Zhu (2020). "APC gene 3′ UTR SNPs and interactions with environmental factors are correlated with risk of colorectal cancer in Chinese Han population." *Bioscience Reports* **40**(3): BSR20192429.

Young, J. and J. R. Jass (2006). "The case for a genetic predisposition to serrated neoplasia in the colorectum: hypothesis and review of the literature." *Cancer Epidemiology Biomarkers & Prevention* **15**(10): 1778–1784.

Yu, L., X. Yang, W. Guan, D. Zhang, S. Ren, Y. Xing, D. An, J. Zhang, Y. Zhu and A. Zhu (2022). "Analysis of key genes for slow transit constipation based on RNA sequencing." *International Journal of General Medicine* **15**: 7569.

Zeuner, A., M. Todaro, G. Stassi and R. De Maria (2014). "Colorectal cancer stem cells: from the crypt to the clinic." *Cell Stem Cell* **15**(6): 692–705.

Zhang, C., L. Qu, S. Lian, L. Meng, L. Min, J. Liu, Q. Song, L. Shen and C. Shou (2019). "PRL-3 promotes ubiquitination and degradation of AURKA and colorectal cancer progression via dephosphorylation of FZR1." *Cancer Research* **79**(5): 928–940.

Zhang, J., G. S. Raju, D. W. Chang, S. H. Lin, Z. Chen and X. Wu (2018a). "Global and targeted circulating microRNA profiling of colorectal adenoma and colorectal cancer." *Cancer* **124**(4): 785–796.

Zhang, Y., X.-l. Wang, M. Zhou, C. Kang, H.-d. Lang, M.-t. Chen, S.-c. Hui, B. Wang and M.-t. Mi (2018b). "Crosstalk between gut microbiota and Sirtuin-3 in colonic inflammation and tumorigenesis." *Experimental & Molecular Medicine* **50**(4): 1–11.

Zhao, Q., F. Wang, Y.-X. Chen, S. Chen, Y.-C. Yao, Z.-L. Zeng, T.-J. Jiang, Y.-N. Wang, C.-Y. Wu and Y. Jing (2022). "Comprehensive profiling of 1015 patients' exomes reveals genomic-clinical associations in colorectal cancer." *Nature Communications* **13**(1): 2342.

Zheng, J., R. Dou, X. Zhang, B. Zhong, C. Fang, Q. Xu, Z. Di, S. Huang, Z. Lin and J. Song (2023). "LINC00543 promotes colorectal cancer metastasis by driving EMT and inducing the M2 polarization of tumor associated macrophages." *Journal of Translational Medicine* **21**(1): 153.

Zheng, Z., C. Bian, H. Wang, J. Su, L. Meng, Y. Xin and X. Jiang (2022). "Prediction of immunotherapy efficacy and immunomodulatory role of hypoxia in colorectal cancer." *Therapeutic Advances in Medical Oncology* **14**: 17588359221138383.

Zhong, C., T. Xie, L. Chen, X. Zhong, X. Li, X. Cai, K. Chen and S. Lan (2022). "Immune depletion of the methylated phenotype of colon cancer is closely related to resistance to immune checkpoint inhibitors." *Frontiers in Immunology* **13**: 983636.

Zhou, G., L. Noordam, D. Sprengers, M. Doukas, P. P. C. Boor, A. A. van Beek, R. Erkens, S. Mancham, D. Grünhagen and A. G. Menon (2018). "Blockade of LAG3 enhances responses of tumor-infiltrating T cells in mismatch repair-proficient liver metastases of colorectal cancer." *Oncoimmunology* **7**(7): e1448332.

4 Colorectal Cancer Diagnosis, Tools, and Techniques

Method of Testing Colorectal Cancer Using Cellular, Molecular, and Genetic Techniques

Khalid Umar Fakhri[1], *Deeksha Mehta*[2], *Prohit Jumnani*[2], *Jasna Shaji*[2], *Moshahid Alam Rizvi*[1], *and Anil Kumar Mavi*[3]

[1]Department of Biosciences, Jamia Millia Islamia, New Delhi, India

[2]Amity Institute of Virology and Immunology, Amity University, Noida, Uttar Pradesh, India

[3]Department of Botany & Life Sciences, Sri Aurobindo College, University of Delhi, Delhi, India

4.1 INTRODUCTION

Colorectal cancer (CRC) forms an important public health problem, especially in advanced countries, and ranks third among the most generally diagnosed cancers worldwide [1]. It is the alternate cause of cancer-related death in men and the third cause in women [2]. The development of CRC occurs by a process known as the adenocarcinoma sequence, which starts with the normal colonic epithelium of the large intestine's inner walls changing into tiny, benign tumors known as adenomatous polyps, which eventually progress to adenocarcinoma [3]. The process of diagnosing a patient is difficult and involves several interactions with the patient, the treating physician, and the healthcare system. The diagnostic accuracy of the CRC prediction models now in use is restricted since they rely only on the assessment of symptoms [4]. There are several CRC screening techniques available, which may be broadly categorized into two groups: more invasive imaging or endoscopic procedures and noninvasive stool or blood tests. There is not a single best CRC screening technique that is accepted globally. This leads to distinct strategies in different nations [5]. The three most widely accepted tests are the sigmoidoscopy, colonoscopy, and fecal occult blood test (FOBT). While there are many screening strategies, and their number is growing due to the potential of emerging technologies in molecular marker application, not all strategies meet the requirements for screening tests in population programs. A colonoscopy is often regarded as the most reliable method for diagnosing colorectal diseases. Even though its usage is unquestionably linked to a decrease in CRC-related fatalities, no findings from randomized controlled clinical studies have been found to support its efficacy. The standard and sensitive guaiac fecal occult blood test (gFOBT) techniques are the two primary varieties, with the former being more effective in detecting peroxidase activity. The tests are qualitative in nature, and as the findings are obtained using a reagent that alters the color of the stool sample, the lab worker must subjectively assess the results. Screening programs have helped make these techniques more accessible to the population, with FOBT, sigmoidoscopy, and colonoscopy representing the most used. Fecal immunochemical test (FIT) is emerging as the best test for screening due to its better acceptance among populations [6] (Figure 4.1). The most

DOI: 10.1201/9781003503842-4

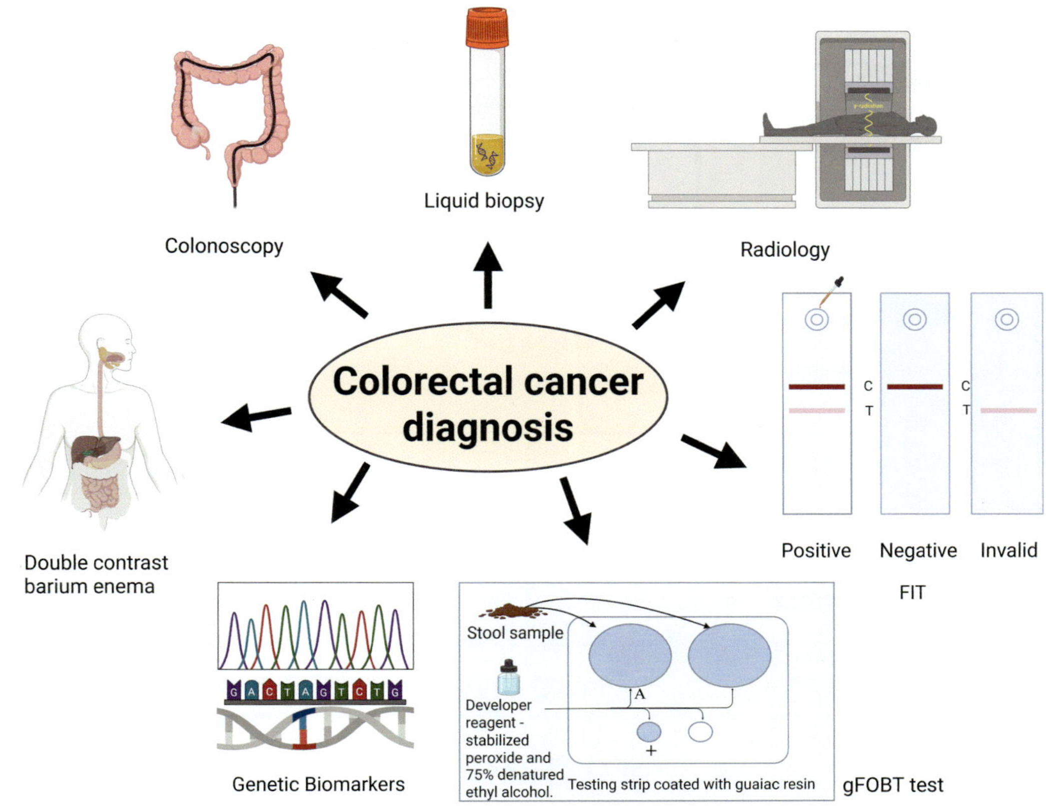

FIGURE 4.1 Diagnostic and screening methods in CRC. CRC, Colorectal cancer.

prevalent genetic instability is chromosome instability (CIN). In addition to the usual chromosomal abnormalities found in CIN tumors, certain oncogenes and tumor suppressor genes show a high accumulation of mutations [7].

There is a growing understanding that the molecular alterations in polyps and CRCs are what primarily cause the distinct behaviors of the tumors currently. Epigenetic changes are also often observed in polyps and CRCs, and they seem to work in tandem with gene mutations to drive the polyp-to-CRC sequence. APC, CTNNB1, KRAS, BRAF, SMAD4, TGFBR2, TP53, PIK3CA, ARID1A, SOX9, FAM123B, and ERBB2 are among the most frequently observed alterations in CRC. These alterations appear to promote colorectal tumorigenesis by interfering with the function of important signaling pathways, such as the TGF-β, EGF-MAPK, PI3K, and WNT-β-catenin signaling pathways, or by influencing genes that control central cell behaviors, such as DNA repair and proliferation [8].

4.2 GUAIAC FECAL OCCULT BLOOD TEST

The predominant noninvasive screening modality employed for the detection of CRC is the gFOBT [9]. This test is employed to identify occult (concealed) blood within stool samples. The presence of hemoglobin in stools may arise from various sources, including mass lesions such as CRC and adenomas, gastrointestinal (GI) infections, inflammation, vascular diseases, and other GI disorders [10]. In the screening process, individuals independently collect a stool specimen at home, which is subsequently submitted to a laboratory for analysis. The gFOBT functions by detecting hemoglobin molecules shed from anomalous epithelial cells, rendering it a fundamental and enduring technique

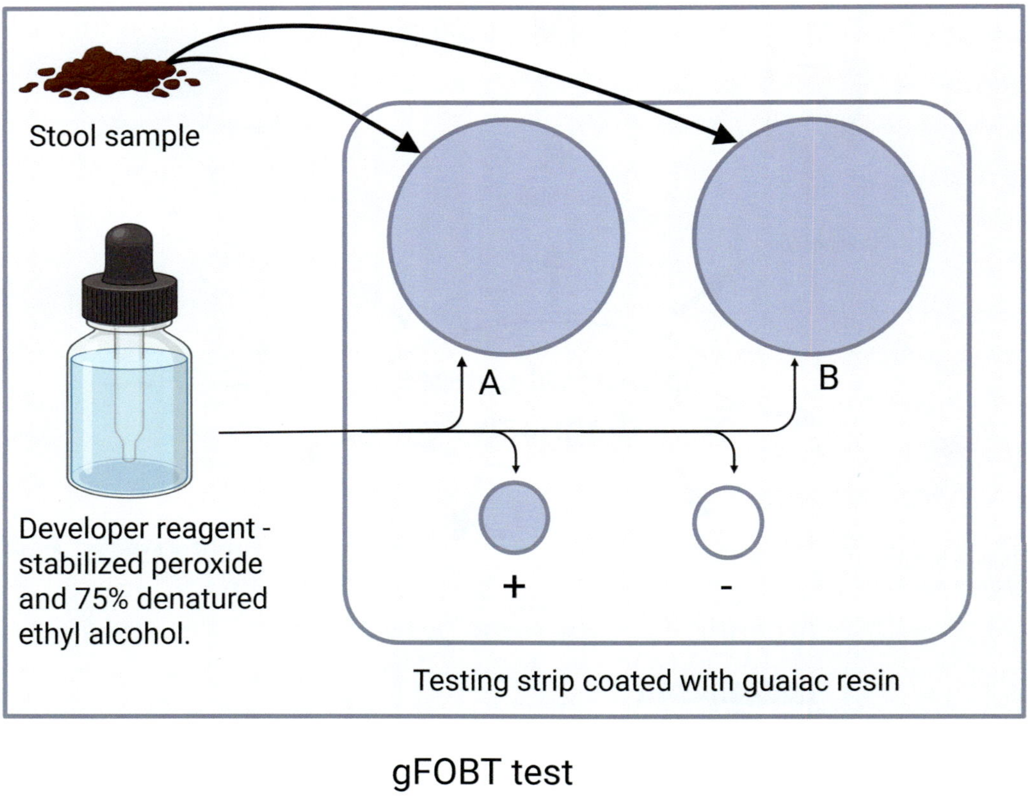

FIGURE 4.2 gFOBT test. The stool sample is placed on a guaiac resin-coated testing strip, and a developer reagent is applied to both the sample and a control area (+, –). Hemoglobin in the sample reacts with peroxide (H_2O_2), releasing oxygen (O_2), causing the guaiac to create a blue result, showing that hemoglobin is present in the stool sample. gFOBT, Guaiac fecal occult blood test.

for CRC screening. The mechanism involves the recognition of color changes during the oxidation of hemoglobin in fecal matter, facilitated by the pseudoperoxidase activity of hemoglobin in the presence of H_2O_2 (Figure 4.2) [11].

The gFOBT, relying on an oxidation reaction, is susceptible to false-positive and false-negative results influenced by dietary peroxidase. False positives, attributed to non-human hemoglobin (e.g., from red meat), drugs (e.g., topical iodine, aspirin), and peroxidase-rich foods, can skew results. Conversely, false negatives may result from vitamin C intake, intermittent bleeding, and hemoglobin degradation by colonic microbiota [10]. To mitigate interference, early recommendations advised dietary restriction to exclude peroxidase-containing substances before gFOBT, optimizing its diagnostic accuracy for CRC screening [12]. The gFOBT is advised annually for individuals aged 50 or older, increasing to thrice yearly at the age of 60 [13]. This screening is coupled with a 5-year interval flexible sigmoidoscopy. Positive gFOBT results prompt additional investigations, including colonoscopy, computed tomography (CT) colonography, or double-contrast barium enema (DCBE) studies [14,15].

The Hemoccult II test, commonly used for guaiac testing, demonstrates approximately 50% sensitivity due to its lower analytical sensitivity and intermittent bleeding in cancers. While the test boasts around 98% specificity, the prevalence of false positives is notable, and rehydration, though enhancing sensitivity, compromises specificity [16]. Five randomized controlled trials (RCTs) with 11- to 30-year follow-ups revealed that biennial gFOBT screening reduced CRC mortality by 22% at 20 years and 22% at 30 years. Annual gFOBT showed similar reductions in CRC mortality at

30 years, with no significant impact on all-cause mortality [17]. The sensitivity of gFOBT for cancer was 37%, doubling to 79% with high-sensitivity gFOBT but requiring a considerable increase in colonoscopies. Rehydration of gFOBT improved sensitivity (92.2%) but decreased specificity (90.4%) [18]. In the United States, Hemoccult Sensa exhibited a CRC sensitivity of 0.50–0.75 and a specificity of 0.96–0.98. Sensitivity for advanced adenomas was lower (0.06–0.17), but specificity remained high (0.96–0.99) [19]. Sensitivity for CRC and advanced neoplasia with gFOBT ranged between 50% and 75%, while specificity was 96%–98% and 96%–99%, respectively [20].

4.3 FECAL IMMUNOCHEMICAL TESTS

The most extensively used biomarker for CRC diagnosis is FIT. This biomarker serves as the foundation for several population screening initiatives aimed at lowering CRC mortality.

The basis for FITs is the response of monoclonal or polyclonal antibodies that are particular to albumin, other fecal blood components, or human fecal hemoglobin (f-Hb) (Figure 4.3). If they don't react with food or blood from the upper digestive system, they don't need to follow any dietary or pharmaceutical restrictions. Its capacity to identify and measure f-Hb concentrations 7–15 times lower than those found by chemical testing is by far its biggest benefit. This raises the sensitivity of advanced adenoma detection and CRC considerably. Additionally, FITs allow for precise and dependable automated analysis, which prevents subjective interpretation [21].

In terms of fecal Hb content per gram of feces, FITs might yield qualitative or quantitative results. The benefit is that the cut-off level in population-based screening may be chosen based on endoscopic and budgetary resources. On the market, there are several FIT brands to choose from.

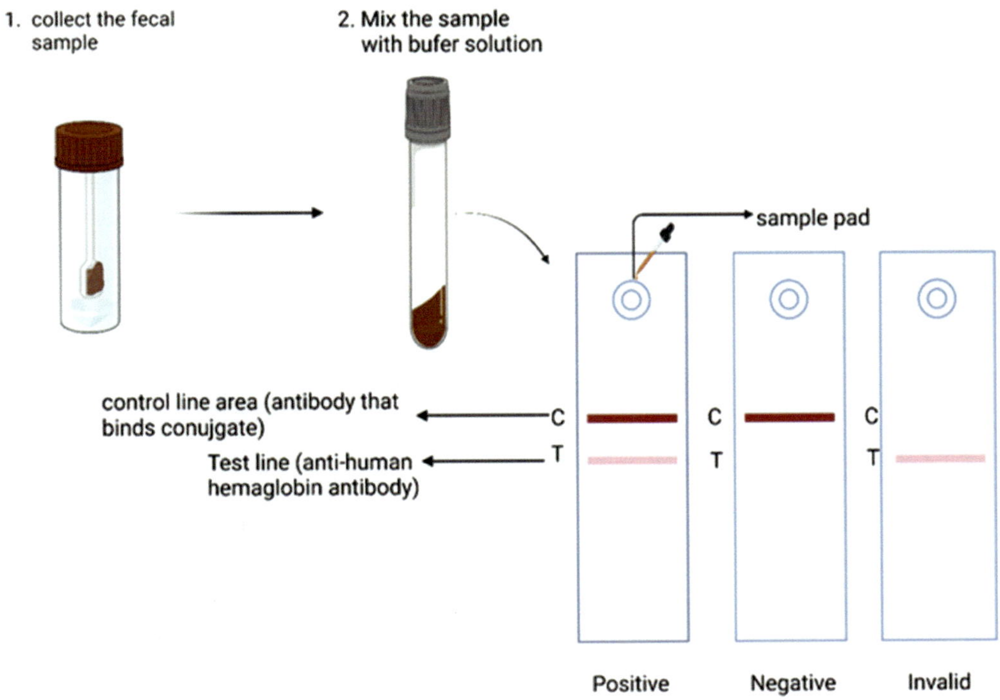

FIGURE 4.3 Illustration of FIT. **(1)** For analysis, collect the fecal sample of the patient with the help of the brush provided in the test kit. The sample taken should be freshly collected on the same day of the test. **(2)** Mix the brush with the buffer solution provided. With the help of the dropper, apply 2–3 drops of the sample into the sample pad of the test card in the kit and interpret the result within 5 minutes. FIT, Fecal immunochemical test.

TABLE 4.1
Comparison between Quantitative Fecal Immunochemical Test and Qualitative Fecal Immunochemical Test

Quantitative Fecal Immunochemical Test	Qualitative Fecal Immunochemical Test
They are based on immunoturbidimetry.	They are based on immunochromatography.
They have higher specificities but lower sensitivities.	They have higher sensitivities but lower specificities.
It is better suited for large-scale screening.	It performs poorly when it comes to large-scale screening.
It allows reproducibility and high-throughput testing.	It has lower reproducibility.
Detection rate is higher.	Detection rate is lower.
Positivity rate is lower compared to qualitative FIT [24].	Positivity rate is higher in qualitative FIT.
It is superior to qualitative FIT as it predicts advanced CRC during screening [25].	It is inferior to quantitative FIT in predicting CRC.

FIT, Fecal immunochemical testing; CRC, colorectal cancer.

These tests employ varying buffer volumes, analytical techniques, reporting units, and fecal material samples. Results are typically shown as the concentration of Hb in nanograms per milliliter of test buffer. It is not possible to compare the quantitative outcomes of several tests one to one because of these variations. In mass screening, different brands of FITs function differently, even when utilizing these standardized Hb concentrations. These variations hold true for both qualitative and quantitative outcomes [5].

Fecal immunochemical tests come in two very distinct formats: quantitative (which estimates the fecal hemoglobin content) and qualitative (where findings are either positive or negative) (Table 4.1).

4.3.1 Qualitative Fecal Immunochemical Tests

The available qualitative fecal immunochemical tests employ card- or tube-based collection devices; however, these tests are not well suited to large-scale organized population-based screening programs when compared to automated analytical approaches. The tests primarily use immunochromatographic test cassettes for analyses. One benefit of qualitative FIT is that, in general, their analysis is straightforward, dependable, and unaffected by food, negating the need for the dietary restrictions typically advised for the more sensitive guaiac-based FOBT. Integrated quality control systems are often included in qualitative FIT to oversee each test's execution [22].

4.3.2 Quantitative Fecal Immunochemical Tests

A distinct fecal sample collecting instrument is used for every quantitative fecal immunochemical assay. Many employ a stick or probe that, depending on the gadget, collects a different amount of feces. The collected fecal waste is subsequently put into a tube containing a buffer; the buffer's amount, content, and preservation properties vary across devices manufactured by various manufacturers. Unexpectedly, fecal hemoglobin concentrations are reported in a range of units, in contrast to other more recent fecal indicators such as calprotectin. These discrepancies render simple data comparisons between publications reporting findings of various quantitative fecal immunochemical assays difficult [22].

Despite the potential benefits of using the fecal hemoglobin concentrations found in individuals, this quantitative test is most frequently used as a qualitative test because results are expressed as either positive or negative without reference to the hemoglobin concentrations found in the subjects studied [23].

4.4 DOUBLE-CONTRAST BARIUM ENEMA

The DCBE is a radiographic method utilized for visualizing the colon and rectum, employing barium sulfate as a contrast medium followed by the introduction of air to enhance contrast. The use of DCBE may be considered in scenarios where colonoscopy poses risks (such as severe respiratory or heart failure), in cases of colon obstruction, or when colonoscopy is declined by the patient. In cases where colonic perforation is suspected, it is imperative to utilize a water-soluble contrast medium instead of barium [26]. DCBE exhibits a sensitivity ranging from 75% to 95% for the detection of CRC, with a specificity of 96% for identifying large adenomas. When supplemented with flexible sigmoidoscopy, DCBE enhances diagnostic efficacy while maintaining a low perforation rate of approximately 1 in 25,000, thus establishing its safety profile for colorectal screening [27].

However, the utilization of DCBE has been progressively diminishing over time, partly due to research indicating its inferiority compared to alternative visualization techniques such as computed tomographic (CT) colonography and flexible sigmoidoscopy. Negative DCBE results have been associated with a 2.8-fold increase in CRC incidence and nearly a 3.3-fold rise in the risk of advanced-stage CRC compared to negative colonoscopy outcomes. Moreover, DCBE exhibits miss rates of 10% for distal cancer (including the sigmoid colon) and 19% for proximal cancer, while its detection rate is lower than that of CT colonography (7.3% for CT colonography compared with 5.6% for barium enema) [28,29]. Nevertheless, a study has suggested that barium enema remains a viable option for detecting massive submucosal invasion in CRC, demonstrating an accuracy of 81% and the ability to predict depths of submucosal invasion ≥1.8 mm [30,31].

4.5 COLONOSCOPY

Colonoscopy is one of the various screening techniques being used in the world. It detects precancerous lesions and cancers at an early stage. Thus, it helps in preventing cancers. It has been a primary tool in screening for CRC. It is highly recommended for high-risk individuals and patients, but the numbers are still low. We will have a look into the reasons for low screening numbers. Colonoscopy is performed by skilled endoscopists. Colonoscopy provides an insight into what stage the cancer has developed. It can be concluded by examining the most advanced neoplasia found. Polyps are examined both inside the colon and, once resected, outside the colon to come to any conclusion regarding the cancer development. Alongside colonoscopy, polypectomy may be performed when the polyps are >10 mm. Once the polyps have been removed successfully, a follow-up colonoscopy needs to be performed every 10 years. Patients with a high probability of developing CRC are also considered eligible patients who are advised to undergo screening. They either have at least two first-degree relatives who were diagnosed with cancer (HNPCC or FAP) or just one first-degree relative with cancer under or over 60 years of age [32]. Colonoscopy alongside polypectomy is a remarkable means to prevent cancer from spreading further. Though colonoscopy seems to be an excellent screening tool, it has its risks. It is an invasive and costly procedure.

4.5.1 Procedure for Colonoscopy

Bowel preparations involve placing the patient on a restrictive diet, combined with the use of cathartic drugs, laxatives, and enemas. A restrictive diet may include foods that leave minimal fecal residue and clear liquids, which may include fruit juices without pulp, tea, and coffee without milk or cream, clear broth, or bouillon. No red-colored juices or food must be consumed, as they might look like blood. No beans or watermelon seeds must be consumed, as they may be confused with polyps by the endoscopists [33]. Cleansing of the colon is important as the time for the procedure is decreased, lowering the risks associated with colonoscopy. Patients with inflammatory bowel syndrome or ones with cardiac disease can be given isosmotic polyethylene glycol-containing electrolyte preparations. These formulations avoid the shifts in the electrolytes in the patient's body [34].

The patient fasted for several hours before the procedure. They lie on their sides. Medications are given intravenously to reduce the pain during the procedure. The use of sedatives is debatable as it makes the procedure resource-intensive. Deep sedation drugs include propofol; moderate sedation includes drugs such as midazolam, spasmolytics, and benzodiazepines, combined with analgesics such as opioids [35].

Air insufflation (AI) is used to distend the colonic folds and improve cecal intubation. The colonoscope is inserted into the rectum via the anus, and guided by the endoscopist, it swiftly and gently moves into the colon. It has a light and tiny camera attached at its tip, which allows colonic walls to be visualized on a monitor. The walls are checked for any abnormal growths or polyps; if any are found, a polypectomy is done. Recovering patients may experience some bloating or cramps, which are normal for successful procedures (Figure 4.4).

4.5.2 Types of Colonoscopies

4.5.2.1 CAP-Assisted Colonoscopy

A transparent rubber cap is attached to the tip of the colonoscope. It is a hollow cylinder of variable length (2–10 mm). Edges are rounded up and do not come into view. Hence, the cap does not cause any hindrance. It helps move the colonic folds, which in turn provides a better adenoma detection rate (ADR) by keeping a proper distance between the mucosa and the lens, which is present at the

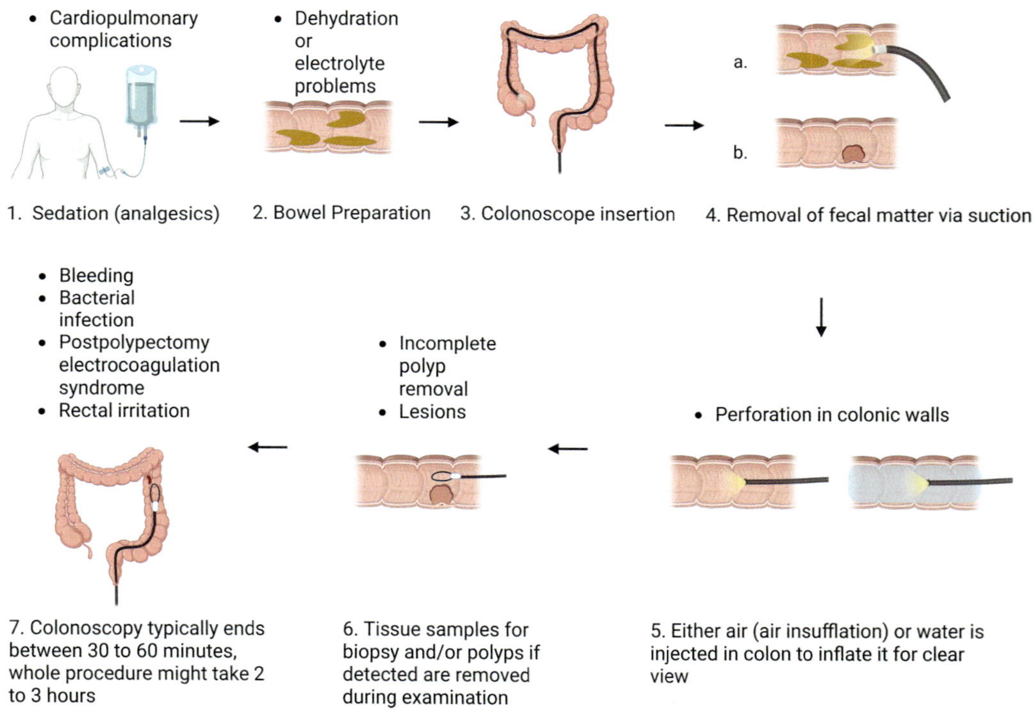

FIGURE 4.4 Steps and risks associated with colonoscopy. **(1)** Sedation in patients depends on their health and choice. The drawbacks are the cardiopulmonary complications generally observed in older patients. **(2)** Bowel preparation is necessary to significantly reduce the fecal matter in the colon. **(3)** A colonoscope is inserted through the anus while the patient lies on their side. **(4)** Fecal matter is removed through suction to expose the polyps. **(5)** AI or water enhances the visibility inside the colon. **(6)** Tissues and polyps detected are removed through polypectomy. They can also be analyzed to stage the cancer. **(7)** Examination time depends on the examiner and their skills. Polypectomy may cause bleeding afterward. Colonic perforation is a major downside of colonoscopy.

tip of the colonoscope. The cap aids in the better visualization of mucosal folds and detection of flat, small, depressed lesions, as well as reducing the number of loops in the colonoscope. Nowadays, the cap has a side hole to relieve any fecal matter or fluid that might get trapped within the cap [36].

4.5.2.2 Water-Assisted Colonoscopy

It is also known as water-aided colonoscopy (WAC). It consists of water immersion or water-infused (WI) colonoscopy and water exchange (WE) colonoscopy. The latter has more advantages but is a tedious procedure. Rather than AI, water is used to distend the colon for enhanced visualization. AI causes pain, and colonic perforation is one of the serious drawbacks. Patients are generally sedated to reduce the pain, but it also poses a risk. Costs and complications posed by sedation can be avoided to an extent by using WAC as both WI and WE, reducing the insertion pain shown in various RCT. WI and WE both include the use of water, but the difference lies in the removal of the infused water. WE prove advantageous as it cleans the colon, removes fecal matter, and thus, enhances the ADR, whereas WI has no significant role in cleaning the colon [37].

4.5.2.3 Retroscope (Retroflexion in Colonoscopy)

Retroflexion is a U-turn that the lens makes. The viewing lens, which is now facing backward, captures the image of the insertion tube as well. It is often used to observe the rectum walls and fully inspect the anus. The rectum being spacious makes the retroflexion easier, whereas, in the left colon or sigmoid colon, it is difficult. In the rectum, where the retroflexion is attempted, if any resistance is found, another site is chosen for the retroflexion. It is not attempted if the patient experiences pain or if the rectum has contracted because of radiotherapy or inflammatory bowel disease. It is avoided in these cases to reduce the instances of rectal perforations. TER is an instrument that provides a way to view both the proximal and distal ends of colonic folds at once. It is a mini endoscope, thin enough to fit inside any standard or pediatric colonoscope. The tip containing the camera comes out of the front end of the colonoscope and bends at 180°. The U-shape formed due to bending is just 7.5 mm; thus, it can be pulled back along with the colonoscope even through the thinnest bends of the sigmoid colon. In the ascending and sigmoid colon, an upper endoscope is used to minimize the chances of perforation during unwinding. If a polyp is seen in a forward view, the instrument can be moved forward until enough space is found to perform retroflexion. Followed by retroflexion comes the withdrawal of the colonoscope. During the withdrawal, the same polyp can be viewed from the other side, exposing the entire polyp. Entire resection may be performed during retroflexion or resection in forward view.

While the rectum and anus can be analyzed well, the detection rates of neoplasia in the proximal colon, especially the ascending colon, are also enhanced. Polypectomy can be performed on the lesions that were not visible earlier [38]. It is also considered one of the best methods for rectal colonoscopy (Figure 4.5).

4.5.3 Limitations

Not all the eligible candidates are screened even though many diagnostic tools and techniques are available. Thus, patients in advanced cancer stages become frequent. Lack of finances, healthcare insurances, costly procedures, repulsion to bowel testing, and apprehension toward invasive procedures are some of the obstacles. People's own beliefs, like homophobia, sexual trauma, dreading the pain, acquiring AIDS, past negative experiences with testing, and shame, have led to people avoiding the testing. Fear of being diagnosed with cancer and distrust in medical facilities lead patients to avoid the examination, and the patient's preference plays a major role when moving forward with the examination. Some serious complications are associated with colonoscopy, such as colonic wall perforation, hemorrhage, effect on the heart, and post-polypectomy syndrome [39]. Bleeding does occur during polypectomy, but it becomes worrisome if it happens post-polypectomy. Post-polypectomy bleeding depends on factors such as the number and size of polyps, histology, and anticoagulant used.

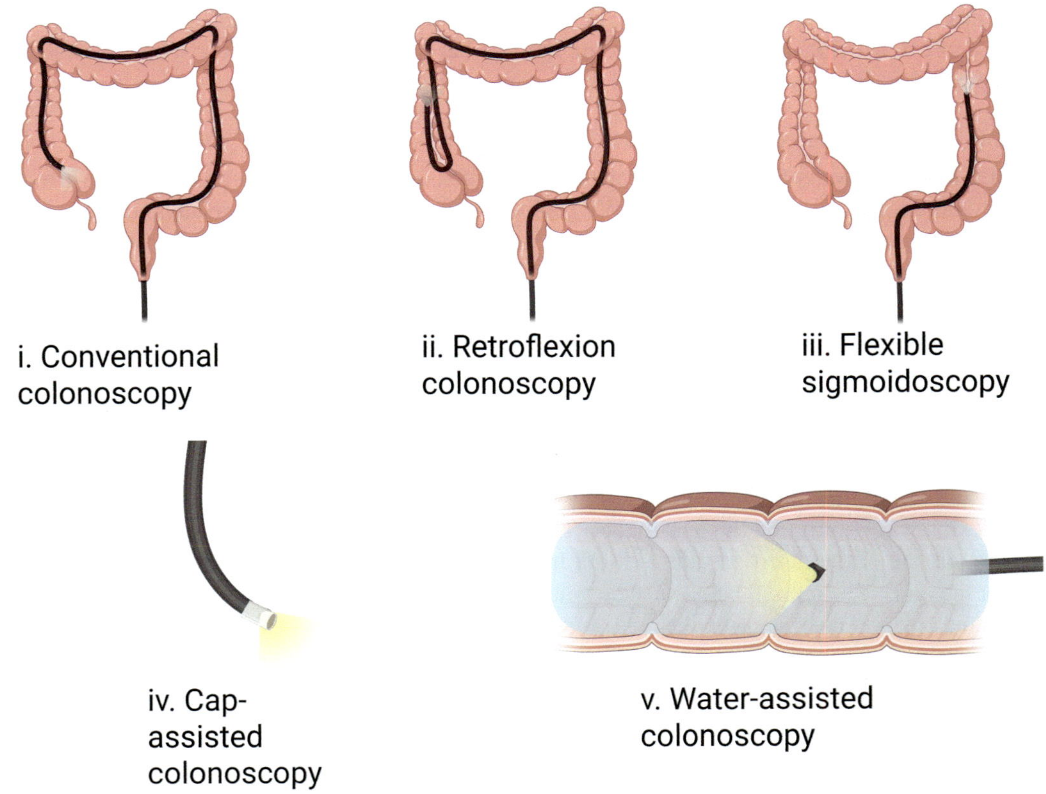

FIGURE 4.5 Types of colonoscopies. Caps in cap-assisted colonoscopy can be flexible, soft, or rigid. Depending on the use, they aid in reducing friction with colonic mucosa and facilitate easier movement. Either air insufflation (AI) or water is used to inflate the colon. Water-assisted colonoscopy can overcome the problem of colonic perforation and pain.

Bowel preparation, if not done correctly, may lead to ineffective examination, as the fecal matter stuck in the colonic walls decreases the visibility of polyps that may be present. The endoscopist's expertise is important as well due to the polyp miss rate or adenoma miss rate (ADR). Detection of precancerous lesions is difficult in the right colon. If ADR is high, it means insufficient detection capability, which will lead to patients coming in for frequent follow-ups and with advanced neoplasia conditions where the treatment becomes more difficult. Negative colonoscopy results do not always mean that all is well, as it could be an indication that the testing might have been ineffective. To avoid the inaccuracy of examination procedures, cecum intubation and its photodocumentation are important. Cecum intubation, when performed well, reduces the number of follow-ups required. The study showed that cecal intubation was easily achievable in males compared to female patients [32]. The withdrawal time of colonoscopy must be less than 6 minutes to minimize the risks attached to colonoscopy, such as colonic perforations. To avoid colonic perforations, endoscopists avoid applying force when any resistance is felt inside the colon while trying to insert the colonoscope. Audible timers may be used to keep track of withdrawal time and enhance inspection techniques.

4.6 FLEXIBLE SIGMOIDOSCOPY

4.6.1 Procedure

Before performing this technique, various steps and bowel preparation are involved. Healthcare providers must begin by describing the rationale for this screening, informing the patients about the bowel preparation method, and communicating during and after the procedure to reassure them and help them remain calm. Patients must be informed about the possible side effects of the sedatives being used, the discomfort caused by the sigmoidoscope, and the risks of the procedure. Bowel preparation sets up the colon so that it can be properly visualized, with no stool matter hindering the path or hiding the polyps. Liquids can be suctioned out during the examination, but the solid stool cannot be removed; if residue remains, the examiner might ask for the bowel preparation to be repeated. Sedatives may be given, which depends on the patient's requirements, but it does add to the risks. Iron-containing sedatives are stopped a week before the procedure, as tannic acid in the iron may lead to darkening and sticking of the stools to the colonic walls.

During insertion of the scope, its tip can be supported with the index finger and then directed toward the anus. Once entered, air may be used to inflate the colon for better movement of the scope inside the anal canal and the colon. But inflation causes pain and might damage the colonic walls if not controlled. The scope is well lubricated, and with inflation, it can move around in the tight turns of the colon. Experienced endoscopists with expertise take care of the withdrawal, rotation, and reinsertion of the scope and regulate the amount of air and water inside the colon by guiding the assistant [40]. Once the scope is fully inserted, the colon is inspected during retraction. The examiner looks for lesions and polyps. Retroflexion can be in flexible sigmoidoscopy, where the scope can retroflex upon withdrawal and give a view of the rectum. The rectum, sigmoid colon, and descending colon can be visualized. Biopsies can be done while screening. A full inspection with good bowel preparation may be completed in less than 10 minutes. Mucosal biopsies with an endoscopic forceps may be done. Polypectomy is not performed without a complete colon inspection.

Under the examination, the colon condition is judged. Normal glistening mucosa indicates a healthy colon. An unhealthy colon may show polyps and/or colonic diverticulum. Polyps include hyperplastic polyps (small, less than 5 mm in diameter), adenomatous polyps (cancerous or precancerous), and large polyps (greater than 2 cm in diameter). Colonic diverticulum is a small, sac-like pouch that shows an area of weakness in the muscular layer of the colon. Flexible sigmoidoscopy may be advised over colonoscopy due to less bleeding, reduced risk of perforation, the potential to avoid sedation or anesthesia, and cost-effectiveness. A limitation is that only a part of the colon is examined [19].

4.7 RADIOLOGY

The role of conventional CT scans and conventional MR imaging in assessing patients with colorectal tumors is now well established. Like other tumors, CRC has a staging-dependent prognosis, with a 40%–60% 5-year survival rate [41]. Although they have demonstrated encouraging results, recently developed methods like perfusion CT and magnetic resonance imaging (MRI) spectroscopy, which provide insights into tumor biology, are not yet validated for use in clinical practice [42]. The gold-standard method for the early diagnosis of CRC is still optical colonoscopy (OC) because of its great diagnostic performance. Low-density X-ray attenuation serves as the foundation for CT colonography image generation. Minimal radiation dosage is obtained from the diagnostic CT colonography pictures, which can be obtained with low X-ray energy. Usually, the CT acquisition is carried out twice: once while supine and once while prone (or vice versa) [43]. In addition to separating polyps that may be adhered to the intestinal walls from fluid and/or fecal remnants, this is done to maximize the distention of the various colonic segments based on gravitational compression by the surrounding abdominal structures [44].

The globally recognized standard for CRC staging is the American Joint Committee on Cancer's tumor node metastasis classification. Because MRI has a high degree of accuracy in defining localization, determining total extension, and establishing the link between the tumor and peritoneal reflection, it is the preferred modality for initial staging [45]. Tumor location and proximity to mesorectal fascia (MRF) and sphincters, tumor size, amount of extramural dissemination (T stage), peritoneal reflection, extramural vascular invasion, lymph nodes, and bone metastases are all evaluated during the MRI staging process for rectal cancer. Since most treating doctors prefer a structured synoptic MRI report, it is considered superior to the descriptive report form since it guarantees that all necessary characteristics are included and are clothed objectively. For tumor localization, the lowest part of the tumor's distance from the anal verge is measured. When the most caudal boundary of a rectal tumor is less than 5 cm from the anal verge, 5–10 cm from the anal verge, or more than 10 cm from the anal verge, correspondingly, the tumor is classified as low, intermediate, or high. MRF circumferentially binds the mesorectal fat that surrounds the rectum in the lowest portions of the rectum. However, the anterior section of the mesorectal fat is covered by the peritoneum in the upper regions up to a point known as the "anterior peritoneal reflection." The peritoneum eventually encircles the rectosigmoid by extending progressively posteriorly from the anterior peritoneal reflection upward. The peritoneal reflection, which attaches to the anterior side of the rectum and may be seen in both the axial and sagittal planes on T2-weighted imaging, appears as a thin (1–2 mm) hypointense line. The peritoneal reflection can be seen on sagittal images near the uterocervical angle in women and above the tip of the seminal vesicles in males [46].

Metastases are present at the time of initial diagnosis in 25% of patients with colon cancer and 18% of patients with rectal cancer. US, CT, MRI, and PET/CT are the imaging modalities most employed to identify CRC metastases. According to a meta-analysis, the combined sensitivity and specificity for identifying tumor invasion in T1/T2 vs T3/T4 patients using multidetector CT methods were 86% and 78%, respectively.

A tumor is only considered to have a poor prognosis if its extension extends five millimeters past the muscularis propria. When classifying tumors for colon cancers using the TNM classification method, the following criteria are used: T1/T2, T3a, and T3b tumors have favorable prognoses (>80% 3-year DFS), but T3c, T3d, and T4 tumors have much worse DFS. The following is the definition of T1 and T2 tumors in accordance with the TNM staging system: tumor T1, restricted to the mucosa; tumor T2, including the submucosa but not the muscularis propria. T1 tumors cause an intraluminal projection or focal mass on CT images; the layers of the gut wall are not visibly distorted. T2 tumors are tumors with greater asymmetrical thickening projecting intraluminally but with preservation of smooth muscle coat [47,48].

The assessment and staging of CRC are pivotal for treatment decisions and prognostic predictions. While conventional CT scans and MRI imaging play established roles, newer methods like perfusion CT and MRI spectroscopy await clinical validation. Despite the encouraging results of advanced imaging techniques, OC remains the gold standard for early CRC diagnosis due to its diagnostic performance [43]. For CRC staging, MRI stands out for its accuracy in defining tumor localization, assessing extension, and linking to peritoneal reflection. Specific MRI evaluations encompass tumor characteristics, lymph nodes, and metastases, guiding treatment planning. The TNM classification method aids in prognostication, with T3c, T3d, and T4 tumors showing worse outcomes compared to T1/T2. Imaging modalities like MRI and CT hold immense value in identifying CRC metastases and informing patient care decisions.

4.8 BIOPSY

4.8.1 Liquid Biopsy

It refers to pathological samples collected from non-solid resources, such as blood, urine, or other bodily fluids, such as cerebrospinal fluid and saliva. They are examined for the presence of circulating tumor cells (CTCs), tumor-educated platelets, cell-free nucleic acids released by cells, such as cell-free RNA, cell-free DNA, circulating tumor DNA (ctDNA), and exospores in peripheral blood [49]. They are a source of genetic and transcriptomic information on the cancer tissue. Screening them helps in early diagnosis and preventing the severity of the disease. The term liquid biopsy is given due to its similarity with conventional tumor tissue biopsy.

The tissue collected during colonoscopy when undergoing histopathological examination is used for staging and clinical assessment. However, clinical scientists are always on the lookout for a better prognostic individual biomarker. A huge focus has been on ctDNA and CTCs. The staging and whole tumor burden are approximated by the amount of ctDNA and CTCs present in the blood. However, the ctDNA has a short life span in circulation. Scientists have tried to find the correlation between ctDNA,

CTCs, and tumor staging with little success [50]. When a tumor is resected, there are different levels of resection. R0, R1, and R2 represent no tumor tissue left, microscopic tumor tissue left, and macroscopic tumor tissue left, respectively. In R0 resection, it is expected to see no ctDNA and CTCs in circulation. Some scientists in their studies have reported a negative correlation between the resections and the ctDNA and CTC levels [51]. Many researchers have reported a negative presence of CTCs and ctDNA after a successful removal of a tumor [52]. Therefore, researchers are trying to utilize them as a prognostic marker. The presence of CTCs and ctDNA may be used as a predictive marker in chemotherapy. EGFR (anti-epidermal growth factor receptor) therapy does not react well to Ras mutations in the cancer cells. Thus, by examining the levels of ctDNA and CTCs, doctors can assess whether a treatment is working effectively against the tumor. Follow-up after examinations and treatment is mandatory, as CRC can relapse after the resection and chemotherapy. Other methods of examination may be costly and labor-intensive. Performing liquid biopsy may become one of the superior methods in the future.

Though liquid biopsy has its advantages, there are limitations to it. The few number of CTCs present in blood makes their detection difficult. Even in metastatic conditions, CTCs have not been found in small amounts of blood. Its detection is possible in a higher amount of blood, which is impractical as it poses a threat to the patient. In this scenario, ctDNA has proved to be a better option [53]. For the blood to be examined, there are preparation steps. The probes used are blood that must reach the laboratory in cold conditions or be processed quickly, which makes it inconvenient. There is no standardization for these tests. Techniques such as next-generation sequencing (NGS) and digital polymerase chain reaction have given a solution to issues in identifying the genetic heterogeneity in populations of tumor cells, but they need to develop further for them to be used as standard methods of detection (Figure 4.6).

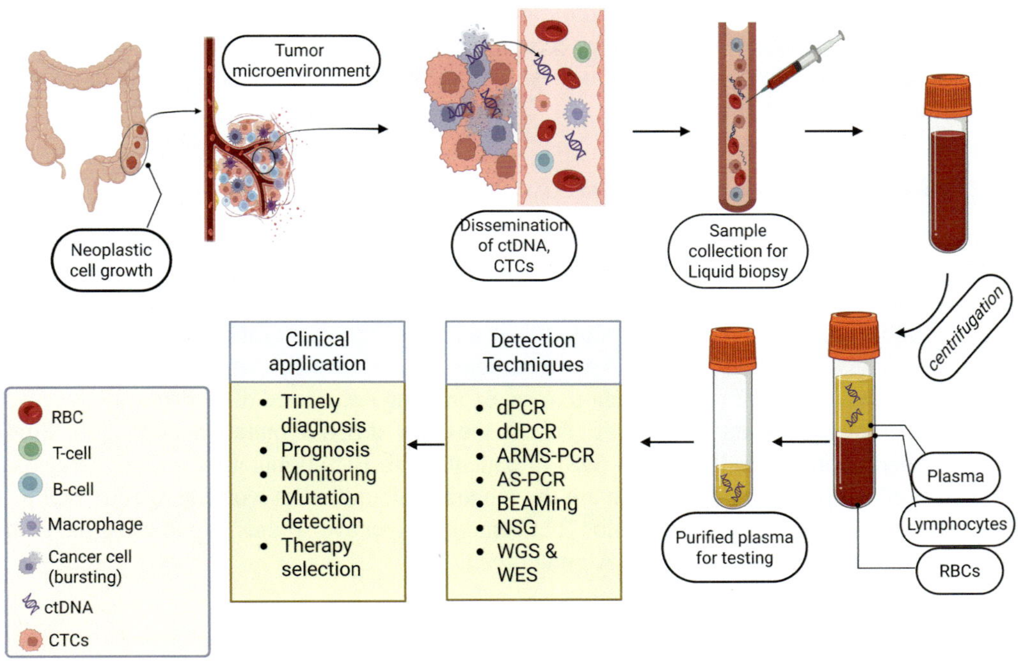

FIGURE 4.6 Liquid biopsy. Neoplastic cells proliferate in the intestine, resulting in the creation of tumors inside the tumor microenvironment. As the tumor grows, it releases circulating tumor DNA (ctDNA) and circulating tumor cells (CTCs) into the bloodstream. To capture these indicators, a patient's blood sample is collected for liquid biopsy. The blood is then centrifuged to separate the components, resulting in purified plasma ready for testing. dPCR, digital PCR; ddPCR, droplet digital PCR; ARMS-PCR, amplification refractory mutation system PCR; AS-PCR, allele-specific PCR; BEAMing, bead emulsification amplification and magnetics; NGS, next-generation sequencing; WES, whole exon sequencing; WGS, whole genome sequencing.

4.8.2 Tissue Biopsy

The most reliable method for identifying tumors is still tissue biopsy. However, a significant issue is that, because of recurrent damage and low patient compliance, it is challenging to track the progression of the illness by repeated biopsies. Furthermore, a single biopsy's limitations include poor sensitivity and accuracy, high procedural costs, the location of tissue excision, and it is typically not representative of a patient's heterogeneity or the constantly shifting whole cancer gene expression profile. Finding a less invasive technique to screen the high-risk population and identify CRC prevalence in asymptomatic individuals at an early and curable stage is, in fact, desperately needed [54].

It is not always possible to do regular biopsies. A single biopsy can provide a restricted picture of the tumor, but it is never able to capture its temporal heterogeneity. This conundrum has created a new diagnostic pathway [55].

4.9 GENETIC MARKERS – PROGNOSTIC AND PREDICTIVE MARKERS

4.9.1 KRAS Gene (v-Ki-ras2 Kirsten Rat Sarcoma)

It is a proto-oncogene located at 12p 12.1. With the approximate size of 38 kb, it encodes a 188-amino acid residue, 21.6 kDa, small guanosine triphosphate/guanosine diphosphate binding protein called p21ras. It has a role in the signal transduction cascade where epidermal growth factor, hepatocyte growth factor, and insulin-like growth factor bind to their receptors [56].

RAS/RAF/MAPK is a signal transduction pathway that mediates cellular responses regulating cell growth, survival, and differentiation. It is activated temporarily in a tightly regulated manner. Mutations in the KRAS gene cause its constitutive activation, which leads to unregulated growth and impaired differentiation. The most common KRAS mutations are G > A transitions and G > T transitions, point mutations. Drugs such as cetuximab or panitumumab work against this signaling pathway, but mutant KRAS is resistant to these drugs. Not all KRAS mutations lead to resistance to this immunotherapy [57].

4.9.2 BRAF Gene (V-raf Murine Sarcoma Viral Oncogene Homolog B1)

It is one of the RAF family proteins, which are activated by RAS small guanidine triphosphate. Mutations in BRAF cause 8%–12% of metastatic CRC (mCRC) [58]. Valine 600 is converted into glutamic acid (V600E) inside the kinase activation domain of BRAF protein. This causes 80% of the mutation in BRAF. BRAF V600E mutations lead to cancer cells proliferating without a functional RAS. Thus, those cells with KRAS mutations do not have BRAF V600E mutations [59]. BRAF mutations have been more evident in men, expressing sex-related differences.

Chemotherapy has proven to be less effective in the cases of BRAF mutation mCRC [60]. Studies have indicated poorer overall survival (OS) in patients with BRAF mutation mCRC. Drugs such as cetuximab and panitumumab, which are the monoclonal antibodies working against EGFR, do show efficacy in the RAS-WT population, but some retrospective studies hypothesize the conferring of resistance to these drugs by BRAF mutation [7].

4.9.3 APC Gene (Adenomatous Polyposis Coli)

Mutations in the APC gene are reported in FAP and sporadic CRC. It encodes a protein in the Wnt/APC/CTNNB1 (beta-catenin) signaling pathway. APC hypermethylation and polymorphisms (D1822V, E1317Q, and I1307K) are the causes of mutations. The I1307K variant poses a significant risk of CRC, whereas the E1317Q variant is associated with high adenoma risk [61]. APC mutations lead to c-myc and cyclin D1, unregulated transcription, enhancing the chances of cancer formation. This mutation also shows poor OS in advanced-stage cancer.

4.9.4 Epidermal Growth Factor Receptor

It is a 170-kDa transmembrane tyrosine kinase receptor. Only 1% of patients are affected by EGFR amplification resulting in mCRC. Mutations in EGFR can lead to increased copy number and altered protein expression. In CRC, high levels of ligands, such as AREG and EREG, can lead to excess activation of EGFR. Its constitutive activation and gene amplification, and their prognostic significance is still uncertain. EGFR kinase inhibitors work against this receptor. Cetuximab or panitumumab are the anti-EGFR therapies. However, the rare numbers of patients affected with EGFR-amplified mCRC have not helped much in developing more personalized therapies.

4.9.5 PI3K Gene (Phosphatidylinositide-3-Kinases)

Mutations in the PIK3CA gene encode for a catalytic subunit of PI3K called p110 α. It is the PI3K hyperactivation that is observed in inappropriate activation of PI3K/AKT signaling. Hyperactivation leads to growth stimulation, increased proliferation, and reduced apoptosis. Aberration in the PI3K gene can overlap with KRAS and BRAF mutations. Overlapping with BRAF mutation has shown severity in CRC [62].

It is a tumor suppressor gene located in the 10q23 region of chromosome 10. It encodes a 47 kDa protein with lipid and phosphatase activity.

4.9.6 PTEN (Phosphatase and TENsin Homolog deleted on chromosome 10)

It dephosphorylates PIP 3 to PIP 2, thus regulating the activity of PI3K. Even though PTEN mutations are less frequent in sporadic CRC, their inactivation can have a deleterious effect. Loss of PTEN is often observed in KRAS, BRAF, and PI3K mutations [63].

4.9.7 TP53

It is a tumor suppressor gene encoding protein that regulates cell division, growth, and apoptosis. It is a significant prognostic marker, as patients responding to chemotherapy show elevated expression of wild-type TP53 and indicate their higher chances of survival.

4.9.8 NDST4 (N-Deacetylase/N-Sulfotransferase)

It is a tumor suppressor gene located at 4q26. Its downregulation is detected in CRC. It is a part of heparan sulfate biosynthesis to form heparan sulfate proteoglycans (HSPGs). They make up the tissue structure. Impaired functioning of NDST4 results in deformed or negative modification of HS chains of HSPGs, leading to invasive tumor cells. NDST4 genetic loss is a good prognostic biomarker (Table 4.2).

4.9.9 Chromosome Instability

CIN in CRC refers to a characteristic of cancer cells where there are frequent changes or abnormalities in the number or structure of chromosomes. CIN is the most common genomic instability, which is observed in ~80% of the CRC and adenoma [81]. Tumors arising from this pathway are typically classified as non-hypermutated due to the limited number of mutations occurring in coding sequences. The mechanisms driving CIN often involve abnormalities in chromosomal segregation, including failures in sister chromatid separation, disruptions in cell senescence triggered by telomere shortening leading to genomic rearrangements, malfunctioning DNA damage-response systems, and the loss of heterozygosity in tumor suppressor genes [82]. Alongside the typical chromosomal abnormalities observed in CIN tumors, there is a distinct accumulation of mutations in

TABLE 4.2
Genetic Biomarkers for Colorectal Cancer and Techniques Used for Screening

Biomarker/Mutation	Chromosome Position	Techniques	References
KRAS and NRAS mutations	KARS: 12p12.1 NRAS: 1p13.2	• Allelic-specific PCR • Pyrosequencing • Multigene assays • NSG	• [64,65]
BRAF mutation	7p34	• PNA clamp real-time PCR. • F-SSCP	• [66,67]
MLH1 promoter methylation	3p21.3	• Immunohistochemistry staining • NSG	• [68,69]
Loss of MLH1 and PMS2	MLH1: 3p21.3 PMS2: 7q22.1	• Immunohistochemistry staining • NSG • PCR	• [70,71]
Loss of MSH2	2p22.2	• EPCAM deletions testing	• [72]
APC mutations (FAP and AFAP)	5q22.2	• PCR for gene polymorphism • NSG	• [73,74]
MUTYH mutations (MUTYH-associated polyposis)	1p34.1	• PCR/RE digestion-based techniques. • Denaturing high-pressure liquid chromatography. • Pyrosequencing. • Sanger sequencing as well as next-generation sequencing. • Allele-specific PCR.	• [75]
POLE and POLD1 mutations (polymerase proofreading-associated polyposis)	POLE: 12q24.33 Pold1: 19q13.33	• NSG	• [76,77]
STK11 mutations (Hamartomatous polyposis syndromes)	19p13.3	• PCR • HRM	• [78]
BMPR1A and SMAD4 mutations (Juvenile polyposis syndrome)	BMPR1A: 12p12.1 SMAD4: 18q21.2	• NSG • Sequence analysis for single nucleotide variation. • Detection of copy number variants.	• [79]
TP53 mutation	17p13.1	• Sanger sequencing	• [80]

PCR, Polymerization chain reaction; NSG, next-generation sequencing; PNA, peptide nucleic acid; F-SSCP, fluorescence – single-strand conformation polymorphism; EPCAM, epithelial cell adhesion molecule; HRM, high-resolution melting.

particular tumor suppressor genes (APC, TP53, SMAD4, SMAD2, DCC) and oncogenes (KRAS, CTNNB1, PIK3CA) [83]. These genes play a crucial role in CRC development. However, the CRC cells' capacity to exploit alternate cancer-specific molecular pathways, either via acquired mutations in other pivotal genes or through epigenetic modifications, adds considerable intricacy to the adenoma-carcinoma sequence. This complexity also presents hurdles in devising targeted therapeutic strategies. Particularly, the genes primarily associated with CIN are predominantly linked to the Wnt signaling pathway, underscoring its significance in the early stages of CRC pathogenesis. The Wnt signaling pathway is pivotal in CRC, with alterations observed in most tumors. Wnt protein binds to Frizzled receptors, causing β-catenin to accumulate in the nucleus and activate transcription. Dysregulation, such as APC inactivation or CTNNB1 activation, results in unchecked β-catenin nuclear accumulation, promoting cell proliferation. Additionally, the MAPK pathway, frequently dysfunctional in sporadic CRC, is activated by receptor tyrosine kinases, contributing further to CRC pathogenesis [84,85].

4.9.10 Microsatellite Instability

Microsatellites are short repetitive sequences, such as dinucleotide, trinucleotide, tetranucleotide, or pentanucleotide repeats, dispersed randomly throughout the human genome, numbering over 60,000. While exhibiting a high propensity for polymorphic variation, they ideally maintain a consistent profile within an individual's cells [12]. The mismatch repair (MMR) system, consisting of proteins like MSH2, MSH3, MSH6, MLH1, PMS1, and PMS2, corrects DNA mismatches, insertions/deletions, and damage. Recognition of mismatched DNA by MutSa (MSH2-MSH6) and MutSb (MSH2-MSH3) complexes recruits MutLa (MLH1-PMS2) for excision and repair, involving DNA polymerase III and DNA ligase [86]. Microsatellite instability (MSI) in tumor DNA involves different-sized repetitive sequences compared to those in germline DNA, stemming from a malfunctioning DNA MMR system. MSI can result from mutations in MMR genes like MLH1, MSH2, MSH6, or PMS2, or from hypermethylation-induced silencing of the MLH1 promoter. This condition often indicates Lynch syndrome, the most common hereditary colon cancer syndrome, characterized by MSI [85]. Accepted classifications for MSI in colon tumors include MSI-H (high level), MSI-L (low level), and MS-S (stable). Tumors classified as MSI-H display instability in over 30% of the microsatellite markers evaluated [14]. The laboratory test for MSI-H/dMMR (mismatch repair deficiency) CRCs has expanded beyond Lynch syndrome diagnosis, aiding in categorizing sporadic cases and guiding clinical management. Techniques such as immunohistochemical staining for MMR protein loss, PCR-based microsatellite testing (standard), and NGS for comprehensive genetic profiling alongside MSI status are utilized [86].

4.9.11 The CpG Island Methylator Phenotype Pathway

The CpG Island Methylator Phenotype (CIMP) Pathway is distinguished by the hypermethylation or hypomethylation of CpG islands in the promoter region of a gene or histone protein modification. In cancer cells, hypermethylation silences tumor suppressor genes (e.g., hMLH1, MGMT), whereas hypomethylation upregulates oncogenes (e.g., GRHL2) [87]. Based on methylation profiles, CIMP-high cancers can be classified as CIMP-H1 or CIMP-H2, with KRAS mutations significantly associated with CIMP-H2 but not CIMP-H1, indicating an age progression correlated with increased methylation, distinct molecular subtypes, gene body methylation in CIMP cancers, susceptibility of oncogenes to gene body methylation, and cluster-specific mutations in chromatin remodeling genes such as SWI/SNF and CHD families [88]. A study also found a substantial correlation between BARF gene mutation and CIMP [89]. In patients with stage I–II CRC, the methylation-based prognostic epigenetic classifier with FAT3, DUSP3, KAZN, and TLE4 methylation levels successfully predicts cancer recurrence following surgery. This increased predictive value outperforms existing clinicopathological and molecular subtyping approaches, particularly in patients with variable microsatellite status [90]. Within CIMP tumors, the demethylated phenotype shows increased ATP production efficiency in mitochondria, activation of the Myc target pathway linked to tumor immune suppression, downregulation of critical immune-related pathways, and a tumor microenvironment marked by reduced immunity and increased fibroblast infiltration, implying increased tumor aggressiveness and immune modulation [91]. Some long non-coding RNAs (LncRNAs) have been shown to act as oncogenes or tumor-suppressing genes in cancer. TSLC8, a tumor-suppressing LncRNA, was found to be hypermethylated in CRC [92], and LINC00152, which is involved in the activation of oncogenic pathways (PI3K/Akt, Ras, WNT, TP53, Notch, and ErbB), was found to be hypomethylated in CRC [93].

4.10 FUTURE PERSPECTIVES

Early diagnosis of CRC increases the chances of the patient's survival multi-fold. Hence, major focus has been on developing techniques that will enhance the prognosis efficiency, leading to better chances of decreasing mortality and gaining more knowledge about the disease. Multi-omics systems and biomarkers are two fields gaining the most attention in the current scenario as future diagnostic tools. Their accuracy as a diagnostic tool can aid in developing a personalized approach to patient care. Two types of multi-omics systems are being considered, which are lipidomics and microbiomics. Microbiomics focuses on identifying microbial metabolites and changes in gut microbiota and metabolites over the course of developing CRC. The contradiction is that the gut microbiota of an individual heavily depends upon their ethnicity, eating habits, geographical locations, and lifestyles. Hence, future research would have to include several factors to develop a protocol for an individual's therapy. Lipidomics is used to analyze the structure and function of lipid molecules. Changes in lipid metabolism can also be observed throughout CRC development. Glycero-, glycerophospho-, and sphingolipids are mainly being studied as they exhibit significant alterations during cancer. Lipidomics integrated with other omics, such as metabolomics, microbiomics, proteomics, etc., may prove to be an efficient tool to diagnose and develop therapy [94].

Biomarkers are being extensively studied. Various DNA, RNA, and protein-based biomarkers for cancer have been identified that can be observed through biological samples such as urine, stool, blood, serum, and tissues. However, they have failed when it comes to high levels of sensitivities and specificities. *PTV1* is an oncogenic lncRNA. It has been found to be associated with a high risk of tumor invasion, advancement of cancer, and poor OS rate of patients. It also plays its part in chemically resisting the drugs used to treat CRC [95]. Its ability to remain stable against endogenous RNases exhibits its capability to be utilized as an effective biomarker. Alterations in *PVT1* through CRC development are not yet determined. Hence, it must be studied carefully to be used as a diagnostic and prognostic biomarker.

The genes involved in hereditary colorectal cancer syndromes have been identified. They can be diagnosed and studied in the patients and their relatives. Extensive personalized care, preventive strategies, and therapeutics can be developed. However, the familial side of CRC does not sit well with the quite characterized colon cancer. Hence, epidemiological studies are required to be conducted and reviewed to identify the polymorphisms in CRC in different populations [96].

REFERENCES

1. N. Mehla, S. Verma, P. Bhan, and K. Vashist, "Recent advances in Monoclonal antibodies based formulation used in the management of CRC," *J. Pharm. Negat.*, vol. 13, no. 6, pp. 324–338, 2022, doi: 10.47750/pnr.2022.13.S06.047.
2. M. M. Center, A. Jemal, R. A. Smith, and E. Ward, "Worldwide variations in colorectal cancer," Diseases of the Colon and Rectum, vol. 53, no. 7. p. 1099, 2010. doi: 10.1007/DCR.0b013e3181d60a51.
3. A. S. Aghabozorgi et al., "SC," *Biochimie*, 2018, doi: 10.1016/j.biochi.2018.11.003.
4. P. Vega, F. Valentín, and J. Cubiella, "Colorectal cancer diagnosis: Pitfalls and opportunities," *World J. Gastroenterol. Oncol.*, vol. 7, no. 12, pp. 422–433, 2015, doi: 10.4251/wjgo.v7.i12.422.
5. E. H. Schreuders, E. J. Grobbee, M. C. W. Spaander, and E. J. Kuipers, "Advances in fecal tests for colorectal cancer screening," *Curr. Treat. Options Gastroenterol.*, vol. 14, no. 1, pp. 152–162, 2016, doi: 10.1007/s11938-016-0076-0.
6. G. Binefa, F. Rodríguez-Moranta, À. Teule, and M. Medina-Hayas, "Colorectal cancer: From prevention to personalized medicine," *World J. Gastroenterol.*, vol. 20, no. 22, pp. 6786–6808, 2014, doi: 10.3748/wjg.v20.i22.6786.
7. F. Di Nicolantonio et al., "Wild-type BRAF is required for response to panitumumab or cetuximab in metastatic colorectal cancer," *J. Clin. Oncol. Off. J. Am. Soc. Clin. Oncol.*, vol. 26, no. 35, pp. 5705–5712, Dec. 2008, doi: 10.1200/JCO.2008.18.0786.
8. B. T. Dickinson, J. Kisiel, D. A. Ahlquist, and W. M. Grady, "Molecular markers for colorectal cancer screening," *Gut*, vol. 64, no. 9, pp. 1485–1494, 2015, doi: 10.1136/gutjnl-2014-308075.

9. J. N. Li and S. Y. Yuan, "Fecal occult blood test in colorectal cancer screening," *J. Dig. Dis.*, vol. 20, no. 2, pp. 62–64, 2019, doi: 10.1111/1751-2980.12712.
10. S. M. Beg M, "Occult gastro – intestinal bleeding: Detection, interpretation, and evaluation," *J Indian Acad Clin. Med.*, vol. 2, no. 3, pp. 153–158, 2002.
11. S. M. Baik and R. A. Lee, "National cancer screening program for colorectal cancer in Korea," *Ann. Surg. Treat. Res.*, vol. 105, no. 6, pp. 333–340, 2023, doi: 10.4174/astr.2023.105.6.333.
12. J. O. Armitage, M. B. Kastan, J. H. Doroshow, and J. E. Tepper, *Abeloff's Clinical Oncology*. Elsevier, 2020.
13. A. Waldmann, P. Borchers, and A. Katalinic, "Temporal trends in age- and stage-specific incidence of colorectal adenocarcinomas in Germany," *BMC Cancer*, vol. 23, no. 1, pp. 1–15, 2023, doi: 10.1186/s12885-023-11660-1.
14. U. Bhalerao et al., "Diagnosis of Colorectal Cancer Using Molecular Techniques," in Colon Cancer Diagnosis and Therapy, Cham: Springer International Publishing, pp. 143–170, 2021. doi: 10.1007/978-3-030-64668-4_7.
15. L. Sali, G. Grazzini, and M. Mascalchi, "CT colonography: Role in FOBT-based screening programs for colorectal cancer," *Clin. J. Gastroenterol.*, vol. 10, no. 4, pp. 312–319, 2017, doi: 10.1007/s12328-017-0744-1.
16. Robert JC Steele & Paula McDonald, "Screening for colorectal cancer," in *Colorectal cancer: Diagnosis and clinical management*, John Wiley & Sons, 2014, pp. 29–30.
17. A. Qaseem et al., "Screening for colorectal cancer in asymptomatic average-risk adults: A guidance statement from the American college of physicians (Version 2)," *Ann. Intern. Med.*, vol. 176, no. 8, pp. 1092–1100, 2023, doi: 10.7326/M23-0779.
18. G. P. Young et al., "Advances in fecal occult blood tests: The FIT revolution," *Dig. Dis. Sci.*, vol. 60, no. 3, pp. 609–622, 2015, doi: 10.1007/s10620-014-3445-3.
19. S. Jain, J. Maque, A. Galoosian, A. Osuna-Garcia, and F. P. May, "Optimal strategies for colorectal cancer screening," *Curr. Treat. Options Oncol.*, vol. 23, no. 4, pp. 474–493, 2022, doi: 10.1007/s11864-022-00962-4.
20. J. F. Toth, M. Trivedi, and S. Gupta, "Screening for colorectal cancer: The role of clinical laboratories," *Clin. Chem.*, vol. 70, no. 1, pp. 150–164, 2024, doi: 10.1093/clinchem/hvad198.
21. N. Pin-Vieito, M. Puga, D. Fernández-De-Castro, and J. Cubiella, "Faecal immunochemical test outside colorectal cancer screening?," *World J. Gastroenterol.*, vol. 27, no. 38, pp. 6415–6429, 2021, doi: 10.3748/wjg.v27.i38.6415.
22. C. G. Fraser, J. E. Allison, S. P. Halloran, and G. P. Young, "A proposal to standardize reporting units for fecal immunochemical tests for hemoglobin," *J. Natl. Cancer Inst.*, vol. 104, no. 11, pp. 810–814, 2012, doi: 10.1093/jnci/djs190.
23. R. M. Ayling, "New faecal tests in gastroenterology," *Ann. Clin. Biochem.*, vol. 49, no. 1, pp. 44–54, 2012, doi: 10.1258/acb.2011.011150.
24. M. J. Park, K. S. Choi, Y. K. Lee, J. K. Jun, and H. Y. Lee, "A comparison of qualitative and quantitative fecal immunochemical tests in the Korean national colorectal cancer screening program," *Scand. J. Gastroenterol.*, vol. 47, no. 4, pp. 461–466, 2012, doi: 10.3109/00365521.2012.668930.
25. Y. Huang, Q. Li, W. Ge, S. Cai, S. Zhang, and S. Zheng, "Predictive power of quantitative and qualitative fecal immunochemical tests for hemoglobin in population screening for colorectal neoplasm," *Eur. J. Cancer Prev.*, vol. 23, no. 1, pp. 27–34, 2014, doi: 10.1097/CEJ.0b013e328364f229.
26. Chahira Mazouzi et al., "Colorectal cancer screening 'literature review,'" *EPH - Int. J. Med. Heal. Sci.*, vol. 10, no. 1, pp. 1–16, Feb. 2024, doi: 10.53555/eijmhs.v10i1.191.
27. J. Yee et al., "ACR appropriateness criteria colorectal cancer screening," *J. Am. Coll. Radiol.*, vol. 11, no. 6, pp. 543–551, 2014, doi: 10.1016/j.jacr.2014.02.006.
28. S. Halligan et al., "Computed tomographic colonography versus barium enema for diagnosis of colorectal cancer or large polyps in symptomatic patients (SIGGAR): A multicentre randomised trial," *Lancet*, vol. 381, no. 9873, pp. 1185–1193, 2013, doi: 10.1016/S0140-6736(12)62124-2.
29. S. Halligan et al., "Computed tomographic colonography compared with colonoscopy or barium enema for diagnosis of colorectal cancer in older symptomatic patients: Two multicentre randomised trials with economic evaluation (the SIGGAR trials)," *Health Technol.* Assess. *(*Rockv*)*., vol. 19, no. 54, pp. 1–134, 2015, doi: 10.3310/hta19540.
30. K. Kawasaki et al., "Is barium enema examination negligible for the management of colorectal cancer? Comparison with conventional colonoscopy and magnifying colonoscopy," *Jpn. J. Radiol.*, vol. 39, no. 12, pp. 1159–1167, 2021, doi: 10.1007/s11604-021-01157-x.

31. K. Kawasaki et al., "Role of barium enema examination for the diagnosis of submucosal invasion depth in T1 colorectal cancers," *Cancer Imaging*, vol. 21, no. 1, pp. 1–11, 2021, doi: 10.1186/s40644-021-00437-z.
32. J. Regula et al., "Colonoscopy in colorectal-cancer screening for detection of advanced neoplasia," *N. Engl. J. Med.*, vol. 355, no. 18, pp. 1863–1872, 2006, doi: 10.1056/nejmoa054967.
33. K. E. Occhipinti and J. A. Di Palma, "How to choose the best preparation for colonoscopy," *Nat. Rev. Gastroenterol. Hepatol.*, vol. 6, no. 5, pp. 279–286, 2009, doi: 10.1038/nrgastro.2009.42.
34. V. O. Millien and N. M. Mansour, "Bowel preparation for colonoscopy in 2020: A look at the past, present, and future," Curr. Gastroenterol. Rep., vol. 2, pp. 1–9, 2020.
35. S. Eberl, B. Preckel, P. Fockens, and M. W. Hollmann, "Analgesia without sedatives during colonoscopies: Worth considering?," *Tech. Coloproctol.*, vol. 16, no. 4, pp. 271–276, 2012, doi: 10.1007/s10151-012-0834-5.
36. A. Rastogi, "Cap-assisted colonoscopy," *Gastroenterol. Clin. North Am.*, vol. 42, no. 3, pp. 479–489, 2013, doi: 10.1016/j.gtc.2013.05.008.
37. S. Cadoni and F. W. Leung, "Water-assisted colonoscopy," *Curr. Treat. Options Gastroenterol.*, vol. 15, no. 1, pp. 135–154, 2017, doi: 10.1007/s11938-017-0119-1.
38. J. D. Waye, "Retroview colonoscopy," *Gastroenterol. Clin. North Am.*, vol. 42, no. 3, pp. 491–505, 2013, doi: 10.1016/j.gtc.2013.05.013.
39. P. E. Young and C. M. Womeldorph, "Colonoscopy for colorectal cancer screening," *J. Cancer*, vol. 4, no. 3, pp. 217–226, 2013, doi: 10.7150/jca.5829.
40. O. S. Ashley, M. Nadel, and D. F. Ransohoff, "Achieving quality in flexible sigmoidoscopy screening for colorectal cancer," *Am. J. Med.*, vol. 111, no. 8, pp. 643–653, 2001, doi: 10.1016/S0002-9343(01)00959-7.
41. M. A. Barish, J. A. Soto, and J. T. Ferrucci, "Consensus on current clinical practice of virtual colonoscopy," *Am. J. Roentgenol.*, vol. 184, no. March, pp. 786–792, 2005.
42. A. Engledow et al., "Integrated 18F-FDG PET/CT and perfusion CT of primary colorectal cancer: Effect of inter- and intraobserver agreement on metabolic-vascular parameters," *Am. J. Roentgenol.*, vol. 199, no. 5, November, pp. 1003–1009, 2012, doi: 10.2214/AJR.11.7823.
43. M. Kekelidze, L. D'Errico, M. Pansini, A. Tyndall, and J. Hohmann, "Colorectal cancer: Current imaging methods and future perspectives for the diagnosis, staging and therapeutic response evaluation," *World J. Gastroenterol.*, vol. 19, no. 46, pp. 8502–8514, 2013, doi: 10.3748/wjg.v19.i46.8502.
44. M. L. Puckett, H. A. Hildebrandt, R. K. Wong, P. A. Nugent, P. A. Mysliwiec, and W. R. Schindler, "Computed tomographic virtual colonoscopy to screen for colorectal neoplasia in asymptomatic adults," *N. Engl. J. Med.*, vol. 349, no. 23, pp. 2191–2200, 2003.
45. S. B. Edge and C. C. Compton, "The American joint committee on cancer: The 7th edition of the AJCC cancer staging manual and the future of TNM," *Ann. Surg. Oncol.*, vol. 17, no. 6, pp. 1471–1474, 2010, doi: 10.1245/s10434-010-0985-4.
46. K. S. Jhaveri and H. Hosseini-nik, "MRI of rectal cancer: An overview and update on recent advances," *Am. J. Roentgenol.*, vo. 205, no. 1, July, pp. 42–55, 2015, doi: 10.2214/AJR.14.14201.
47. D. Towey and L. Hill, "Data Acquisition, Reconstruction and Transfer," in PET/CT in Radiotherapy Planning, Cham: Springer International Publishing, 2017, pp. 23–27. doi: 10.1007/978-3-319-54744-2_4.
48. N. Bees, Y. Barbachano, and G. Brown, "Preoperative computed tomography staging of nonmetastatic colon cancer predicts outcome: Implications for clinical trials," *Br. J. Cancer*, vol 96, no. 7, pp. 1030–1036, 2007, doi: 10.1038/sj.bjc.6603646.
49. S. Najafi, J. Majidpoor, and K. Mortezaee, "Liquid biopsy in colorectal cancer," *Clin. Chim. Acta*, vol. 553, p. 117674, 2024, doi: 10.1016/j.cca.2023.117674.
50. J.-K. Lin et al., "Clinical relevance of alterations in quantity and quality of plasma DNA in colorectal cancer patients: Based on the mutation spectra detected in primary tumors," *Ann. Surg. Oncol.*, vol. 21 Suppl 4, pp. S680–6, Dec. 2014, doi: 10.1245/s10434-014-3804-5.
51. C. Wittekind, C. C. Compton, F. L. Greene, and L. H. Sobin, "TNM residual tumor classification revisited," *Cancer*, vol. 94, no. 9, pp. 2511–2516, May 2002, doi: 10.1002/cncr.10492.
52. J. Tie et al., "Circulating tumor DNA analysis detects minimal residual disease and predicts recurrence in patients with stage II colon cancer," *Sci. Transl. Med.*, vol. 8, no. 346, p. 346ra92, Jul. 2016, doi: 10.1126/scitranslmed.aaf6219.
53. A. Esposito et al., "Monitoring tumor-derived cell-free DNA in patients with solid tumors: Clinical perspectives and research opportunities," *Cancer Treat. Rev.*, vol. 40, no. 5, pp. 648–655, Jun. 2014, doi: 10.1016/j.ctrv.2013.10.003.
54. H. Zhou et al., "Liquid biopsy at the frontier of detection, prognosis and progression monitoring in colorectal cancer," *Mol. Cancer*, vol. 21, no. 1, pp. 1–21, 2022, doi: 10.1186/s12943-022-01556-2.

55. K. Pantel and C. Alix-Panabières, "Circulating tumour cells in cancer patients: Challenges and perspectives," *Trends Mol. Med.*, vol. 16, no. 9, pp. 398–406, 2010, doi: 10.1016/j.molmed.2010.07.001.
56. D. Dinu et al., "Prognostic significance of KRAS gene mutations in colorectal cancer--preliminary study," *J. Med. Life*, vol. 7, no. 4, pp. 581–587, 2014.
57. F. Coppedè, A. Lopomo, R. Spisni, and L. Migliore, "Genetic and epigenetic biomarkers for diagnosis, prognosis and treatment of colorectal cancer," *World J. Gastroenterol.*, vol. 20, no. 4, pp. 943–956, 2014, doi: 10.3748/wjg.v20.i4.943.
58. T. Yokota et al., "BRAF mutation is a powerful prognostic factor in advanced and recurrent colorectal cancer," *Br. J. Cancer*, vol. 104, no. 5, pp. 856–862, Mar. 2011, doi: 10.1038/bjc.2011.19.
59. W. De Roock et al., "Association of KRAS p.G13D mutation with outcome in patients with chemotherapy-refractory metastatic colorectal cancer treated with cetuximab," *JAMA*, vol. 304, no. 16, pp. 1812–1820, Oct. 2010, doi: 10.1001/jama.2010.1535.
60. A. Fariña-Sarasqueta et al., "The BRAF V600E mutation is an independent prognostic factor for survival in stage II and stage III colon cancer patients," *Ann. Oncol. Off. J. Eur. Soc. Med. Oncol.*, vol. 21, no. 12, pp. 2396–2402, Dec. 2010, doi: 10.1093/annonc/mdq258.
61. J. Liang et al., "APC polymorphisms and the risk of colorectal neoplasia: A HuGE review and meta-analysis," *Am. J. Epidemiol.*, vol. 177, no. 11, pp. 1169–1179, Jun. 2013, doi: 10.1093/aje/kws382.
62. C. Rosty et al., "PIK3CA activating mutation in colorectal carcinoma: Associations with molecular features and survival," *PLoS One*, vol. 8, no. 6, p. e65479, 2013, doi: 10.1371/journal.pone.0065479.
63. A. Sartore-Bianchi et al., "Multi-determinants analysis of molecular alterations for predicting clinical benefit to EGFR-targeted monoclonal antibodies in colorectal cancer," *PLoS One*, vol. 4, no. 10, p. e7287, Oct. 2009, doi: 10.1371/journal.pone.0007287.
64. T. Nagasaka et al., "Colorectal cancer with mutation in BRAF, KRAS, and wild-type with respect to both oncogenes showing different patterns of DNA methylation," *J. Clin. Oncol.*, vol. 22, no. 22, pp. 4584–4594, 2004, doi: 10.1200/JCO.2004.02.154.
65. S. Ouerhani, K. Bougatef, I. Soltani, A. B. A. Elgaaied, S. Abbes, and S. Menif, "The prevalence and prognostic significance of KRAS mutation in bladder cancer, chronic myeloid leukemia and colorectal cancer," *Mol. Biol. Rep.*, vol. 40, no. 6, pp. 4109–4114, 2013, doi: 10.1007/s11033-013-2512-8.
66. W. Q. Li, K. Kawakami, A. Ruszkiewicz, G. Bennett, J. Moore, and B. Iacopetta, "BRAF mutations are associated with distinctive clinical, pathological and molecular features of colorectal cancer independently of microsatellite instability status," *Mol. Cancer*, vol. 5, pp. 2–7, 2006, doi: 10.1186/1476-4598-5-2.
67. T. S. Ahn et al., "The BRAF mutation is associated with the prognosis in colorectal cancer," *J. Cancer Res. Clin. Oncol.*, vol. 140, no. 11, pp. 1863–1871, 2014, doi: 10.1007/s00432-014-1735-y.
68. Z. Dayang et al., "Characteristics and clinical analysis of MLH1 c.463dupC gene mutation in a Lynch syndrome family," *Natl. Med. J. China*, vol. 104, no. 7, pp. 547–551, 2024, doi: 10.3760/cma.j.cn112137-20231122-01170.
69. M. Tarancón-Diez, R. Büttner, and N. Friedrichs, "Enhanced tumoral MLH1-expression in MLH1-/PMS2-deficient colon cancer is indicative of sporadic colon cancer and not HNPCC," *Pathol. Oncol. Res.*, vol. 26, no. 3, pp. 1435–1439, 2020, doi: 10.1007/s12253-018-00571-3.
70. S. W. Bajwa - ten Broeke et al., "The coding microsatellite mutation profile of PMS2-deficient colorectal cancer," *Exp. Mol. Pathol.*, vol. 122, no. June, p. 104668, 2021, doi: 10.1016/j.yexmp.2021.104668.
71. P. Bryant, S. Walton Bernstedt, J. Thutkawkorapin, A. S. Backman, A. Lindblom, and K. Lagerstedt-Robinson, "Exome sequencing in a Swedish family with PMS2 mutation with varying penetrance of colorectal cancer: Investigating the presence of genetic risk modifiers in colorectal cancer risk," *Eur. J. Cancer Prev.*, vol. 32, no. 2, pp. 113–118, 2023, doi: 10.1097/CEJ.0000000000000769.
72. V. J. Forster et al., "Biallelic EPCAM deletions induce tissue-specific DNA repair deficiency and cancer predisposition," *NPJ Precis. Oncol.*, vol. 8, no. 1, 2024, doi: 10.1038/s41698-024-00537-6.
73. C. Wang, C. Ouyang, J. S. Sandhu, M. Kahn, and M. Fakih, "Wild-type APC and prognosis in metastatic colorectal cancer," *J. Clin. Oncol.*, vol. 38, no. 4_suppl, pp. 223–223, 2020, doi: 10.1200/jco.2020.38.4_suppl.223.
74. Z. N. Jawad, "Association between APC gene SNPs and the risk of occurrence of colorectal cancer," *AIP Conf. Proc.*, vol. 2414, no. 1, p. 20021, 2023, doi: 10.1063/5.0114675.
75. M. Hegde, M. Ferber, R. Mao, W. Samowitz, and A. Ganguly, "ACMG technical standards and guidelines for genetic testing for inherited colorectal cancer (Lynch syndrome, familial adenomatous polyposis, and MYH-associated polyposis)," *Genet. Med.*, vol. 16, no. 1, pp. 101–116, 2014, doi: 10.1038/gim.2013.166.

76. C. Palles et al., "Germline mutations affecting the proofreading domains of POLE and POLD1 predispose to colorectal adenomas and carcinomas," *Nat. Genet.*, vol. 45, no. 2, pp. 136–143, 2013, doi: 10.1038/ng.2503.
77. Y. Cui et al., "Combined immunodeficiency caused by a loss-of-function mutation in DNA polymerase delta 1," *J. Allergy Clin.* Immunol., vol. 145, no. 1, pp. 391–401.e8, 2020, doi: 10.1016/j.jaci.2019.10.004.
78. S. J. Hosseini et al., "Using High Resolution Melting Method for Detecting Nucleotide Polymorphisms of STK11 Gene Germline in Patients with Digestive System Cancers," *Asian Pacific J. Cancer Biol.*, vol. 1, no. 2, pp. 47–50, 2016, doi: 10.31557/APJCB.2016.1.2.47.
79. A. M. Jelsig et al., "Whole genome sequencing and disease pattern in patients with juvenile polyposis syndrome: A nationwide study," *Fam. Cancer*, vol. 22, no. 4, pp. 429–436, 2023, doi: 10.1007/s10689-023-00338-z.
80. F. Manirakiza et al., "TP53 mutations in Romanian patients with colorectal cancer," *Genes Environ.*, vol. 45, no. 1, pp. 1–8, 2023, doi: 10.1186/s41021-023-00277-2.
81. R. E. YS and I. Rey, "Molecular diagnostics in colorectal cancer," *Indones. J. Gastroenterol. Hepatol. Dig. Endosc.*, vol. 16, no. 1, pp. 26–33, Apr. 2015.
82. L. H. Nguyen, A. Goel, and D. C. Chung, "Pathways of colorectal carcinogenesis," *Gastroenterology*, vol. 158, no. 2, pp. 291–302, 2020, doi: 10.1053/j.gastro.2019.08.059.
83. M. S. Pino and D. C. Chung, "The chromosomal instability pathway in colon cancer," *Gastroenterology*, vol. 138, no. 6, pp. 2059–2072, 2010, doi: 10.1053/j.gastro.2009.12.065.
84. M. De Rosa et al., "Genetics, diagnosis and management of colorectal cancer (Review)," *Oncol. Rep.*, vol. 34, no. 3, pp. 1087–1096, 2015, doi: 10.3892/or.2015.4108.
85. C. Woischke, M. Michl, and J. Neumann, "Molecular pathology of colorectal cancer," *Pathologie*, vol. 44, no. 5, pp. 279–286, 2023, doi: 10.1007/s00292-023-01201-9.
86. B. L. Sun, "Current microsatellite instability testing in management of colorectal cancer," Clin. Colorectal Cancer, vol. 20, no. 1, pp. e12–e20, 2021, doi: 10.1016/j.clcc.2020.08.001.
87. V. Vymetalkova, P. Vodicka, S. Vodenkova, S. Alonso, and R. Schneider-Stock, "DNA methylation and chromatin modifiers in colorectal cancer," *Mol. Aspects Med.*, vol. 69, no. January, pp. 73–92, 2019, doi: 10.1016/j.mam.2019.04.002.
88. L. Fennell et al., "Integrative genome-scale DNA methylation analysis of a large and unselected cohort reveals 5 distinct subtypes of colorectal adenocarcinomas," *CMGH*, vol. 8, no. 2, pp. 269–290, 2019, doi: 10.1016/j.jcmgh.2019.04.002.
89. Y. Luo, B. Cheng, S. Liu, L. Xu, X. Wu, and S. Sun, "Relationship between CpG island methylation phenotype, microsatellite instability phenotype and mutation of KRAS, NRAS, and BRAF genes in colorectal cancer," *Int. J. Clin. Exp. Pathol.*, vol. 12, no. 3, pp. 1101–1107, 2019.
90. H. Yu et al., "DNA methylation profile in CpG-depleted regions uncovers a high-risk subtype of early-stage colorectal cancer," *J. Natl. Cancer Inst.*, vol. 115, no. 1, pp. 52–61, 2023, doi: 10.1093/jnci/djac183.
91. H. Huang et al., "DNA hypomethylation patterns and their impact on the tumor microenvironment in colorectal cancer," *Cell. Oncol.*, 2024, doi: 10.1007/s13402-024-00933-x.
92. Z. Du, T. Yu, M. Sun, Y. Chu, and G. Liu, "The long non-coding RNA TSLC8 inhibits colorectal cancer by stabilizing puma," *Cell Cycle*, vol. 19, no. 23, pp. 3317–3328, 2020, doi: 10.1080/15384101.2020.1843773.
93. O. Galamb et al., "Promoter hypomethylation and increased expression of the long non-coding RNA LINC00152 support colorectal carcinogenesis," *Pathol. Oncol. Res.*, vol. 26, no. 4, pp. 2209–2223, 2020, doi: 10.1007/s12253-020-00800-8.
94. I. Ullah et al., "Multi-omics approaches in colorectal cancer screening and diagnosis, recent updates and future perspectives," *Cancers (Basel)*, vol. 14, no. 22, 2022, doi: 10.3390/cancers14225545.
95. O. O. Ogunwobi, F. Mahmood, and A. Akingboye, "Biomarkers in colorectal cancer: Current research and future prospects," *Int. J. Mol. Sci.*, vol. 21, no. 15, pp. 1–20, 2020, doi: 10.3390/ijms21155311.
96. I. Mármol, C. Sánchez-de-Diego, A. P. Dieste, E. Cerrada, and M. J. R. Yoldi, "Colorectal carcinoma: A general overview and future perspectives in colorectal cancer," *Int. J. Mol. Sci.*, vol. 18, no. 1, 2017, doi: 10.3390/ijms18010197.

5 Multi-Faceted Surgical Intervention Using Robotic Technology for Better Colorectal Cancer Management

Nazia Afroze, Ritu Raina, and Arif Hussain
School of Life Sciences, Manipal Academy of Higher Education, Dubai Campus, United Arab Emirates

5.1 INTRODUCTION

Colorectal cancer is ranked as the fourth most lethal cancer globally, and it causes around 700,000 deaths annually [1]. Even the first world nations are observing a sharp rise in cancer morbidity and mortality [2]. Amid recent advancements in the medical field, various novel surgical techniques, such as laparoscopy, transanal minimally invasive surgery (TAMIS), and robotics surgery, are used widely for the treatment of this critical health condition [3–6]. The latest clinical trials have focused on comparing outcomes between open surgery and laparoscopy, laparoscopy and robotics, laparoscopy and TAMIS, etc., as well as finding the pre- and postoperative efficacy of these techniques, which have consequently helped in better treatment options for colorectal cancer patients [7–9]. Although surgical options such as colectomy and hemicolectomy, as well as laparoscopy, which is considered the gold standard technique, are well established, however, evolving novel techniques, including TAMIS, laparoscopic lateral pelvic lymph node (LN) dissection, and robotic surgery are attaining prominent reputation [10].

Although these techniques are novel and gaining much more prominence than conventional surgery, surgeons must be vigilant in selecting surgery types for their patients, and remain thoughtful regarding the inapt use and associated risks of these novel techniques, and be careful about the pre- and postoperative therapies (chemotherapy, targeted therapy, chemoradiotherapy, etc.) to have better outcomes [7]. In colorectal cancer surgery, the development of robotic surgery symbolizes a critical evolution that offers better precision and comfort in executing traditionally complex procedures [11]. Therefore, these developments in technology highlight how surgical procedures are evolving in the industry. The aim of this book chapter is to explicate current trends around the globe and updated surgical treatment options along with recent advancements in robotic surgery for colorectal cancer.

5.2 SURGICAL TREATMENT MODALITIES IN COLON CANCER

The only treatment for confined colon cancer is surgical removal. For certain patients with minimal spread and possibly manageable liver and lung metastases, surgery is recommended. The type of surgery you require will depend on the size, location, and severity of the colon cancer. The type of surgery will depend on the size of the colon cancer and how far it has gone [12].

DOI: 10.1201/9781003503842-5

Local resection is a surgical procedure sometimes used to treat very early-stage colon tumors. Using a colonoscope, the surgeon performs this procedure to remove the cancer from the colon's lining. This implies that there is no need for the surgeon to make an incision in your abdomen [13,14]. After the cancer has been removed, a pathologist studies it to determine its grade. To remove more colon tissue, the surgeon can advise undergoing a second procedure. This is done when the cancer is not eradicated in the first instance.

5.2.1 Eradicating Part or Complete Colon

The surgical removal of all or part of the colon is known as an excision or colectomy. The amount of cancer removed depends on the colon's dimensions and the cancer's site. Once part of the colon has been eliminated, the two sides that remain visible are joined. It may also be necessary to create a stoma. Usually, the LNs encircling the colon must also be tested and removed [15].

5.2.2 Degree of Removal

The amount of the tumor that must be removed depends on the blood supply that flushes the intestinal segment that contains it. There should be a minimum of 5 cm between the proximal and distal margins. The mesentery of the affected segment should be removed in one piece along with the main feeding vessel(s) and those that rely on lymphatic drainage [15]. Any LNs from regions beyond the resection that seem cancerous should be excised if at all possible. The American Society of Colon and Rectal Surgeons (ASCRS) advocates for the standard shutdown of the feeding vessel at the origin. Furthermore, when strong vascular ligation is used for Stages II and III (i.e., Dukes B and C) colorectal cancers, a higher overall survival and a decreased recurrence rate are reported. At least 12 LNs from the excised colon cancer material should be identified and evaluated to facilitate accurate staging [16]. If fewer than 12 LNs are initially discovered, additional sample analysis should be carried out to identify more nodes, according to the College of American Pathologists. Adjuvant chemotherapy should be administered to patients with higher-risk Stage II cancer who have fewer than 12 LNs examined but N0 disease. Complete mesocolic excision with central vascular ligation is a technique used to resect the colon and its lymphovascular supply by removing the colon and mesocolon along with an intact envelope of visceral peritoneum. This concept is like that of total mesorectal excision for rectal cancer [15]. The method involves splitting the feeding arteries centrally at the origin of the vessels and separating the visceral plane from the parietal one. This allows for a greater harvesting of LNs and resection of more mesocolic tissue. Although stage migration may be possible due to the higher LN yield, no randomized research has been conducted to verify the better results [16].

5.2.3 Evaluating the Stages of Cancer

The American Joint Committee on Cancer/TNM approach for colon cancer staging should be observed, which shows that tumor depth, nodal metastasis, and distant metastases are significant prognostic factors and should be considered. Additionally, it has been demonstrated that histologic grade is a significant outcome predictor and is typically considered when making therapy recommendations. The residual tumor code "R" indicates if the surgical resection was complete, and this should be evaluated [17].

- R0: total tumor removal with histologically negative margins on all sides.
- R1: partial tumor resection with involvement of the microscopic surgical resection margins (grossly uninvolved).
- R2: partial resection of the tumor accompanied by a substantial residual tumor (primary tumor, regional nodes, and macroscopic margin involvement) that was not resected [18].

5.2.4 Several Types of Colectomies

Complete colon excision: the entire colon is removed.

Hemicolectomy: the colon's left half is removed [17].

A left hemicolectomy removes the left half of the transverse colon and descending colon.

A right hemicolectomy involves the removal of the ascending colon and the right half of the transverse colon.

A right-sided hemicolectomy begins with the right half having access to the retroperitoneum. There are four methods for doing this [19]: Cutting the colon's lateral peritoneal attachments is the first step in the traditional lateral-to-medial mobilization.

Second, the retroperitoneum is penetrated by a posterior approach. A posterior approach enters the retroperitoneum by projecting the small bowel to the right upper quadrant and cutting the peritoneum beneath the small bowel mesentery from the fourth section of the duodenum to the cecum [19]. Third, the smaller sac is opened by cutting the peritoneum near the hepatic flexure. **A right hemicolectomy** is typically used to treat ascending colon and cecum cancers. The distal ileum is where the operation begins, and it may continue to the mid-transverse colon (Figure 5.1). The ileocolic, right colic, and perhaps the right or hepatic branch of the middle colic artery and vein must be isolated. When malignant tumors are found close to the mid-transverse colon or at the hepatic flexure, a right extended colectomy is recommended. It entails ligating the ileocolic, right colic, and middle colic vessels in addition to resecting the distal transverse colon and occasionally the splenic flexure. There are two methods for colon mobilization. The first starts laterally and moves medially, mobilizing the mesentery before vascular division occurs. This makes it possible to precisely identify the right ureter, the retroperitoneal duodenum, and the mesentery root before the veins are divided [19].

The second method, called medial-to-lateral, starts medially by locating the lymphovascular bundle and moves laterally. The latter has been applied mostly in laparoscopic surgery, where arterial ligation before mobilization maximizes the mesenteric structures' traction and exposure. Nonetheless, this method can also be applied in open environments. The optimum course of action is determined by the surgeon's choices and knowledge. The anastomosis can be sewn by hand or using a stapler device, and it can be completed as an end-to-side or side-to-side ileocolic anastomosis to restore bowel continuity. The best procedure is determined by the surgeon's inclination, experience, and the availability of equipment.

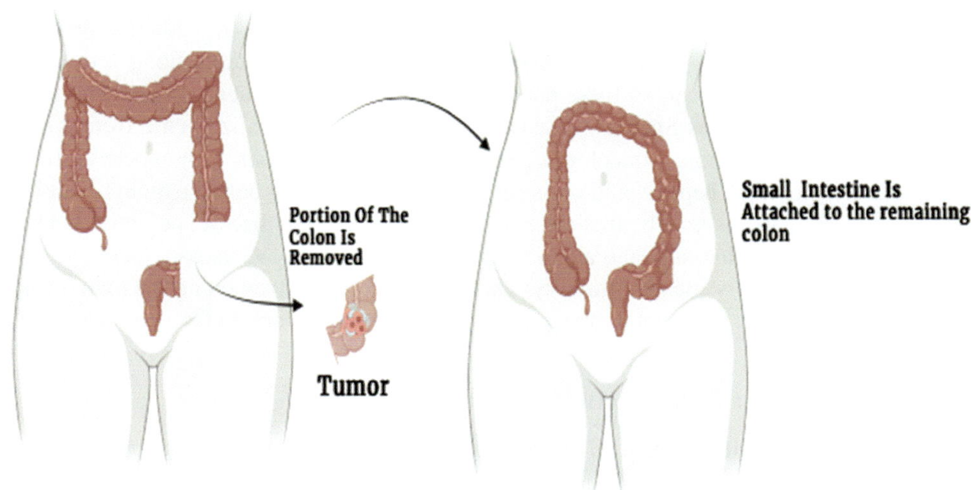

FIGURE 5.1 Colectomy of colorectal cancer.

Left Hemicolectomy: The preferred surgery for splenic flexure and descending colon cancers is left hemicolectomy. The excision reaches approximately 2–3 cm above the sacral promontory, from the distal transverse colon to the sigmoid colon or upper rectum. During resection, vessels such as the first sigmoid branch and the left colic artery pedicle are ligated. Usually, the anastomosis occurs intraperitoneally. There are two ways to mobilize the left colon: medial-to-lateral or lateral-to-medial [17,20]. To protect the spleen, the splenic flexure should be mobilized with caution. Finding the ureter is crucial for left-sided colon resections, ideally before the colon is transected. Intestinal continuity is reestablishing through colonic or colorectal anastomosis, which is hand-sewn, stapled, or sewed end to end or side to end. The anastomosis can be checked for leaks by immersing it in liquid and introducing air through a rigid or flexible endoscope. If a leak is found, the assessment procedure is repeated until no leak is found, at which point the staple line is inverted using nonabsorbable sutures [17].

A sigmoid colectomy is indicated in cases of sigmoid colon cancer. The sigmoid colon cancer is a prevalent form of colorectal cancer that often spreads to the liver [21]. This surgery starts with the inferior mesenteric artery being separated at its origin; the dissection continues toward the pelvis, just below the superior rectal arteries, until sufficient margins are achieved [22].

For localized colon cancer, laparoscopic and open colectomy yield comparable oncological results. The specific features of the tumor and the patient, as well as the surgeon's familiarity with advanced laparoscopic procedures, should determine the adoption of the laparoscopic approach [23]. Laparoscopic and open surgical resection of localized colon cancer have similar overall and recurrence-free survival rates, according to many large multi-institutional randomized trials undertaken in the United States and overseas, but in these experiments, the cancers in the rectum or transverse colon were excluded. Short-term outcomes from laparoscopic colectomy are better; these include less discomfort, less scarring, shorter hospital stays, and a faster return to normal activities [24].

For metastatic colon cancers: Currently, the treatment options for metastatic colon cancer consist of local surgery or radiation and chemotherapy, targeted therapy, and immunotherapy, which are often successful in reducing the mass of metastatic tumors [25]. Invasions into adjacent organs or inflammatory adhesions involving surrounding structures occur in about 10% of colon cancer patients. Prior to surgical exploration, preoperative imaging with a computed tomography scan or magnetic resonance imaging typically makes it easier to identify nearby organ involvement. This allows for appropriate planning and the formation of a multidisciplinary team. It is typically impossible to differentiate between malignant and inflammatory adhesions after surgery. It has been demonstrated that malignant cells are present in the peritumoral adhesions in over 40% of instances. Thus, the optimal course of treatment involves removing the tumor in its entirety together with any nearby tissues that are affected, and if at all feasible, avoiding peritumoral adhesion division. Getting a negative margin and removing a sufficient sample of LNs are crucial to the resection process [26].

Twenty to thirty percent of patients will first show lung or liver metastases along with colon cancer. Twenty to twenty-five percent of these patients will have diseases that could be treated. Patients who present with synchronous Stage IV disease should receive personalized care from a multidisciplinary team including medical oncologists, radiologists, and surgeons with expertise in specific diseases, such as colorectal, hepatic, and/or thoracic surgery. Regarding both the main tumor location and any metastases, patients may be first classed as having resectable disease, potentially resectable disease, or unresectable illness. When a patient presents with Stage IV disease, there are different ways to address the primary location initially. These consist of either removing the main tumor and any metastases upfront or, if practical, using neoadjuvant **chemotherapy before resection** [27].

5.2.5 Management of Patients with Obstruction

When a patient has concurrent, possibly treatable solid organ metastatic disease and a symptomatic primary tumor (bleeding, blockage, perforation, etc.), the initial tumor should be removed first. A one-stage combination resection may be carried out if the patient has a little or easily treatable metastatic illness. However, just the main tumor should be removed at first, followed by treatment, and then a metastatic reevaluation for resection if a metastasectomy would significantly reduce the risk of sequelae [28]. Palliative therapies may be taken into consideration for symptomatic individuals who are not candidates for initial tumor excision. These consist of laser ablation for non-obstructing tumors, surgical bypass, endoluminal stenting, and diverting ostomy.

Individualized treatment is necessary for individuals with obstructed malignancies, depending on the location of the blockage and whether proximal colonic distention with fecal load is present. Colostomy or endoluminal stenting to relieve the blockage and fecal burden are treatment possibilities, as are resection (e.g., Hartmann resection) with or without anastomosis or excision of the distended colon (e.g., subtotal/total colectomy). Because their tumors are more advanced than those without obstruction, people with obstructing cancers may have a worse prognosis. This does not, however, rule out the possibility of curative resection. Tumor-directed resection of the right or transverse colon eliminates the enlarged colonic segment. In most cases, low bacterial counts in the right and transverse colon, together with the patient's overall health at the time of resection and the lack of other indications that a protective stoma is necessary, allow for a safe enterocolonic anastomosis.

Therapeutic resection should adhere to the oncological resection recommendations. There are a number of surgical options available to patients who present with a malignant obstruction in their left colon. Appropriate surgical procedures include subtotal colectomy with ileorectal anastomosis, resection with primary anastomosis, and resection with end colostomy and rectal pouch (Hartmann's surgery) [19]. In a retrospective study of 243 consecutive patients who underwent emergent surgery for obstructing colon cancer at Queen Mary's Hospital in Hong Kong, patients who underwent resection and primary anastomosis did not show significant differences in terms of hospital mortality or anastomotic leak rates, regardless of whether the tumor was located on the right or left side. Nevertheless, a temporary proximal diverting colostomy is a common procedure used by many surgeons to remove an obstructing left-sided colon malignancy. A primary anastomosis should be checked for integrity and patency prior to the stoma being closed. If the entire colon was not evaluated prior to the formation of the stoma due to an obstructing tumor or an emergency, a colonoscopy is advised prior to ostomy closure [19,28]. Preoperative stent insertion may be effective in some circumstances.

5.2.6 Asymptomatic Colon Cancer With or Without Metastases

The treatment of asymptomatic primary colon cancer depends on whether the metastatic disease can be treated [28]. If the metastases are possibly resectable, an aggressive surgical strategy is required to cure the patient in both the primary and metastatic regions. The quantity of liver metastases used to be a key factor in assessing resectability. The most helpful instruments in the present period, however, are residual liver volume and tumor location. Patients with bilobar metastatic tumors or those whose locations make them marginally resectable may benefit more from upfront treatment followed by a reevaluation for delayed resection. The chance that a tumor may develop to the point where it is incurable does not rise with postponed metastatic resection. However, chemotherapy degrades the liver's quality. Unless major hepatic resection (three or more segments) is required, most patients clearly prefer simultaneous resection of the primary and metastatic disease. Several studies and meta-analyses have demonstrated equivalent survival and morbidity for patients who undergo a single-stage procedure compared with delayed (staged) hepatic resection [28,29]. The choice between single-stage surgery and a staged approach is influenced by various factors such as the expected intricacy of the hepatectomy and colectomy, the size of the remaining liver, the possibility of significant blood loss or prolonged ischemia durations in the liver, and the comorbidities of the patient [29].

5.2.7 Patients with Recurring Colon Cancer

According to recent estimates, 2%–5% of people with primary colorectal cancer also have concurrent primary colorectal malignancies. The best course of action for identical colon malignancies is still up for debate and frequently comes down to the surgeon's opinions. Subtotal colectomy or numerous segmental resections are available surgical procedures. Individuals who have synchronous tumors should be assessed for other underlying colonic diseases, such as chronic ulcerative colitis or hereditary non-polyposis colorectal cancer syndrome, or for a genetic colorectal cancer syndrome that may relate to them. This could have an impact on surgical decision-making about the degree of resection needed to guarantee the best possible care for the underlying condition [30].

5.2.8 Patients with Symptomatic Colon or Rectal Cancer

The following methods can be used for surgical palliation in patients with symptomatic colon or rectal cancer who have incurable metastatic disease: primary stoma diversion, primary cancer resection, and primary anastomosis [31].

5.2.8.1 Get Around the Process

Resection with primary anastomosis is the best palliative technique for those who can withstand an intra-abdominal operation [31]. However, severe comorbid diseases or widespread local diseases that may affect other structures may make excision or primary anastomosis impractical. The preferred surgery in these situations, particularly for individuals with distal colonic malignancies, is a diverting colostomy. Although an end colostomy is favored due to its ease of management and lower risk of long-term problems, a loop colostomy can also be employed [32].

A bypass procedure between the small intestine and the colon distal to the obstruction may be used for individuals with obstructive malignancies that are incurable. The patient may have distention of the bypassed segment and the eventual development of closed-loop obstruction owing to the accumulation of secretions in the bypassed colon, even if they have a functional ileocecal valve. Seromuscular sutures can be used in these situations to secure the cecum to the abdominal wall, and surgical clips can be inserted to identify the location of a future cecostomy tube placement if necessary. The laparoscopic approach is recommended to reduce the risk of postoperative problems, regardless of the surgical palliation technique used. In that context, randomized experiments have been performed to check the advantages and disadvantages of the two.

5.2.9 Natural Orifice Transluminal Endoscopic Surgery

The advancement of laparoscopic surgery without leaving abdominal scars was introduced with the introduction of natural orifice transluminal endoscopic surgery (NOTES). This approach suggests entering the peritoneal cavity through natural openings like the mouth (transgastric), urethra (transvesical), vagina (transvaginal), and anus (transanal), utilizing flexible endoscopic or rigid laparoscopic equipment. As NOTES is also referred to as "scarless" surgery, that can hypothetically reduce the uneasiness and consequences associated with wounds. Transanal NOTES has gained acceptance in the field of colorectal surgery as both a pure access method for resecting a rectal or colonic specimen and a hybrid approach aided by laparoscopy. Transvaginal or transrectal NOTES colectomy is safe and possible, according to a recent analysis of the German NOTES registry's first 139 colonic NOTES surgeries.

5.3 LAPAROSCOPY SURGERY

The cornerstone of curative therapy for colorectal cancer is surgery. An improved outcome of this technique was seen in other less invasive abdominal surgery procedures, sparking interest in laparoscopic colectomy. In the early 1990s, the first narrative of the technical viability of laparoscopic surgery for colonic malignancies appeared [33,34]. Despite initial skepticism, largely due to the technical difficulty of the procedure, laparoscopic surgery for colorectal cancer has seen substantial advancements over the past two decades. These developments have refined the techniques and successfully integrated them into modern clinical practice, improving patient outcomes and procedural efficiency [33,34].

One of the key advantages is the small incision size in laparoscopic procedures. The mean incision length for laparoscopic procedures ranges from 4.5 to 10.6 cm, significantly smaller compared to laparotomy incisions, which measure between 17.4 and 22 cm, with a maximum difference of up to 17.5 cm [35–40] (Figure 5.2). This reduction in incision size correlates with decreased pain and lower analgesic requirements. A systematic review of randomized controlled trials by Tjandra and Chan reported a reduction in analgesic consumption by up to 30% in the laparoscopic group [3]. Additionally, several studies have demonstrated notable improvements in pain scores. Milsom et al. further investigated postoperative pulmonary function in patients undergoing laparoscopic surgery. The study revealed a faster reappearance of normal lung function in patients [38,39,41,42].

Laparoscopic procedures also minimize the exposure of visceral organs and reduce retraction of the bowel with limited bowel handling. These factors contribute to reduced surgical trauma and promote faster recovery of gastrointestinal motility. It is validated in various studies that the laparoscopy procedure causes early restoration of bowel function in patients undergoing laparoscopic surgery [37,41–44]. This significant improvement in gastrointestinal recovery was further substantiated in a meta-analysis conducted by Tjandra and Chan [3].

The evidence consistently suggests that laparoscopic surgery, in comparison to open surgery, offers substantial benefits in terms of earlier return to normal activities and faster recovery [41]. Patients undergoing laparoscopic procedures experience quicker independent mobility, earlier resumption of household activities, and shorter times to regain physical, social, and work-related functions [42]. These findings, supported by various clinical trials, also indicate significantly

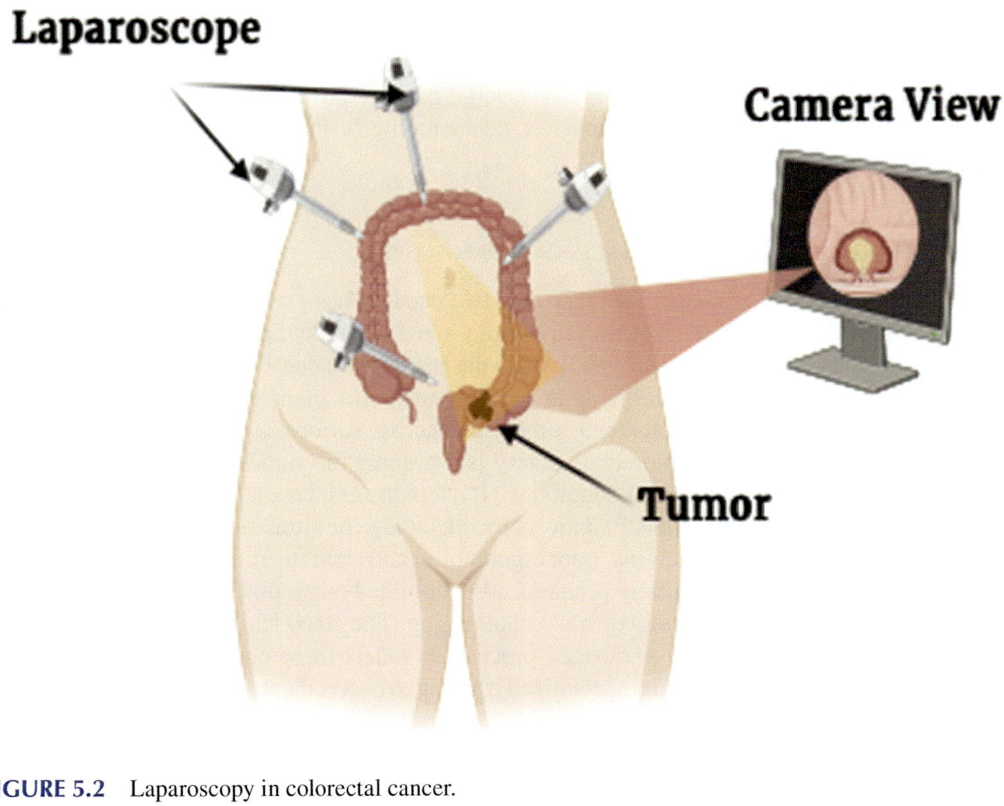

FIGURE 5.2 Laparoscopy in colorectal cancer.

reduced hospital stays, underscoring the advantages of minimally invasive techniques in promoting quicker recovery and enhancing patient outcomes. While cultural and patient-specific factors may influence these results, the overall trend demonstrates that laparoscopic surgery is a favorable option for reducing postoperative recovery times, further solidifying its role in modern colorectal surgical practices [3,45].

5.3.1 Conversion

Owing to the high conversion rates (10% and 25%) from laparoscopic procedure to open surgery, colorectal patient faces both short- and long-term complications (higher rates of intra- and postoperative complications, extended hospital stays, and poorer survival) [3,46]. Most of these conversions are usually done because of technical challenges, including the invasive nature of the tumor or difficulty in getting visuals of the tumor [47].

Markedly, the CLASICC trial has demonstrated that the learning curve can play a significant role in reducing conversion rates, as conversion rates were found to be reduced with augmented surgical experience [36,48,49]. Thus, emphasizing surgeon skills may lower conversion rates and enhance patient outcomes, especially at high-volume centers [50]. Further research is needed to clarify the underlying causes of poorer prognosis in converted cases and to optimize strategies for reducing conversions in laparoscopic colorectal surgery [48,49].

5.3.2 Mortality and Morbidity

Laparoscopic surgery for colon cancer has demonstrated significant advantages, particularly in reducing postoperative complications. Meta-analyses confirm that decreased morbidity in laparoscopic procedures is largely attributed to fewer wound-related issues and a lower incidence of postoperative ileus [35,36,43]. These improvements are linked to quicker bowel recovery, smaller incisions, less tissue retraction, and the use of wound protectors. However, higher conversion rates may obscure these benefits in certain studies, highlighting the importance of surgical experience. Therefore, while laparoscopic surgery offers specific postoperative benefits, mortality outcomes remain comparable across surgical approaches, necessitating further research to optimize patient selection and minimize conversion rates.

5.3.3 Advantages and Disadvantages of Laparoscopy

The evolution in the field of medical surgery over two decades has given various advantages, such as shorter hospitalizations, faster return to day-to-day activities, reduction of postoperative distress, and enhanced cosmetic outcomes [35,36,43]. Over the span of two decades, the field of general surgery has been revolutionized by minimally invasive surgery, with many advantages. These benefits have been widely recognized in colorectal surgery and across various surgical disciplines. Despite early uncertainty, laparoscopic colorectal surgery has become the standard approach for treating both benign and malignant colorectal conditions [4]. Nonetheless, its adoption is hampered due to the technical challenges of traditional laparoscopy, including the limitations of 2D imaging, reliance on an assistant for camera stability, poor ergonomics, and rigid instruments that lack flexibility [51]. These difficulties are particularly pronounced in confined anatomical regions like the pelvis, which make laparoscopic rectal surgery more challenging. The steep learning curve, high conversion rates, and limited number of performed procedures reflect these obstacles, underscoring the need for continued advancements and specialized training to overcome them [52]. Table 5.1 explains the advantages and disadvantages of laparoscopic surgery.

TABLE 5.1
Current Main Advantages and Disadvantages of Robotic Surgery

	Laparoscopic Surgery		Transanal Minimally Invasive Surgery	
Aspect	**Advantage**	**Disadvantage**	**Advantage**	**Disadvantage**
Minimally invasive	Smaller incisions and less tissue trauma.	Highly specialized skills required, may not be available in all medical centers.	Provides access to the rectum without external incisions and reduces tissue trauma.	Limited to tumors located in the rectum.
Postoperative recovery	Faster recovery, shorter hospital stay, and quicker return to normal activities.	Longer operating time in some cases, due to the complexity of the procedure.	Faster recovery time, shorter hospital stay, and quicker return to normal activities.	May require complementary procedures like chemotherapy or radiation for more advanced cancers.
Pain and scarring	Reduced postoperative pain, minimal scarring.	Conversion to open surgery may be required in case of complications.	Less postoperative pain and no visible scars.	Not suitable for large tumors or tumors located higher in the colon.
Blood loss	Less intraoperative blood loss compared to open surgery.	Increased risk of complications in patients with previous abdominal surgeries.	Reduced intraoperative blood loss compared to open surgery.	Risk of rectal bleeding or postoperative complications such as infections and leaks.
Complications	Lower risk of wound infections and fewer respiratory complications.	May have a learning curve for surgeons and can lead to increased complication rates early in the training.	Lower risk of surgical site infections and fewer complications.	Can lead to bowel control issues or leakage, especially in complex cases.
Tumor specialization	Offers excellent magnified views of the surgical site, improving tumor identification.	Limited tactile feedback.	Offers enhanced visualization of rectal tumors through the use of specialized cameras and instruments.	May not provide sufficient access or view for larger invasive tumors.
Oncological outcomes	Comparable long-term oncological outcomes to open surgery.	Potential for port-site metastasis if not managed properly.	Effective for treating early-stage rectal cancers or precancerous lesions.	Limited oncological application suited for early-stage or benign tumors.
Cost	Can be cost-effective due to shorter hospital stays and quick recovery.	Higher upfront costs due to training requirements and specialized equipment.	Potentially lower overall costs due to fewer complications and shorter hospital stays.	Requires specialized equipment and surgeon training, which increases upfront costs.
Patient suitability	Beneficial for selected patients, particularly for those who are younger or in early cancer stages.	May not be suitable for patients with locally advanced tumors or those with complex anatomical challenges.	Ideal for early-stage rectal cancers or benign tumors close to the anal verge, improving the patient's comfort.	May not be suitable for advanced cancer or patients with complex anatomy or pre-existing conditions.

5.4 TRANSANAL MINIMALLY INVASIVE SURGERY

Surgical treatment has been found to be appropriate because of the growing use of rectal cancer screening and greater rates of full clinical response to chemoradiotherapy for both early- and late-stage rectal cancers. Despite being the gold standard for treating rectal cancer, complete mesorectal excision (TME) is linked to substantial morbidity and longstanding deficits in urinary, anorectal, and sexual function [53,54]. Although around 19% of patients do not get their temporary stomas reversed, others may require permanent stomas [55]. The proctectomy-correlated challenges have impelled both doctors and patients to reassess the compulsion of improvisation and personalized treatment strategies. Hence, there is an increasing movement toward organ-preserving approaches, particularly the local excision of early rectal tumors in selected patients [56,57]. Since local excision does not adequately target local LNs, it is currently only a viable curative option for early-stage cancers with a minimal chance of LN metastasis. For the local excision of rectal tumors, available surgical techniques include conventional transanal excision, transanal endoscopic microsurgery (TEM), TAMIS, and colonoscopic resection. Transanal excision is performed under direct visibility using standard open surgical instruments. Except for the most distant rectal lesions, this technique provides little exposure, which results in high specimen fragmentation rates and positive margins [58].

TEM, which utilizes a fixed rigid resectoscope with insufflation to maintain a pneumorectum, was first introduced in 1984, providing a steady, enlarged view through either a digital or binocular system. Specialized instruments are then used via two working ports. TEM provides a superior resection quality in comparison to standard transanal excision, as it gives a microscopic view, reduced fragmentation of the specimen, and a much-reduced rate of relapse, although exhibiting similar postoperative difficulties [59]. Additionally, TEM is found to be superior to conventional advanced colonoscopic techniques in averting lesion relapse [33,60].

Despite its advantages, adoption of the TEM technique is slow, mainly due to the high cost of the equipment and a steep learning curve [61]. Therefore, the shortcomings of the TEM technique and the combined need for a cost-effective and better oncological outcome have led to the advancement of TAMIS [62]. In the TAMIS technique, rather than a specialized platform, a standard laparoscopic instrument is used with single-port access, which leads to a reduction in cost for this procedure. A fundamental significance of using TAMIS over the TEM technique is its ability to offer a 360° vision of the rectal lumen, enabling surgery to be performed in any quadrant without moving the patient or platform [63]. Therefore, the TAMIS technique, which was introduced in 2010 and offers similar benefits at a lower cost, has been materialized as a viable alternative to TEM [64]. The advantages and disadvantages of the TAMIS technique are listed in Table 5.1. Initially, TAMIS was only employed for benign and malignant rectal tumors. Nevertheless, its use has been extended to a wider range of clinical consequences, including the management of rectal prolapse, fistula repair, treatment of foreign body removal, and anastomotic impediments [65]. Furthermore, transanal total mesorectal excision is an evolved platform of TAMIS [66].

5.4.1 Complications

Generally minor complications are linked with TAMIS, of which the post-procedural hemorrhage is the most common, which might appear right after surgery or can be late. These cases can be managed successfully through surgical repair [40–43] or endoscopic intervention. Other problems, i.e., scrotal emphysema, can be resolved without intervention [67], while antibiotics can be used for transient fever, and for urinary retention, catheterization can resolve the issue [68].

One of the more serious complications, inadvertent intraperitoneal entry, requires careful consideration [67]. Resections of larger, anteriorly positioned lesions—especially those that are more than 10 cm from the anal verge—carry a higher risk. Peritoneal entrance is reported to occur 2% of the time as per the published TAMIS dataset [67,69]. The defect size and patient-specific factors should be factored in while managing this kind of defect either transanally or through a joint abdominal

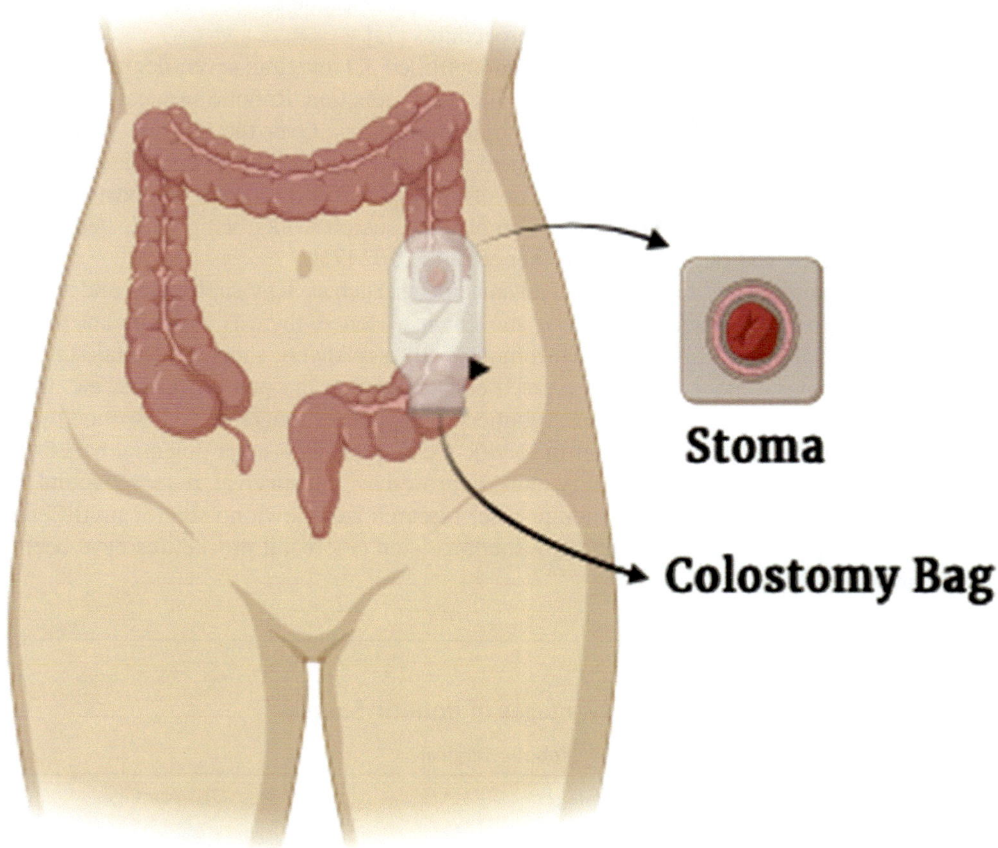

FIGURE 5.3 Illustration of colostomy for treating colorectal cancer.

approach. A personalized approach should be used before deciding to create a temporary diverting ostomy, factoring in the risk of anastomotic leak depending on the patient's ability to handle it. Figure 5.3 illustrates the technique of colostomy in colorectal cancer. In TEM procedures, the incidence of ostomy creation is reported to range between 0% and 14%.

5.4.2 Oncologic Attribute

Being a new surgical technique, TAMIS has limited oncological follow-up data in the short term. A systematic review involving 259 patients post-TAMIS surgery showed the examined recurrence rates of benign and malignant rectal neoplasms to be 2.7% with an average follow-up period of 7.1 months [70]. A study reported that 60% of the surgeons chose to perform full-thickness excisions, while 30% chose full-thickness or submucosal excision depending on the pathology and the site of tumor location. These reported outcomes suggest that TAMIS is effective in managing rectal neoplasms, although long-term data is required to fully evaluate its oncological safety [70].

5.5 ROBOTIC SURGICAL

To address the limitations of laparoscopic techniques, robotic surgical systems were developed that provide stable, surgeon-controlled 3D imaging; angulated instruments with seven degrees of freedom; and superior engineering, with a reduction in tremor. Innovative robotic surgeries have led to the widespread acceptance of this technique over the past decade across various surgical disciplines, predominantly in rectal and colorectal procedures [71].

Robotic surgical systems offer steady, surgeon-controlled 3D imaging; seven-degree-of-freedom angulated instruments; enhanced ergonomics; and tremor reduction. Robotic surgical systems were created to overcome the drawbacks of laparoscopic procedures. Over the past ten years, robotic surgery has become widely used in many surgical specialties, especially in colon and rectal procedures, thanks to these advancements [72]. The efficacy of this technique is demonstrated by the growing body of studies, with an increasing number of research works being published since the first robotic colectomy was performed by Weber et al. in 2002 [71].

The introduction of various technological advancements, such as 3D visualization and angulated instruments, aids in complicated and precise movements. Hence, there is a significant reduction in intraoperative blood loss through this technique with a relatively shorter hospital stay, faster bowel function recovery, along with a decreased conversion rate to open procedure, etc. In comparison to laparoscopic and open surgeries, robotic surgery causes longer distal resection margins. Additionally, the learning curve is shorter in robotic techniques, with other potential benefits such as lower perioperative and 30-day mortality rates, improved overall survival, reduced wound infections, and better functional outcomes, though other research has shown no significant differences (Table 5.2). It is reported that a total of 1,635 robot-assisted colorectal procedures have been performed worldwide [72].

TABLE 5.2
Current Main Advantages and Disadvantages of Robotic Surgery

	Robotic Surgical	
Aspects	**Advantages**	**Disadvantages**
Precision and control	Provides enhanced precision and control with highly articulated robotic arms.	Limited tactile feedback for the surgeon.
Visualization	Offers a 3D, high-definition view of the surgical field, improving tumor identification and safety.	Requires specialized equipment and a learning curve for the surgeon to master the technology.
Minimally invasive	Small incisions and less tissue trauma compared to open surgery.	Higher upfront costs due to expensive equipment and maintenance.
Postoperative recovery	Shorter hospital stays and faster recovery with less postoperative pain.	Longer setup time for surgery due to complexity of the robotic system.
Blood loss	Reduced intraoperative blood loss compared to open surgery.	In cases of emergency or complications, converting to open surgery can be challenging and time-consuming.
Complications	Reduced risk of postoperative complications such as infections and wound issues.	May still have a risk of other complications, like anastomotic leakage or nerve damage in some cases.
Surgeon fatigue	Reduces surgeon fatigue by providing ergonomically comfortable seating positions for long surgeries.	Prolonged operating time, especially during the surgeon's early experience with robotic systems.
Oncological outcomes	Comparable long-term oncological outcomes to laparoscopic and open surgery, with potential for fewer positive margins.	No proven superior oncological outcomes compared to other minimally invasive techniques.
Cost	Potentially cost-effective, long-term due to quicker recovery and shorter hospital stays.	High initial and operational costs due to robotic systems and instruments.
Patient suitability	Especially useful for patients with complex anatomy or challenging tumor locations, like the pelvis.	May not be available at all the hospitals or may not be accessible for all the patients due to cost and equipment limitations.

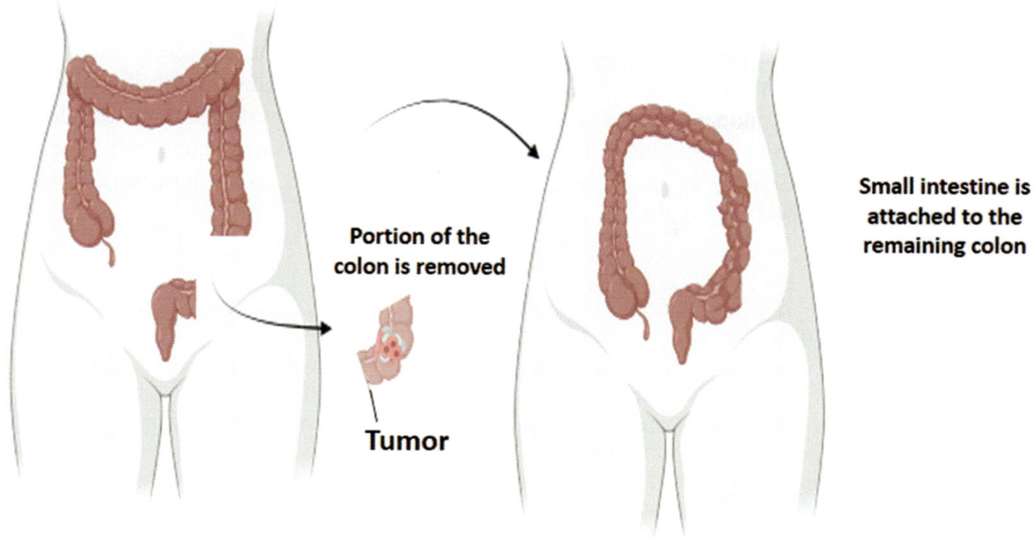

FIGURE 5.4 Robotic surgery for colorectal cancer.

5.5.1 Benefits and Drawbacks of Robotic Surgery Systems

The da Vinci robotic system has been able to successfully address many of the shortcomings of traditional laparoscopic surgery while preserving the benefits of minimally invasive techniques. It provides the lead surgeon with stable, three-dimensional views of the operative field. The flexible, jointed wrist designs of the instruments allow a range of movement that outshines the human hand with a 540° rotation and 180° articulation feature. Furthermore, due to the filtration of the physiological tremors of the surgeon's hands, more accurate movements of the machine will be plausible. Another factor is that the surgeon's physical fatigue experienced during laparoscopy can be reduced (Figure 5.4) [73]. However, some limitations will remain, such as the high investment in the machine with higher maintenance and consumables expenses and lengthier operative time. Although the newer Xi models have been significantly improved and are emerging as the favored option. the technological advancements of robotic colorectal surgery must be supported by validated clinical outcomes, overcoming these financial and logistical barriers [73] (Table 5.2).

5.5.2 Comparison of the Outcomes of Robotic versus Laparoscopic and Open Surgery

When treating colorectal cancer, robotic, laparoscopic, and open operations each have unique benefits and drawbacks. In anatomically challenging areas such as the pelvis, the use of robotic surgery improves control and enhances precision. Robotic systems have been linked to improved functional outcomes, especially in urogenital functions (continence and potency), with a much lower conversion rate to open surgery compared to laparoscopic surgery. Although these functional advantages of robotic surgery are distinguished, these do not translate into having a better oncological outcomes (such as LN yield, circumferential margins, or long-term survival) compared to open or laparoscopic procedures.

Even though laparoscopic surgery is more common and less costly than robots, it still remained constrained due to technical issues such as limited dexterity and the possibility of physician fatigue during lengthier procedures. However, the technique still has advantages over open procedures (including faster recovery, reduced postoperative pain, and shorter hospital stays, but with higher conversion rates in complex cases). In situations where minimally invasive techniques are impractical, open surgery is still a necessary option, despite its higher morbidity and longer recovery periods [74,75].

5.6 CONCLUSION

Therefore, based on the enhanced functional outcomes and a significantly decreased conversion rate, robotic surgery has an edge over laparoscopic and open surgery. Although to fully validate its oncological effectiveness, more validated clinical outcomes data are required, along with efforts to reduce the cost of the procedure. Although robotic procedures have gained superiority over laparoscopic and open surgery, the ultimate choice among the procedures can be best advised by the experienced surgeon based on the clinical scenario.

REFERENCES

1. Brody H. Colorectal cancer. *Nature*. 2015 May 14;521(7551):S1.
2. Arnold M, Sierra MS, Laversanne M, Soerjomataram I, Jemal A, Bray F. Global patterns and trends in colorectal cancer incidence and mortality. *Gut*. 2017 Apr;66(4):683–91.
3. Tjandra JJ, Chan MKY. Systematic review on the short-term outcome of laparoscopic resection for colon and rectosigmoid cancer. *Colorectal Dis*. 2006 Jun;8(5):375–88.
4. Araujo SEA. Robotic surgery for rectal cancer: current immediate clinical and oncological outcomes. *World J Gastroenterol*. 2014;20(39):14359.
5. Devane LA, Burke JP, Kelly JJ, Albert MR. Transanal minimally invasive surgery for rectal cancer. *Ann Gastroenterol Surg*. 2021 Jan 26;5(1):39–45.
6. Kwak JM, Kim SH. Robotic surgery for rectal cancer: an update in 2015. *Cancer Res Treat*. 2016 Apr 15;48(2):427–35.
7. Melstrom KA, Kaiser AM. Role of minimally invasive surgery for rectal cancer. *World J Gastroenterol*. 2020 Aug 14;26(30):4394–414.
8. Reza MM, Blasco JA, Andradas E, Cantero R, Mayol J. Systematic review of laparoscopic *versus* open surgery for colorectal cancer. *Br J Surg*. 2006 Jul 17;93(8):921–8.
9. Ng KT, Tsia AKV, Chong VYL. Robotic versus conventional laparoscopic surgery for colorectal cancer: a systematic review and meta-analysis with trial sequential analysis. *World J Surg*. 2019 Apr 15;43(4):1146–61.
10. Stitzenberg KB, Barnes E. Advances in rectal cancer surgery. *Clin Colorectal Cancer*. 2022 Mar;21(1):55–62.
11. Prete FP, Pezzolla A, Prete F, Testini M, Marzaioli R, Patriti A, et al. Robotic versus laparoscopic minimally invasive surgery for rectal cancer. *Ann Surg*. 2018 Jun;267(6):1034–46.
12. Hohenberger P. Colorectal cancer — what is standard surgery? *Eur J Cancer*. 2001;37:173–87.
13. Chang GJ, Kaiser AM, Mills S, Rafferty JF, Buie WD. Practice parameters for the management of colon cancer. *Dis Colon Rectum*. 2012;55(8).
14. Wiggers T, Jeekel J, Arends JW, Brinkhorst AP, Kluck HM, Luyk CI, et al. No-touch isolation technique in colon cancer: a controlled prospective trial. *Br J Surg*. 1988 May 1;75(5):409–15.
15. Shinji S, Yamada T, Matsuda A, Sonoda H, Ohta R, Iwai T, et al. Recent advances in the treatment of colorectal cancer: a review. *J Nippon Medi Sch*. 2022;89(3):246–54.
16. Tabaja L, Akmal Y, Abbas MA. Colon cancer surgical therapy. *Clinical Decision Making in Colorectal Surgery*. In: Steele SR, Maykel JA, Wexner SD, editors. Cham: Springer International Publishing; 2020. p. 489–501.
17. Ahmed M. Colon cancer: a clinician's perspective in 2019. *Gastroenterology Res*. 2020;13(1):1–10.
18. Shida D, Kobayashi H, Kameyama M, Hase K, Maeda K, Suto T, et al. Factors affecting R0 resection of colorectal cancer with synchronous peritoneal metastases: a multicenter prospective observational study by the Japanese Society for Cancer of the Colon and Rectum. *Int J Clin Oncol*. 2020;25:330–7.
19. Steele SR, Maykel JA, Wexner, SD, (Editors). *Clinical Decision Making in Colorectal Surgery*. Springer, USA, 2020. ISBN-10: 3319659413
20. Cui W, Zhu G, Zhou T, Mao X, Wang X, Chen Y. Laparoscopic and conventional left hemicolectomy in colon cancer. *J BUON*. 2020;25(1):240–7.
21. Shao S. Survival prediction in sigmoid colon cancer patients with liver metastasis : a prospective cohort study. *JNCI Cancer Spectr*. 2024;8(5):pkae080.
22. Furuichi Y, Kumamoto K, Asano E, Kondo A, Uemura J, Suto H, et al. Four cases of laparoscopic colectomy for sigmoid colon and rectal cancer with persistent descending mesocolon. *Surg Case Rep*. 2020;6(1):1–7.

23. Argilés G, Tabernero J, Labianca R, Hochhauser D, Salazar R, Iveson T, et al. Localised colon cancer: ESMO Clinical Practice Guidelines for diagnosis, treatment and follow-up. *Ann Oncol.* 2020;31(10):1291–305.
24. Li C, Wang Q, Jiang KW. What is the best surgical procedure of transverse colon cancer? an evidence map and minireview. *World J Gastrointest Oncol.* 2021;13(5):391–9.
25. Ganesh K, Massagué J. Targeting metastatic cancer. *Nat Med.* 2021;27(1):34–44.
26. Biller LH, Schrag D. Diagnosis and treatment of metastatic colorectal cancer: a review. *JAMA.* 2021 Feb 16;325(7):669–85.
27. Wang J, Li S, Liu Y, Zhang C, Li H, Lai B. Metastatic patterns and survival outcomes in patients with stage IV colon cancer: a population-based analysis. *Cancer Med.* 2020;9(1):361–73.
28. Vogel JD, Felder SI, Bhama AR, Hawkins AT, Langenfeld SJ, Shaffer VO, et al. The American society of colon and rectal surgeons clinical practice guidelines for the management of colon cancer. *Dis Colon Rectum.* 2022;65(2):148–77.
29. Kanemitsu Y, Shitara K, Mizusawa J, Hamaguchi T, Shida D, Komori K, et al. Primary tumor resection plus chemotherapy versus chemotherapy alone for colorectal cancer patients with asymptomatic, synchronous unresectable metastases (JCOG1007; iPACS): a randomized clinical trial. *J Clin Oncol.* 2021;39(10):1098–107.
30. Brown KGM, Koh CE. Surgical management of recurrent colon cancer. *J Gastrointest Oncol.* 2020;11(3):513.
31. Sawicki T, Ruszkowska M, Danielewicz A, Niedźwiedzka E, Arłukowicz T, Przybyłowicz KE. A review of colorectal cancer in terms of epidemiology, risk factors, development, symptoms and diagnosis. *Cancers.* 2021;March 2021:1–23.
32. Bhat BK, Yadav A, Nundy S. The role of palliative surgery in the management of acute intestinal obstruction secondary to peritoneal carcinomatosis. *Chirurgia (Bucur).* 2024;119(2):218–26.
33. Jacobs M, Verdeja JC, Goldstein HS. Minimally invasive colon resection (laparoscopic colectomy). *Surg Laparosc Endosc.* 1991 Sep;1(3):144–50.
34. Phillips EH, Franklin M, Carroll BJ, Fallas MJ, Ramos R, Rosenthal D. Laparoscopic colectomy. *Ann Surg.* 1992 Dec;216(6):703–7.
35. Clinical Outcomes of Surgical Therapy Study Group. A comparison of laparoscopically assisted and open colectomy for colon cancer. *N Engl J Med.* 2004 May 13;350(20):2050–9.
36. Guillou PJ, Quirke P, Thorpe H, Walker J, Jayne DG, Smith AMH, et al. Short-term endpoints of conventional versus laparoscopic-assisted surgery in patients with colorectal cancer (MRC CLASICC trial): multicentre, randomised controlled trial. *Lancet.* 365(9472):1718–26.
37. Hewett PJ, Allardyce RA, Bagshaw PF, Frampton CM, Frizelle FA, Rieger NA, et al. Short-Term outcomes of the Australasian randomized clinical study comparing laparoscopic and conventional open surgical treatments for colon cancer. *Ann Surg.* 2008 Nov;248(5):728–38.
38. Milsom JW, Böhm B, Hammerhofer KA, Fazio V, Steiger E, Elson P. A prospective, randomized trial comparing laparoscopic versus conventional techniques in colorectal cancer surgery: a preliminary report. *J Am Coll Surg.* 1998 Jul;187(1):46–54; discussion 54–5.
39. Liang JT, Huang KC, Lai HS, Lee PH, Jeng YM. Oncologic results of laparoscopic versus conventional open surgery for stage II or III left-sided colon cancers: a randomized controlled trial. *Ann Surg Oncol.* 2007 Jan;14(1):109–17.
40. Braga M, Frasson M, Zuliani W, Vignali A, Pecorelli N, Di Carlo V. Randomized clinical trial of laparoscopic versus open left colonic resection. *Br J Surg.* 2010 Aug;97(8):1180–6.
41. Leung KL, Kwok SPY, Lam SCW, Lee JFY, Yiu RYC, Ng SSM, et al. Laparoscopic resection of rectosigmoid carcinoma: prospective randomised trial. *Lancet.* 2004 Apr 10;363(9416):1187–92.
42. Braga M, Vignali A, Gianotti L, Zuliani W, Radaelli G, Gruarin P, et al. Laparoscopic versus open colorectal surgery: a randomized trial on short-term outcome. *Ann Surg.* 2002 Dec;236(6):759–66; discussion 767.
43. Colon Cancer Laparoscopic or Open Resection Study Group. Laparoscopic surgery versus open surgery for colon cancer: short-term outcomes of a randomised trial. *Lancet Oncol.* 2005 Jul;6(7):477–84.
44. Lacy AM, García-Valdecasas JC, Delgado S, Castells A, Taurá P, Piqué JM, et al. Laparoscopy-assisted colectomy versus open colectomy for treatment of non-metastatic colon cancer: a randomised trial. *Lancet.* 2002 Jun 29;359(9325):2224–9.
45. Ohtani H, Tamamori Y, Arimoto Y, Nishiguchi Y, Maeda K, Hirakawa K. A Meta-analysis of the short- and long-term results of randomized controlled trials that compared laparoscopy-assisted and open colectomy for colon cancer. *J Cancer.* 2012;3:49–57.

46. Abraham NS, Byrne CM, Young JM, Solomon MJ. Meta-analysis of non-randomized comparative studies of the short-term outcomes of laparoscopic resection for colorectal cancer. *ANZ J Surg*. 2007 Jul;77(7):508–16.
47. Chan ACY, Poon JTC, Fan JKM, Lo SH, Law WL. Impact of conversion on the long-term outcome in laparoscopic resection of colorectal cancer. *Surg Endosc*. 2008 Dec;22(12):2625–30.
48. Fleshman J, Sargent DJ, Green E, Anvari M, Stryker SJ, Beart RW, et al. Laparoscopic colectomy for cancer is not inferior to open surgery based on 5-year data from the COST Study Group trial. *Ann Surg*. 2007 Oct;246(4):655–62; discussion 662–4.
49. Jayne DG, Thorpe HC, Copeland J, Quirke P, Brown JM, Guillou PJ. Five-year follow-up of the Medical Research Council CLASICC trial of laparoscopically assisted versus open surgery for colorectal cancer. *Br J Surg*. 2010 Nov;97(11):1638–45.
50. Law WL, Poon JTC, Fan JKM, Lo OSH. Survival following laparoscopic versus open resection for colorectal cancer. *Int J Colorectal Dis*. 2012 Aug;27(8):1077–85.
51. Mak TWC, Lee JFY, Futaba K, Hon SSF, Ngo DKY, Ng SSM. Robotic surgery for rectal cancer: a systematic review of current practice. *World J Gastrointest Oncol*. 2014;6(6):184.
52. Chew MH, Ng KH, Fook-Chong MCS, Eu KW. Redefining conversion in laparoscopic colectomy and its influence on outcomes: analysis of 418 cases from a single institution. *World J Surg*. 2011 Jan;35(1):178–85.
53. Stephens RJ, Thompson LC, Quirke P, Steele R, Grieve R, Couture J, et al. Impact of short-course preoperative radiotherapy for rectal cancer on patients' quality of life: data from the Medical Research Council CR07/National Cancer Institute of Canada Clinical Trials Group C016 randomized clinical trial. *J Clin Oncol*. 2010 Sep 20;28(27):4233–9.
54. Peeters KCMJ, van de Velde CJH, Leer JWH, Martijn H, Junggeburt JMC, Kranenbarg EK, et al. Late side effects of short-course preoperative radiotherapy combined with total mesorectal excision for rectal cancer: increased bowel dysfunction in irradiated patients – a Dutch colorectal cancer group study. *J Clin Oncol*. 2005 Sep 1;23(25):6199–206.
55. Zhou X, Wang B, Li F, Wang J, Fu W. Risk factors associated with nonclosure of defunctioning stomas after sphincter-preserving low anterior resection of rectal cancer: a meta-analysis. *Dis Colon Rectum*. 2017 May;60(5):544–54.
56. Rullier E, Vendrely V, Asselineau J, Rouanet P, Tuech JJ, Valverde A, et al. Organ preservation with chemoradiotherapy plus local excision for rectal cancer: 5-year results of the GRECCAR 2 randomised trial. *Lancet Gastroenterol Hepatol*. 2020 May;5(5):465–74.
57. Garcia-Aguilar J, Renfro LA, Chow OS, Shi Q, Carrero XW, Lynn PB, et al. Organ preservation for clinical T2N0 distal rectal cancer using neoadjuvant chemoradiotherapy and local excision (ACOSOG Z6041): results of an open-label, single-arm, multi-institutional, phase 2 trial. *Lancet Oncol*. 2015 Nov;16(15):1537–46.
58. Paty PB, Nash GM, Baron P, Zakowski M, Minsky BD, Blumberg D, et al. Long-term results of local excision for rectal cancer. *Ann Surg*. 2002 Oct;236(4):522–9; discussion 529–30.
59. Clancy C, Burke JP, Albert MR, O'Connell PR, Winter DC. Transanal endoscopic microsurgery versus standard transanal excision for the removal of rectal neoplasms. *Dis Colon Rectum*. 2015 Feb;58(2):254–61.
60. Arezzo A, Passera R, Saito Y, Sakamoto T, Kobayashi N, Sakamoto N, et al. Systematic review and meta-analysis of endoscopic submucosal dissection versus transanal endoscopic microsurgery for large noninvasive rectal lesions. *Surg Endosc*. 2014 Feb;28(2):427–38.
61. Barendse RM, Dijkgraaf MG, Rolf UR, Bijnen AB, Consten ECJ, Hoff C, et al. Colorectal surgeons' learning curve of transanal endoscopic microsurgery. *Surg Endosc*. 2013 Oct;27(10):3591–602.
62. Maslekar S, Pillinger SH, Sharma A, Taylor A, Monson JRT. Cost analysis of transanal endoscopic microsurgery for rectal tumours. *Colorectal Dis*. 2007 Mar;9(3):229–34.
63. Atallah S, Albert M, Larach S. Transanal minimally invasive surgery: a giant leap forward. *Surg Endosc*. 2010 Sep;24(9):2200–5.
64. Melin AA, Kalaskar S, Taylor L, Thompson JS, Ternent C, Langenfeld SJ. Transanal endoscopic microsurgery and transanal minimally invasive surgery: is one technique superior? *Am J Surg*. 2016 Dec;212(6):1063–7.
65. Keller DS. Applications beyond local excision. In: *Transanal Minimally Invasive Surgery (TAMIS) and Transanal Total Mesorectal Excision (taTME)*. Cham: Springer International Publishing; 2019. pp. 143–52.
66. Atallah S, Albert M, DeBeche-Adams T, Nassif G, Polavarapu H, Larach S. Transanal minimally invasive surgery for total mesorectal excision (TAMIS-TME): a stepwise description of the surgical technique with video demonstration. *Tech Coloproctol*. 2013 Jun;17(3):321–5.

67. Albert MR, Atallah SB, deBeche-Adams TC, Izfar S, Larach SW. Transanal minimally invasive surgery (TAMIS) for local excision of benign neoplasms and early-stage rectal cancer: efficacy and outcomes in the first 50 patients. *Dis Colon Rectum.* 2013 Mar;56(3):301–7.
68. Hompes R, Ris F, Cunningham C, Mortensen NJ, Cahill RA. Transanal glove port is a safe and cost-effective alternative for transanal endoscopic microsurgery. *Br J Surg.* 2012 Oct;99(10):1429–35.
69. Marks JH, Frenkel JL, Greenleaf CE, D'Andrea AP. Transanal endoscopic microsurgery with entrance into the peritoneal cavity: is it safe? *Dis Colon Rectum.* 2014 Oct;57(10):1176–82.
70. Martin-Perez B, Andrade-Ribeiro GD, Hunter L, Atallah S. A systematic review of transanal minimally invasive surgery (TAMIS) from 2010 to 2013. *Tech Coloproctol.* 2014 Sep;18(9):775–88.
71. Weber PA, Merola S, Wasielewski A, Ballantyne GH. Telerobotic-assisted laparoscopic right and sigmoid colectomies for benign disease. *Dis Colon Rectum.* 2002 Dec;45(12):1689–94; discussion 1695–6.
72. Kwak JM, Kim SH. Robotic surgery for rectal cancer: an update in 2015. *Cancer Res Treat.* 2016 Apr 15;48(2):427–35.
73. Aly EH. Robotic colorectal surgery: summary of the current evidence. *Int J Colorectal Dis.* 2014 Jan 1;29(1):1–8.
74. Bonjer HJ, Deijen CL, Abis GA, Cuesta MA, Van Der Pas MHGM, De Lange-De Klerk ESM, et al. A randomized trial of laparoscopic versus open surgery for rectal cancer. *N Engl J Med.* 2015;372(14):1324–32.
75. Cho MS, Baek SJ, Hur H, Min BS, Baik SH, Lee KY, et al. Short and long-term outcomes of robotic versus laparoscopic total mesorectal excision for rectal cancer: a case-matched retrospective study. *Medicine.* 2015;94(11):e522.

6 Current Advancements in Chemotherapy Regimens for Colorectal Cancer Patients

Mohammad Raghibul Hasan[1], *Bashir Ibrahim Adam Omar*[1], *Sheersh Massey*[2], *Atul Kumar*[2], *and Kapil Dev*[2]
[1]Department of Clinical Laboratory Sciences, Alquwayiyah, Shaqra University, Riyadh Province, Saudi Arabia
[2]Department of Biotechnology, Jamia Millia Islamia, New Delhi, India

6.1 INTRODUCTION

Colorectal cancer (CRC) is a malignant tumor that originates from the lining (mucosa) of the large intestine (colon and rectum). Majority of CRCs arise in the colon, while others may develop in the rectum. CRC can be classified as Dukes' stage A, B, or C [1]. Some studies in this chapter have reported staging of CRC as stage I, II, or III. Stage I corresponds to Dukes' stages A–B, stage II corresponds to Dukes' stages B2–B3, and stage III corresponds to Dukes' stage C. In this review, we have included people with Dukes' A, B, and C (stages I, II, and III) and excluded people with distant metastatic disease (Dukes' stage D or stage IV).

CRC ranks as the third most prevalent malignancy in developed nations, with around one-fifth of individuals exhibiting intestinal obstruction or perforation upon presentation. The primary risk factors for CRC are dietary and genetic. The overall 5-year survival rate is approximately 50%, with 50% of individuals who undergo surgery experiencing illness recurrence [2,3].

Adjuvant systemic chemotherapy decreases mortality in individuals with Dukes' C CRC compared to surgery alone. The impact of adjuvant systemic chemotherapy on mortality, in comparison to surgery alone, remains uncertain for individuals with Dukes' B CRC.

Individuals with high-risk Dukes' B may have advantages from adjuvant systemic chemotherapy in addition to surgery alone [4]. Evidence suggests that oral fluoropyrimidines (with or without leucovorin) may be comparably effective to intravenous fluorouracil (with or without leucovorin) regimens in decreasing mortality. Oxaliplatin, when combined with fluorouracil and leucovorin, enhances disease-free survival (DFS) at 3 and 4 years compared to fluorouracil and leucovorin alone in individuals with Dukes' B or C colon cancer. The addition of irinotecan to fluorouracil and leucovorin does not decrease mortality compared to fluorouracil and leucovorin alone in individuals with Dukes' C CRC, while also increasing adverse effects. Preoperative radiation may somewhat diminish local tumor recurrence and mortality in individuals with rectal cancer compared to surgery alone [5,6].

6.2 CHEMOTHERAPY FOR COLORECTAL CANCER

6.2.1 First-Line Chemotherapy for Colorectal Cancer

Patients diagnosed with advanced or metastatic CRC receive first-line chemotherapy as their initial systemic treatment. The primary objective of this treatment is to mitigate symptoms, regulate tumor growth, and extend the patient's lifespan. The selection of therapy is contingent upon a variety of

DOI: 10.1201/9781003503842-6

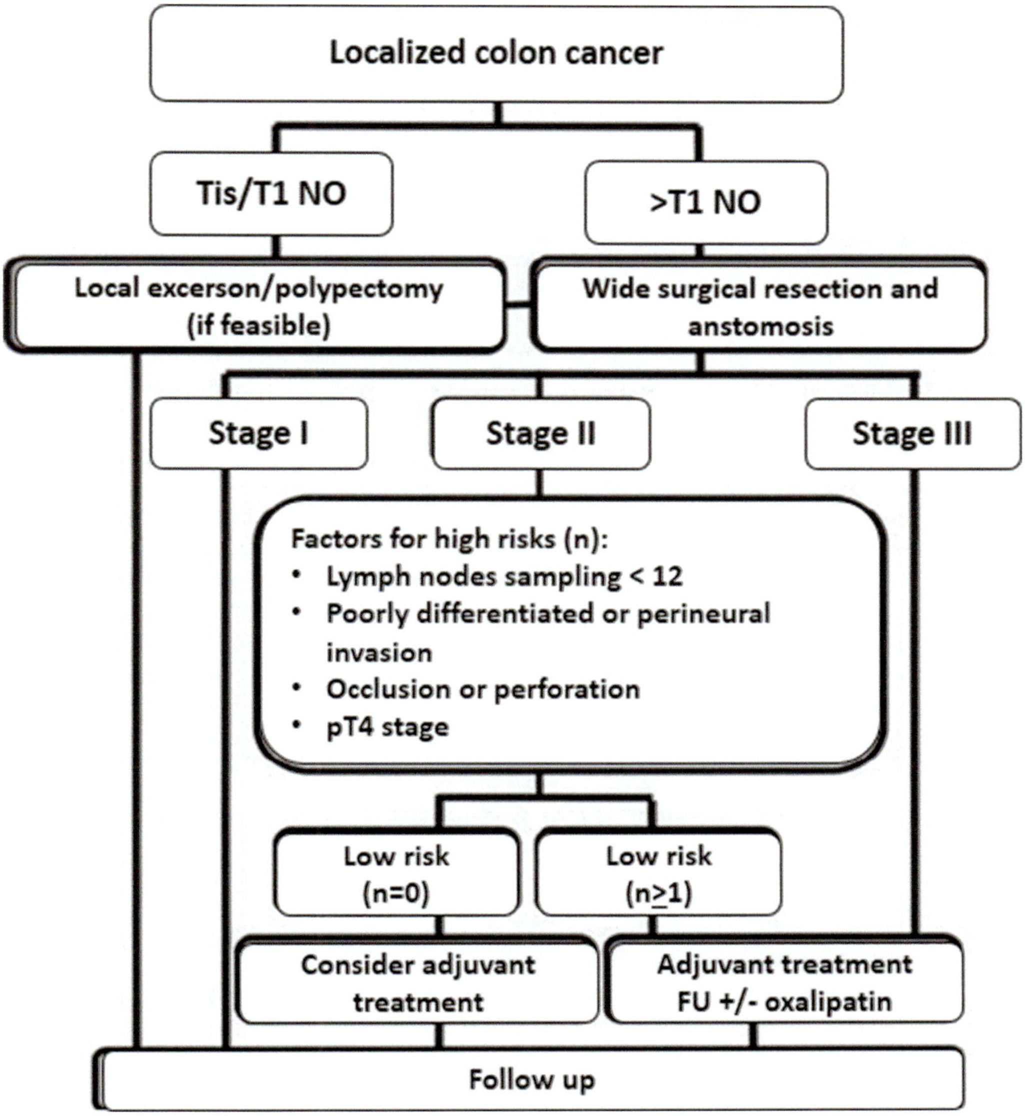

FIGURE 6.1 Different stages and treatment strategies in CRC.

factors, including the stage of the cancer, the patient's overall health, genetic mutations (including KRAS, NRAS, and BRAF), and whether the cancer is localized or has metastasized [7,8] (Figure 6.1).

6.2.2 Factors Affecting First-Line Treatment Choices

- **Tumor Stage and Location**: In contrast to localized, resectable tumors, first-line therapy is more aggressive in cases of metastatic disease.
- **Molecular Markers**: Genetic tests can help find certain mutations (like KRAS/NRAS, BRAF, and microsatellite instability status) that affect how to treat the cancer.
- **Patient Health Status**: Certain routines may not work for everyone because of their age, performance status (as measured by the ECOG or Karnofsky scales), and other health problems.
- **Goals of Treatment**: If the cancer is locally advanced, the goal is to cure it. If the cancer has spread, the goal is to relieve symptoms [8].

6.2.3 Common First-Line Chemotherapy Regimens

Several chemotherapeutic agents and combinations are widely used as first-line treatment for CRC. These include:

1. **5-Fluorouracil (5-FU)**
   Mechanism: 5-FU is a pyrimidine analog that interferes with DNA synthesis, leading to the inhibition of cancer cell replication. It has been an important part of treating CRC for many years. Administration: It is usually given as a continuous drip over a few days, either by itself or with leucovorin to make it work better [9].
2. **Capecitabine**
   The mechanism involves an oral prodrug that, once inside the body, transforms into 5-FU. It provides patients with a more convenient option than intravenous 5-FU and is a useful alternative. For the treatment of CRC, it is typically administered in conjunction with other medications, such as oxaliplatin [10,11].
3. **Oxaliplatin**
   The mechanism involves a platinum-based chemotherapeutic drug that causes DNA crosslinking, which ultimately results in the death of cells. When used in conjunction with 5-FU, it is most effective. In the context of treatment regimens such as FOLFOX, it is frequently coupled with 5-FU and leucovorin [12,13].
4. **Irinotecan**
   Topoisomerase I is an enzyme that is needed for DNA replication. Irinotecan stops this enzyme from working, which kills the cell.
5. **Uses:** It is often used with 5-FU in treatments like FOLFIRI.

6.3 COMBINATION CHEMOTHERAPY REGIMENS

One of the most common approaches utilized in first-line therapy is the combination of different chemotherapeutic drugs. Regimens that are frequently used include:

1. **FOLFOX (5-FU, Leucovorin, Oxaliplatin)**
   Regarding the indication, it is widely utilized as a conventional first-line treatment for advanced CRC, particularly in patients who have a decent performance status. In comparison to 5-FU alone, it has been demonstrated to be effective in improving overall survival (OS) as well as progression-free survival (PFS). Other adverse effects include neuropathy (caused by oxaliplatin), myelosuppression, and problems with the gastrointestinal tract [14].
2. **FOLFIRI (5-FU, Leucovorin, Irinotecan)**
   Indication: A substitute first-line regimen, particularly for patients who are intolerant to oxaliplatin or have a history of neuropathy.
   Efficacy: Equivalent to FOLFOX regarding effectiveness; selection is contingent upon patient-specific considerations.
   Adverse Effects: Diarrhea (due to irinotecan), myelosuppression, and nausea [14].
3. **CAPOX (Capecitabine, Oxaliplatin)**
   Indication: An oral substitute for FOLFOX, providing convenience for patients who favor oral administration over infusion.
   Efficacy: Comparable to FOLFOX regarding results, although dosage modifications may be required due to adverse effects.
   Adverse Effects: Neuropathy, hand-foot syndrome (attributable to capecitabine), and gastrointestinal complications [15].
4. **FOLFOXIRI (5-FU, Leucovorin, Oxaliplatin, Irinotecan)**
   Indication: A rigorous treatment protocol is employed for individuals with significant tumor burden or aggressive pathology.

Effectiveness: Demonstrated superior efficacy in enhancing OS and response rates relative to FOLFIRI, albeit with an increased risk of toxicity.

Adverse Effects: Substantial risk of neutropenia, gastrointestinal complications, and neuropathy [16].

6.4 USE OF TARGETED THERAPIES IN COMBINATION WITH CHEMOTHERAPY

The fifth recommendation is to utilize targeted medicines alongside chemotherapy. In the early stages of CRC treatment, targeted treatments are often employed alongside standard chemotherapeutic agents. This is especially applicable to people with metastatic illness [17]. The following items are included:

a. **Bevacizumab (Avastin)**

   Mechanism: A monoclonal antibody that obstructs vascular endothelial growth factor (VEGF), hence impeding the development of new blood vessels that supply the tumor.

   Combination: Utilized in regimens such as FOLFOX, FOLFIRI, or CAPOX.

   Indication: Enhances OS and PFS when administered in conjunction with chemotherapy in metastatic CRC.

   Adverse Effects: Hypertension, elevated bleeding risk, gastrointestinal perforation [18].

b. **Cetuximab (Erbitux) and Panitumumab (Vectibix)**

   Mechanism: Monoclonal antibodies that specifically target the epidermal growth factor receptor (EGFR), thereby suppressing cellular proliferation.

   Combination: Frequently utilized alongside FOLFIRI or FOLFOX in individuals with wild-type *KRAS/NRAS* genes. Indication: Efficacious in patients with KRAS/NRAS wild-type tumors (i.e., devoid of mutation).

   Adverse Effects: Dermatitis, hypomagnesemia, infusion-related symptoms [19].

6.5 CONSIDERATIONS FOR SPECIAL POPULATIONS

Geriatric Patients: Chronological age should not only dictate eligibility for chemotherapy. Nevertheless, dosage modifications may be required depending on organ functionality and concomitant conditions. Patients with suboptimal performance status may be considered for more tolerated regimens or monotherapy [20,21].

6.5.1 Second-Line Chemotherapy for Colorectal Cancer

For CRC, second-line chemotherapy is given if the disease gets worse or comes back after the first (first-line) treatment. Often, the goal is to stop the tumor from growing, ease the symptoms, and extend the patient's life, even if the original goal of curing the cancer is no longer possible. The choice of second-line therapy relies on the first-line regimen, the genetics of the tumor, and patient factors like their performance status and other health problems they may have [22].

6.5.2 Factors Affecting Second-Line Treatment as Follows

a. **Response to First-Line Therapy**: It is possible that patients who initially responded favorably to the initial treatment could benefit from moving to a different class of medications or from switching to alternative treatment courses.
b. **Duration of Response**: The choice of second-line treatment can be based on how long the response lasts or how long it takes for the disease to get worse.

c. **Molecular and Genetic Factors**: The selection of targeted therapy relies heavily on tumor mutations, such as KRAS/NRAS, BRAF, and microsatellite instability.
d. **Patient Health Status**: The patient's age, overall health, and capacity to accept treatment are all factors that need to be taken into consideration.

6.6 COMMON SECOND-LINE CHEMOTHERAPY REGIMENS

Oncologists usually change the chemotherapy plan if the cancer gets worse during or after the first line of treatment. These are the most popular routines:

a. **Irinotecan-Based Regimens**

Mechanism: Irinotecan is an inhibitor of topoisomerase I that disrupts DNA replication, resulting in cellular apoptosis. It is generally utilized when the initial treatment regimen does not incorporate irinotecan.

Usage: If the patient has previously received an oxaliplatin-based regimen, it is frequently administered as part of the FOLFIRI regimen (5-FU, leucovorin, irinotecan).

Combination with Targeted Therapies: For patients with wild-type KRAS/NRAS, irinotecan may be combined with biologics such as bevacizumab, cetuximab, or panitumumab [6].

b. **Oxaliplatin-Based Regimens**

Mechanism: The platinum-based chemotherapeutic agent oxaliplatin kills cells by crosslinking DNA. If it isn't utilized in the first-line context, it can be added to the second-line.

Usage: Patients who underwent irinotecan-based therapy (such as FOLFIRI) in the initial treatment phase may transition to an oxaliplatin-based regimen like FOLFOX (5-FU, leucovorin, oxaliplatin) or CAPOX (capecitabine, oxaliplatin).

Combination with Targeted Agents: The genetic profile of the tumor may necessitate the addition of bevacizumab or EGFR inhibitors (cetuximab/panitumumab) [12,13].

c. **FOLFOXIRI (5-FU, Leucovorin, Oxaliplatin, Irinotecan)**

Mechanism: Integrates the mechanisms of action of 5-FU, oxaliplatin, and irinotecan to target DNA synthesis, repair, and cell division.

Usage: Despite its increased aggressiveness, this combination may be appropriate for patients with a high tumor burden and a robust performance status. It is occasionally employed as a second-line treatment option when the first-line therapy is less intense (e.g., single-agent 5-FU or capecitabine).

Side Effects: Patients must be carefully selected due to the increased toxicity [14,16,23].

6.7 USE OF TARGETED THERAPIES IN SECOND-LINE TREATMENT

a. **Bevacizumab (Avastin)**

A monoclonal antibody stops VEGF from working, which stops the growth of new blood vessels that tumors need to spread and get bigger. When used with a different chemotherapy regimen (for example, moving from FOLFOX+bevacizumab to FOLFIRI+bevacizumab), it is often kept up in the second-line setting. Studies have shown that keeping bevacizumab in the second-line setting can improve survival without disease progression, even if the disease gets worse during the first-line treatment with bevacizumab [24,25].

b. **Cetuximab (Erbitux) and Panitumumab (Vectibix)**

These are EGFR inhibitors, which stop the EGFR from working. This stops the growth of cancerous cells. They only work on people whose *KRAS* or *NRAS* genes are "wild-type," which means they don't have any changes. For example, it can be mixed with irinotecan or used by itself. For example, FOLFIRI plus cetuximab can be used for people who did not get an EGFR inhibitor in the first place. Benefits in the clinic: Cetuximab and panitumumab work well for people with wild-type KRAS/NRAS cancers that don't respond to chemotherapy [26,27].

c. **Regorafenib (Stivarga)**

A multikinase inhibitor that targets different signaling pathways, such as VEGF receptors, that are involved in tumor growth and angiogenesis. Usage: This is usually used as a last resort, but it may be considered for some people who haven't responded to standard chemotherapy and biologics. Side Effects: It can lead to hand-foot syndrome, tiredness, diarrhea, and high blood pressure [28].

d. **Trifluridine/Tipiracil (TAS-102, Lonsurf)**

The mechanism involves the combination of trifluridine, which disrupts DNA synthesis, and tipiracil, which enhances the bioavailability of trifluridine. It is administered to individuals with advanced disease who have progressed despite conventional treatments. Indication: Authorized for metastatic CRC following progression on fluoropyrimidine, oxaliplatin, irinotecan, and biological therapies. Clinical Benefit: Provides slight enhancements in survival for patients who have depleted alternative therapy alternatives [29].

6.8 ROLE OF MOLECULAR PROFILING IN SECOND-LINE THERAPY

The selection of second-line treatment options can be substantially affected by the genetic profiling of malignancies. For instance:

BRAF V600E Mutations: Patients possessing this mutation may be administered a combination of BRAF inhibitors (e.g., encorafenib) and EGFR inhibitors (e.g., cetuximab).

Microsatellite Instability-High (MSI-H) or Mismatch Repair Deficient (dMMR) Tumors: Immunotherapy, such as pembrolizumab or nivolumab, may serve as an alternative to conventional chemotherapy [30,31].

Preoperative vs. Postoperative Chemotherapy Regimens in Colorectal Cancer

Chemotherapy plays a crucial role in the management of CRC, especially for patients with locally advanced disease. Treatment can be administered either before surgery (preoperative or neoadjuvant chemotherapy) or after surgery (postoperative or adjuvant chemotherapy). Both approaches have their own indications, advantages, and mechanisms of action, and the decision to use one over the other depends on several factors, including the stage of the disease, tumor location, and patient-specific characteristics [32,33].

6.8.1 Preoperative (Neoadjuvant) Chemotherapy

Neoadjuvant chemotherapy is given before surgical resection with the aim of shrinking the tumor, making it easier to remove, and potentially improving surgical outcomes. It is particularly relevant for rectal cancer, where the goal is also to preserve sphincter function and reduce the need for a permanent colostomy [34,35].

6.8.2 Mechanisms and Benefits

- **Tumor Downstaging**: Neoadjuvant chemotherapy can render previously inoperable cancers resectable or enable less invasive surgery by diminishing tumor size.
- **Improved Surgical Outcomes**: Smaller tumors can be more easily and completely resected, leading to higher rates of clear surgical margins (R0 resection).
- **Early Treatment of Micrometastases**: Administering chemotherapy before surgery allows for the early systemic treatment of micrometastatic disease, potentially preventing disease recurrence.
- **Enhanced Radiotherapy Efficacy**: In rectal cancer, chemotherapy is often combined with radiation therapy (chemoradiotherapy) preoperatively, which can enhance the effects of radiation by making cancer cells more susceptible to damage [34,35].

6.8.3 Common Preoperative Regimens

a. **FOLFOX (5-FU, Leucovorin, Oxaliplatin)**

Mechanism: Integrates a DNA synthesis inhibitor (5-FU), a DNA repair disruptor (oxaliplatin), and a folate analog (leucovorin) to enhance the efficacy of 5-FU.

Application: Frequently employed in neoadjuvant contexts for locally advanced rectal carcinoma, typically in conjunction with radiation (chemoradiotherapy).

Objective: Reduce tumor size prior to surgery, enhance the probability of obtaining clear surgical margins, and augment the chances of a favorable surgical result [36].

b. **CAPOX (Capecitabine, Oxaliplatin)**

Mechanism: Comparable to FOLFOX, it employs oral capecitabine in lieu of 5-FU, which is metabolized into 5-FU within the body. The oral administration of capecitabine is advantageous for patients.

Application: Employed similarly to FOLFOX, frequently in conjunction with radiation therapy.

Objective: Augments convenience through oral administration, enhances patient adherence, and attains comparable efficacy to intravenous therapies [36,37].

c. **Capecitabine with Radiotherapy**

Mechanism: Capecitabine functions as a radiosensitizer, enhancing the susceptibility of cancer cells to radiation impacts.

Application: Conventional protocol for locally advanced rectal carcinoma. The objective is to augment the efficacy of radiation therapy, hence facilitating tumor downstaging and improving the likelihood of successful surgical resection [11,38].

6.8.4 Postoperative (Adjuvant) Chemotherapy

Adjuvant chemotherapy is provided following surgical resection, especially in stage II and III CRC, where there is a potential for remaining microscopic disease. The aim is to eliminate any residual cancer cells and avert recurrence.

6.8.5 Mechanisms and Benefits

Elimination of Residual Disease: Surgery may leave behind tiny cancer cells that can cause the disease to come back. To lower the chance of relapse, adjuvant chemotherapy goes after these last few cells. Improvements in Survival: Adjuvant chemotherapy has been shown in clinical studies to make DFS and OS much better in people with stage III CRC and some high-risk stage II patients. Lowering the Risk of Recurrence: Chemotherapy after surgery is very important for lowering the risk of local recurrence, especially in rectal cancer, where the risk of local recurrence is higher without additional treatment [39,40].

6.8.6 Common Postoperative Regimens

a. **FOLFOX (5-FU, Leucovorin, Oxaliplatin)**

The mechanism of FOLFOX, akin to its application in the preoperative context, interferes with DNA replication and repair, resulting in the demise of cancer cells. Application: Recognized as the standard treatment for stage III CRC and high-risk stage II CRC. Aim: Mitigate the likelihood of recurrence and enhance long-term survival [14,36].

b. **CAPOX (Capecitabine, Oxaliplatin)**

The mechanism provides comparable advantages to FOLFOX while facilitating oral delivery of capecitabine, hence enhancing patient compliance with treatment. Indication:

A substitute for FOLFOX, especially for individuals who favor oral administration or are intolerant to intravenous 5-FU. Objective: To achieve efficacy comparable to FOLFOX while providing a more comfortable dosing regimen [15,37].

c. **5-FU/Leucovorin**

The mechanism involves 5-FU's inhibition of DNA synthesis, while leucovorin augments its binding to the target enzyme, hence enhancing potency. Application: Traditionally employed in adjuvant therapy, it is currently predominantly utilized for patients unable to endure oxaliplatin-based regimens due to adverse effects such as neuropathy. Objective: Offers a less hazardous alternative to FOLFOX or CAPOX [14,39,41].

Feature	Adjuvant Chemotherapy	Neoadjuvant Chemotherapy
Timing	After surgery	Before surgery
Goal	Eliminate residual disease, prevent recurrence	Shrink tumor, improve resectability, treat micrometastases early
Benefits	Improved survival, reduced local and distant recurrence	Improved surgical outcomes, potential for less invasive surgery
Challenges	Delayed start due to postoperative recovery, limited by patient health	Requires accurate staging, potential for overtreatment
Common regimens	FOLFOX, CAPOX, 5-FU/LV	FOLFOX, CAPOX, Capecitabine+Radiotherapy
Patient selection	Stage III and high-risk stage II patients	Locally advanced rectal cancer, specific cases of locally advanced colon cancer

6.8.7 Adjuvant vs. Neoadjuvant Chemotherapy Regimens in Colorectal Cancer

6.8.7.1 Introduction

In CRC, chemotherapy can be administered either as an **adjuvant** (postoperative) or **neoadjuvant** (preoperative) treatment. Adjuvant therapy is given after surgical resection to eliminate residual cancer cells and reduce the risk of recurrence, whereas neoadjuvant therapy is used before surgery to shrink the tumor, improve resectability, and potentially treat micrometastases early [42,43].

Both strategies have distinct roles and mechanisms, depending on tumor characteristics, stage, and patient factors.

6.8.7.2 Adjuvant Chemotherapy

Adjuvant chemotherapy is administered after surgery to target any remaining microscopic cancer cells. It is a standard of care for patients with stage III CRC and some patients with high-risk stage II CRC.

6.8.8 Mechanisms and Benefits

- **Eradication of Micrometastatic Disease**: Surgery alone may not remove all cancer cells, especially those that have spread to other parts of the body but are not yet detectable. Adjuvant chemotherapy aims to kill these cells and reduce the risk of relapse.
- **Improvement in Disease-Free and Overall Survival**: Clinical trials have shown that adjuvant chemotherapy improves DFS and OS in patients with stage III colon cancer and selected high-risk stage II cases [44,45].

6.8.9 Common Adjuvant Chemotherapy Regimens

1. **FOLFOX (5-FU, Leucovorin, Oxaliplatin)**
   - **Mechanism**: 5-FU inhibits thymidylate synthase, leading to disrupted DNA synthesis. Leucovorin enhances the binding of 5-FU to its target, increasing its efficacy. Oxaliplatin forms DNA crosslinks, which interfere with DNA replication and transcription.
   - **Usage**: Standard adjuvant treatment for stage III and high-risk stage II CRC. Administered intravenously over several cycles.
   - **Clinical Outcome**: FOLFOX has been shown to improve survival rates and reduce the risk of cancer recurrence in patients after surgery [14,46].
2. **CAPOX (Capecitabine, Oxaliplatin)**
   - **Mechanism**: Capecitabine is an oral prodrug that converts to 5-FU in the body. Combined with oxaliplatin, it offers a similar mechanism to FOLFOX but with the convenience of oral administration.
   - **Usage**: An alternative to FOLFOX, particularly suitable for patients who prefer oral medications or those with logistical barriers to intravenous treatment.
   - **Clinical Outcome**: CAPOX is as effective as FOLFOX in the adjuvant setting for reducing recurrence risk and improving survival [47].
3. **5-FU/Leucovorin (5-FU/LV)**
   - **Mechanism**: The combination targets thymidylate synthase, inhibiting DNA synthesis. Leucovorin boosts the effect of 5-FU.
   - **Usage**: Often used when oxaliplatin is contraindicated (e.g., in patients with preexisting neuropathy).
   - **Clinical Outcome**: While less effective than FOLFOX or CAPOX, it still offers benefits in reducing recurrence rates in patients unable to tolerate oxaliplatin [48].
4. **Neoadjuvant Chemotherapy**

   Neoadjuvant chemotherapy, often combined with radiation therapy (chemoradiotherapy), is given before surgery, primarily for locally advanced rectal cancer. The aim is to shrink the tumor and improve surgical outcomes.

6.8.10 Mechanisms and Benefits

- **Tumor Downstaging**: Reduces tumor size, which can make surgery easier and more effective. For rectal cancer, this can improve the chances of sphincter preservation and reduce the need for permanent colostomy.
- **Early Treatment of Micrometastases**: Initiating systemic chemotherapy before surgery can target microscopic cancer spread that may not yet be clinically detectable.
- **Enhanced Efficacy of Radiation Therapy**: When combined with chemotherapy, radiation can be more effective. Chemotherapy acts as a radiosensitizer, making cancer cells more susceptible to radiation damage [34,49].

6.8.11 Common Neoadjuvant Chemotherapy Regimens

1. **FOLFOX (5-FU, Leucovorin, Oxaliplatin)**
   - **Mechanism**: As described above, FOLFOX interferes with DNA synthesis and repair.
   - **Usage**: In rectal cancer, it is sometimes given as a neoadjuvant regimen, particularly when combined with radiation. It can be used to shrink tumors before surgery.
   - **Clinical Outcome**: Improves surgical resectability and can lead to tumor downstaging, increasing the chances of complete tumor removal [50].

2. **CAPOX (Capecitabine, Oxaliplatin)**
   - **Mechanism**: Similar to FOLFOX but with the advantage of oral capecitabine.
   - **Usage**: Often administered along with radiation in the neoadjuvant setting for rectal cancer.
   - **Clinical Outcome**: Effective in reducing tumor size, with similar outcomes to FOLFOX in preoperative scenarios [36,47].
3. **Capecitabine with Radiotherapy**
   - **Mechanism**: Capecitabine serves as a radiosensitizer, enhancing the effects of radiotherapy by disrupting cancer cell DNA.
   - **Usage**: Standard neoadjuvant regimen for locally advanced rectal cancer. Administered as an oral drug during a course of radiotherapy.
   - **Clinical Outcome**: Helps reduce tumor size, improve surgical margins, and increase the possibility of preserving sphincter function [11,51].
4. **Adjuvant vs. Neoadjuvant Approaches: Pros and Cons**

Feature	Adjuvant Chemotherapy	Neoadjuvant Chemotherapy
Timing	After surgery	Before surgery
Goal	Eliminate residual disease, prevent recurrence	Shrink tumor, improve resectability, treat micrometastases early
Benefits	Improved survival, reduced local and distant recurrence	Improved surgical outcomes, potential for less invasive surgery
Challenges	Delayed start due to postoperative recovery, limited by patient health	Requires accurate staging, potential for overtreatment
Common regimens	FOLFOX, CAPOX, 5-FU/LV	FOLFOX, CAPOX, Capecitabine + Radiotherapy
Patient selection	Stage III and high-risk stage II patients	Locally advanced rectal cancer, specific cases of locally advanced colon cancer

6.8.12 Chemotherapy Regimens for Metastatic and Recurrent Colorectal Cancer

6.8.12.1 Introduction

Metastatic and recurrent CRC typically indicates that the disease has spread beyond the colon or rectum to distant organs, such as the liver, lungs, or peritoneum, or has recurred after initial treatment. The management of metastatic or recurrent CRC often involves systemic chemotherapy, sometimes combined with targeted therapies. The choice of chemotherapy regimens depends on various factors, including the patient's prior treatment, tumor genetics, and overall health status [52].

6.8.12.2 Goals of Chemotherapy in Metastatic and Recurrent CRC

- **Control of Disease Progression**: Since metastatic CRC is often not curable, chemotherapy aims to control disease spread, reduce tumor size, and manage symptoms.
- **Prolongation of Survival**: Effective chemotherapy can extend OS and improve quality of life.
- **Downstaging for Surgery**: In some cases, chemotherapy can shrink metastases enough to make them surgically resectable, offering a chance for longer DFS [42,52].

6.8.12.3 Common Chemotherapy Regimens for Metastatic and Recurrent CRC

Chemotherapy for metastatic CRC can be categorized into first line, second line, and beyond. Here are some of the most commonly used regimens:

6.8.12.3.1 First-Line Regimens

1. **FOLFOX (5-FU, Leucovorin, Oxaliplatin)**
   - **Mechanism**:

5-FU: Inhibits thymidylate synthase, disrupting DNA synthesis and repair.
Leucovorin: Enhances the activity of 5-FU.
Oxaliplatin: Forms DNA crosslinks, leading to impaired DNA replication and cell death.

   - **Usage**: Standard first-line treatment for metastatic CRC, especially for patients who are fit and can tolerate oxaliplatin's side effects.
   - **Outcome**: FOLFOX has been shown to extend survival and improve response rates, making it a foundational treatment in metastatic CRC [53,54].

2. **FOLFIRI (5-FU, Leucovorin, Irinotecan)**
   - **Mechanism**:

Irinotecan: Inhibits topoisomerase I, preventing the unwinding of DNA necessary for replication and transcription.
5-FU and Leucovorin: Work in the same manner as described in FOLFOX.

   - **Usage**: Another first-line regimen, often chosen for patients with prior exposure to oxaliplatin or for those who cannot tolerate it.
   - **Outcome**: Like FOLFOX, FOLFIRI has shown significant benefits in extending survival in patients with metastatic CRC [53,55].

3. **CAPOX (Capecitabine, Oxaliplatin)**
   - **Mechanism**:

Capecitabine: An oral prodrug that converts to 5-FU, offering a similar mechanism of action.
Oxaliplatin: Forms DNA crosslinks.

   - **Usage**: Alternative to FOLFOX, especially for patients preferring oral administration of chemotherapy.
   - **Outcome**: Offers comparable efficacy to FOLFOX, with the convenience of oral administration [36,47,56].

4. **FOLFOXIRI (5-FU, Leucovorin, Oxaliplatin, Irinotecan)**
   - **Mechanism**: Combines the mechanisms of both FOLFOX and FOLFIRI, providing a more aggressive approach.
   - **Usage**: Used primarily in patients who are fit enough to tolerate intensive chemotherapy and often in cases where tumor shrinkage is necessary for potential surgical resection.
   - **Outcome**: Shown to improve response rates, PFS, and OS, but with increased toxicity [57,58].

6.8.12.3.2 Second-Line Regimens

1. **FOLFIRI + Bevacizumab (Avastin)**
   - **Mechanism**:

Bevacizumab: A monoclonal antibody targeting VEGF, which inhibits angiogenesis, thereby reducing the tumor's ability to develop new blood vessels.
FOLFIRI: Chemotherapy regimen as described above.

   - **Usage**: Often used after disease progression on FOLFOX or CAPOX.
   - **Outcome**: The addition of bevacizumab can improve survival and disease control [59,60].

2. **FOLFOX + Bevacizumab**
   - **Mechanism**: Combines anti-angiogenic effects of bevacizumab with chemotherapy.
   - **Usage**: Typically used as a second-line regimen after initial failure of FOLFIRI-based therapy.
   - **Outcome**: Helps in controlling disease progression and extending survival [61,62].

3. **Cetuximab or Panitumumab with Chemotherapy (KRAS/NRAS Wild-Type)**
   - **Mechanism**:

Cetuximab/Panitumumab: Monoclonal antibodies targeting the EGFR, inhibiting cell proliferation in tumors expressing wild-type KRAS/NRAS.
Combination with FOLFIRI or FOLFOX: Enhances the efficacy of the chemotherapy regimen.
- **Usage**: Reserved for patients with wild-type KRAS/NRAS metastatic CRC who have progressed on previous chemotherapy.
- **Outcome**: Improves response rates and OS compared to chemotherapy alone [63,64].

6.8.12.3.3 Beyond Second-Line (Third-Line and Subsequent Therapy)

1. **Regorafenib (Stivarga)**
   - **Mechanism**: A multikinase inhibitor that targets angiogenic, stromal, and oncogenic kinases, leading to reduced tumor growth and angiogenesis.
   - **Usage**: Used as a third-line or beyond for patients who have failed previous standard chemotherapy and targeted therapy regimens.
   - **Outcome**: Can offer modest survival benefits, though with notable side effects [65].
2. **Trifluridine/Tipiracil (TAS-102)**
   - **Mechanism**:

Trifluridine: A nucleoside analog that incorporates into DNA, inhibiting cell proliferation.
Tipiracil: Inhibits the degradation of trifluridine, increasing its efficacy.
   - **Usage**: Approved for patients with metastatic CRC who have been treated with or are not candidates for other standard therapies.
   - **Outcome**: Shown to provide a modest improvement in OS, making it a viable option for late-stage treatment [66,67].

6.8.12.4 Combination of Chemotherapy with Targeted Therapy

Targeted therapies, such as anti-EGFR (cetuximab, panitumumab) and anti-VEGF (bevacizumab) agents, have transformed the management of metastatic CRC. The combination of these agents with standard chemotherapy regimens can improve outcomes in patients with specific molecular profiles:

- **Anti-EGFR Therapy**: Beneficial for patients with wild-type *KRAS/NRAS* genes. Cetuximab and panitumumab are effective in combination with chemotherapy or as monotherapy in later-line settings.
- **Anti-VEGF Therapy**: Bevacizumab, when combined with chemotherapy, can improve PFS and OS in metastatic CRC [68,69].

6.8.12.5 Conclusion

The treatment of metastatic and recurrent CRC involves a tailored approach, taking into account previous treatments, genetic mutations, and patient health. First-line treatments usually involve combinations like FOLFOX, FOLFIRI, or CAPOX, often paired with targeted therapies. Upon progression, second-line and beyond treatments are selected based on the patient's response to prior regimens, incorporating newer agents like regorafenib and TAS-102. Personalized medicine, utilizing genetic profiling, is essential to optimize outcomes in this setting [65,68,70,71].

6.8.13 Current Advancements in Chemotherapy Regimens for CRC Patients

6.8.13.1 Introduction

Advancements in chemotherapy regimens for CRC have significantly evolved over the past decade, focusing on enhancing efficacy, reducing toxicity, and personalizing treatment based on tumor genetics. Innovations in combining conventional chemotherapy with novel agents, immunotherapy, and precision medicine approaches have offered new hope for CRC patients, especially those with metastatic and recurrent disease [72].

6.8.13.2 Recent Developments in Chemotherapy Regimens

6.8.13.2.1 Immunotherapy and Chemotherapy Combinations

The integration of immunotherapy with chemotherapy has been a promising area of research for treating CRC, especially for patients with certain genetic profiles, such as MSI-H or dMMR tumors.

1. **Pembrolizumab (Keytruda)**
   - **Mechanism**: Pembrolizumab is a PD-1 inhibitor that works by blocking the interaction between PD-1 and PD-L1/PD-L2, allowing the immune system to recognize and attack cancer cells.
   - **Usage**: Approved for first-line treatment of patients with MSI-H or dMMR metastatic CRC. This approval was significant because it marked the first time a non-chemotherapy regimen was given priority for a subset of CRC patients.
   - **Clinical Outcome**: Studies like KEYNOTE-177 showed that pembrolizumab significantly improved PFS compared to standard chemotherapy for patients with MSI-H CRC, with fewer side effects [73–75].
2. **Nivolumab (Opdivo) and Ipilimumab (Yervoy)**
   - **Mechanism**: Nivolumab is another PD-1 inhibitor, while ipilimumab targets CTLA-4, another checkpoint protein. The combination enhances the immune response against cancer cells.
   - **Usage**: Approved for patients with dMMR/MSI-H CRC who have progressed on prior therapy. The combination provides a more robust immune response compared to PD-1 inhibitors alone.
   - **Clinical Outcome**: In the CheckMate-142 trial, the combination led to durable responses and extended OS in previously treated patients with MSI-H CRC [76].

6.8.13.2.2 Chemotherapy Enhancers: Trifluridine/Tipiracil (TAS-102)

1. **Mechanism**:
   - **Trifluridine**: A nucleoside analog that incorporates into DNA, disrupting cell proliferation.
   - **Tipiracil**: Inhibits trifluridine metabolism, increasing its bioavailability and efficacy.
2. **Usage**: Approved for patients with metastatic CRC who have been treated with or are not candidates for other standard therapies, including fluoropyrimidine, oxaliplatin, and irinotecan.
3. **Clinical Outcome**: The RECOURSE trial demonstrated that TAS-102 improved OS compared to placebo in patients with heavily pretreated metastatic CRC [77].

6.8.13.2.3 Personalized Chemotherapy Approaches Based on Tumor Profiling

1. **RAS Mutation Testing**
   - **Advancement**: Understanding the *KRAS* and *NRAS* gene mutations has revolutionized CRC treatment. Anti-EGFR therapies, such as cetuximab and panitumumab, are only effective in patients with wild-type *RAS* genes.
   - **Usage**: This personalized approach has allowed clinicians to select therapies that are more likely to be effective and avoid those that may be ineffective or cause unnecessary side effects in patients with *RAS* mutations.
   - **Clinical Outcome**: This stratification has led to improved response rates and OS for patients receiving anti-EGFR therapies [78].
2. ***BRAF* V600E Mutation**
   - **Advancement**: Approximately 8%–10% of metastatic CRC patients have *BRAF* V600E mutations, which are associated with a poor prognosis. Recent advances include targeted therapies that inhibit the BRAF pathway.

- **Usage**: A combination of encorafenib (BRAF inhibitor), cetuximab (EGFR inhibitor), and binimetinib (MEK inhibitor) was approved for BRAF-mutant metastatic CRC, offering a new line of targeted treatment.
- **Clinical Outcome**: The BEACON CRC trial showed that this combination significantly improved OS compared to standard chemotherapy alone [79].

6.8.13.2.4 Circulating Tumor DNA for Treatment Monitoring

1. **Advancement**: Circulating tumor DNA (ctDNA) testing allows for real-time monitoring of tumor response to therapy, early detection of resistance, and assessment of minimal residual disease after treatment. This technique is being used to tailor ongoing treatment strategies.
2. **Clinical Outcome**: ctDNA testing has been shown to help predict relapse earlier than conventional imaging, enabling earlier interventions and adjustments in therapy [80].

6.8.13.2.5 Enhancing Efficacy with Maintenance Therapy

1. **Advancement**: Recent studies have focused on the concept of maintenance therapy—administering a less intensive regimen after an initial robust response to control disease progression without the full burden of standard chemotherapy toxicity.
2. **Usage**: A typical example is the use of capecitabine plus bevacizumab as maintenance after an initial period of treatment with a more aggressive regimen like FOLFOX plus bevacizumab.
3. **Clinical Outcome**: Trials have shown that maintenance therapy can prolong PFS with a lower side effect profile, thus improving the quality of life for patients [81].

6.8.13.3 Conclusion

The landscape of chemotherapy for CRC continues to evolve, with significant advancements in integrating immunotherapy, targeted therapy, and personalized medicine. These approaches are redefining the standard of care for CRC patients, leading to more tailored, effective, and less toxic treatments. Ongoing research is vital to further understanding resistance mechanisms and identifying new targets, which could lead to even better outcomes for CRC patients in the future.

6.8.14 Colorectal Cancer Staging: Dukes' Stages A, B, and C

6.8.14.1 Introduction to Dukes' Staging System

The Dukes' staging system is one of the traditional ways to classify CRC based on the extent of tumor spread. Although modern TNM staging has largely replaced it, Dukes' staging remains a useful and historical reference, especially in clinical settings and literature.

- **Dukes' Stage A**: Cancer is confined to the innermost layers of the colon or rectum (mucosa and submucosa).
- **Dukes' Stage B**: Cancer has spread through the muscular layer of the colon or rectum but has not yet reached the lymph nodes.
- **Dukes' Stage C**: Cancer has spread to the regional lymph nodes but has not metastasized to distant organs.

Each stage has different implications for treatment strategy, including the use of chemotherapy, which varies based on the likelihood of recurrence and the extent of disease spread [82].

6.8.14.2 Chemotherapy Regimens by Dukes' Stages

6.8.14.2.1 Chemotherapy for Dukes' Stage A

- **Clinical Approach**:
    - Surgery is the primary treatment for Dukes' stage A CRC, as the tumor is usually localized and has not spread beyond the inner layers of the bowel wall.
    - **Chemotherapy**: Typically not required for most patients with Dukes' stage A, given the high likelihood of complete cure with surgery alone.
- **Rationale**:
    - Since the cancer is localized, the prognosis is excellent, and additional treatment does not provide significant benefits [83].

6.8.14.2.2 Chemotherapy for Dukes' Stage B

- **Clinical Approach**:
    - Surgery is also the main treatment, but there is a consideration for adjuvant chemotherapy (postoperative chemotherapy) in some patients, especially those with high-risk features (e.g., poor differentiation, lymphovascular invasion, or obstruction).
- **Chemotherapy Regimens**:
    1. **Capecitabine (Xeloda)**
        - **Mechanism**: Oral prodrug of 5-FU, which inhibits thymidylate synthase, leading to disruption of DNA synthesis and cell death.
        - **Usage**: Often used as a monotherapy in adjuvant settings [84].
    2. **FOLFOX (5-FU, Leucovorin, Oxaliplatin)**
        - **Mechanism**:
        - **5-FU**: Inhibits thymidylate synthase, impairing DNA synthesis.
        - **Leucovorin**: Enhances the binding of 5-FU to its target enzyme, increasing its efficacy.
        - **Oxaliplatin**: A platinum-based compound that causes DNA crosslinking, leading to cancer cell death.
        - **Usage**: Considered in high-risk stage B patients who might benefit from additional chemotherapy to reduce the risk of recurrence [85].

6.8.14.2.3 Chemotherapy for Dukes' Stage C

- **Clinical Approach**:
    - Surgery remains the initial treatment to remove the primary tumor and affected lymph nodes. However, because the disease has spread to regional lymph nodes, there is a higher risk of recurrence, necessitating adjuvant chemotherapy.
- **Chemotherapy Regimens**:
    1. **FOLFOX (5-FU, Leucovorin, Oxaliplatin)**
        - **Mechanism**: Combines the effects of 5-FU, leucovorin, and oxaliplatin to inhibit DNA synthesis and induce cell death.
        - **Usage**: Standard of care for adjuvant chemotherapy in Dukes' stage C patients, reducing the risk of cancer recurrence and improving OS.
        - **Clinical Outcome**: The MOSAIC trial demonstrated that FOLFOX significantly improves DFS compared to 5-FU and leucovorin alone [86].
    2. **Capecitabine or CAPOX (Capecitabine + Oxaliplatin)**
        - **Mechanism**:
        - **Capecitabine**: Oral agent converted to 5-FU in the body, leading to inhibition of DNA synthesis.
        - **Oxaliplatin**: Induces DNA crosslinking.

   - **Usage**: As effective as FOLFOX, but with the convenience of oral administration (capecitabine). Used in adjuvant settings for patients who have undergone surgery for stage C CRC.
   - **Clinical Outcome**: Comparable DFS rates to FOLFOX with possibly less frequent administration [84].
3. **FOLFIRI (5-FU, Leucovorin, Irinotecan)**
   - **Mechanism**:
   - **Irinotecan**: Inhibits topoisomerase I, preventing DNA unwinding and replication, leading to cell death.
   - **5-FU/Leucovorin**: Enhances 5-FU activity against cancer cells.
   - **Usage**: Not typically a first-line adjuvant treatment but may be considered in cases where oxaliplatin cannot be used or in patients with recurrent disease [87].

6.8.14.2.4 Emerging Approaches for Stage B and C Patients

1. **Personalized Medicine and Genetic Profiling**
   - **Advancement**: Genomic profiling of tumors can help tailor chemotherapy based on genetic mutations (e.g., MSI status, *KRAS* mutations) [88].
2. **ctDNA for Monitoring Recurrence Risk**
   - **Advancement**: ctDNA testing is becoming a tool for detecting minimal residual disease and deciding on the need for adjuvant therapy [80].

6.8.14.3 Conclusion

The Dukes' staging system offers a straightforward way to understand the progression of CRC, with treatment regimens varying significantly between stages. Chemotherapy is generally not needed for stage A but becomes more important as the disease advances to stage B (especially high-risk cases) and stage C, where the goal is to prevent recurrence and improve survival outcomes. Recent advancements are aimed at optimizing these regimens and exploring new biomarkers to guide treatment decisions.

6.8.15 Chemotherapy for Dukes' Stage D Colorectal Cancer

6.8.15.1 Introduction to Dukes' Stage D

Dukes' stage D CRC represents the most advanced stage of the disease, where the cancer has metastasized to distant organs, such as the liver, lungs, or other sites. The goal of chemotherapy in stage D is typically palliative, aiming to control the spread of cancer, prolong survival, and improve the quality of life. Unlike earlier stages, curative surgery is often not possible, and systemic therapies play a central role in management.

6.8.15.2 Chemotherapy Regimens for Dukes' Stage D

6.8.15.2.1 First-Line Chemotherapy Options

1. **FOLFOX (5-FU, Leucovorin, Oxaliplatin)**
   - **Mechanism**:
     - **5-FU**: Antimetabolite that inhibits thymidylate synthase, blocking DNA synthesis and leading to cell death.
     - **Leucovorin**: Enhances the binding of 5-FU to its target enzyme, increasing its effectiveness.
     - **Oxaliplatin**: A platinum compound that forms DNA crosslinks, preventing DNA replication and causing cancer cell apoptosis.
   - **Usage**: Commonly used as a first-line therapy for metastatic CRC, often in combination with targeted agents like bevacizumab.

- **Clinical Outcome**: The addition of oxaliplatin to 5-FU and leucovorin has been shown to improve response rates and prolong PFS [89].

2. **FOLFIRI (5-FU, Leucovorin, Irinotecan)**
   - **Mechanism**:
     - **Irinotecan**: A topoisomerase I inhibitor that prevents DNA unwinding, leading to breaks in the DNA strands and subsequent cell death.
     - **5-FU and Leucovorin**: Work synergistically to inhibit DNA synthesis.
   - **Usage**: An alternative to FOLFOX for first-line treatment, especially in patients who may not tolerate oxaliplatin. It is often used with targeted agents.
   - **Clinical Outcome**: Demonstrated improved survival outcomes when compared to 5-FU/leucovorin alone [90].
3. **CAPOX (Capecitabine + Oxaliplatin)**
   - **Mechanism**:
     - **Capecitabine**: An oral prodrug of 5-FU that is selectively activated in tumor tissues, leading to DNA synthesis inhibition.
     - **Oxaliplatin**: Causes DNA crosslinking and cancer cell death.
   - **Usage**: Often used as an alternative to FOLFOX, providing the convenience of an oral chemotherapeutic (capecitabine).
   - **Clinical Outcome**: Comparable efficacy to FOLFOX with similar survival benefits [53].

6.8.15.2.2 Targeted Therapy Combinations

1. **Bevacizumab (Avastin)**
   - **Mechanism**: A monoclonal antibody that inhibits VEGF, reducing the blood supply to tumors and limiting their growth.
   - **Usage**: Often added to FOLFOX or FOLFIRI regimens to enhance the effectiveness of chemotherapy by starving the tumor of nutrients. Approved for first-line treatment of metastatic CRC.
   - **Clinical Outcome**: The combination of bevacizumab with chemotherapy improves PFS and OS in patients with metastatic disease [91].
2. **Cetuximab (Erbitux) and Panitumumab (Vectibix)**
   - **Mechanism**: Both are monoclonal antibodies that target the EGFR, inhibiting signaling pathways responsible for tumor cell proliferation.
   - **Usage**: Effective in patients with RAS wild-type metastatic CRC. They are typically used in combination with chemotherapy (e.g., FOLFIRI) or as monotherapy for patients who cannot tolerate more aggressive treatment.
   - **Clinical Outcome**: Improved response rates and survival when combined with chemotherapy in patients with RAS wild-type tumors [92].

6.8.15.2.3 Second-Line Chemotherapy Options

- **FOLFOXIRI (5-FU, Leucovorin, Oxaliplatin, Irinotecan)**
  - **Mechanism**: Combines the mechanisms of FOLFOX and FOLFIRI, offering a more aggressive approach by targeting DNA synthesis, replication, and repair through multiple pathways.
  - **Usage**: Often used as second-line therapy after progression on first-line treatment. May be administered with or without targeted agents.
  - **Clinical Outcome**: Demonstrated improved OS, particularly when combined with bevacizumab, as shown in the TRIBE trial [93].

6.8.15.2.4 Third-Line Therapy and Beyond

1. **Trifluridine/Tipiracil (TAS-102)**
   - **Mechanism**:
     - **Trifluridine**: An antimetabolite that incorporates into DNA and disrupts its function.
     - **Tipiracil**: Inhibits the degradation of trifluridine, increasing its availability and efficacy.
   - **Usage**: Approved for patients with metastatic CRC who have been treated with or are not candidates for other standard therapies.
   - **Clinical Outcome**: The RECOURSE trial showed that TAS-102 improves OS in patients with refractory metastatic CRC [77].
2. **Regorafenib (Stivarga)**
   - **Mechanism**: A multikinase inhibitor that targets several enzymes involved in tumor growth, angiogenesis, and the tumor microenvironment.
   - **Usage**: Used for patients with metastatic CRC who have previously received standard chemotherapy regimens and targeted agents.
   - **Clinical Outcome**: Demonstrated modest improvement in OS, with a manageable safety profile [94] (Tables 6.1 and 6.2).

TABLE 6.1
Summary of Chemotherapy Regimens for Colorectal Cancer across the Dukes' Stages (A, B, C, and D), Including Mechanisms of Action and References

Dukes' Stage	Chemotherapy Regimen	Mechanism of Action	References
Stage A	None	Not typically treated with chemotherapy; surgery is the main approach.	N/A
Stage B	Capecitabine	Oral prodrug of 5-FU; inhibits thymidylate synthase, blocking DNA synthesis.	[84]
	FOLFOX	5-FU inhibits thymidylate synthase; Oxaliplatin causes DNA crosslinking.	[89]
	FOLFIRI	5-FU inhibits thymidylate synthase; Irinotecan inhibits topoisomerase I.	[90]
Stage C	FOLFOX	5-FU inhibits thymidylate synthase; Oxaliplatin causes DNA crosslinking.	[89]
	CAPOX	Capecitabine (5-FU prodrug)+Oxaliplatin.	[95]
	FOLFIRI	5-FU inhibits thymidylate synthase; Irinotecan inhibits topoisomerase I.	[90]
Stage D	FOLFOX	5-FU inhibits thymidylate synthase; Oxaliplatin causes DNA crosslinking.	[89]
	FOLFIRI	5-FU inhibits thymidylate synthase; Irinotecan inhibits topoisomerase I.	[90]
	CAPOX	Capecitabine (5-FU prodrug)+Oxaliplatin.	[95]
	Bevacizumab	Inhibits VEGF, reducing tumor blood supply.	[91]
	Cetuximab	Targets EGFR, inhibiting tumor growth.	[92]
	Regorafenib	Multikinase inhibitor targeting tumor growth and angiogenesis.	[94]
	Trifluridine/Tipiracil (TAS-102)	Antimetabolite disrupting DNA synthesis.	[77]

TABLE 6.2
Summary of Chemotherapy Regimens for Colorectal Cancer

Category	Chemotherapy Regimen	Mechanism of Action	References
Adjuvant	FOLFOX	5-FU inhibits thymidylate synthase; Oxaliplatin causes DNA crosslinking.	[89]
	FOLFIRI	5-FU inhibits thymidylate synthase; Irinotecan inhibits topoisomerase I.	[90]
	CAPOX	Capecitabine (5-FU prodrug) + Oxaliplatin.	[36]
	5-FU + Leucovorin	5-FU inhibits thymidylate synthase; Leucovorin enhances 5-FU efficacy.	[96]
Non-adjuvant	Observation/surgery only	No chemotherapy; only surgical intervention may be employed.	N/A
Preoperative (neoadjuvant)	FOLFOX	Same as above.	[89]
	FOLFIRI	Same as above.	[90]
	CAPOX	Same as above.	[36]
	5-FU + Radiotherapy	5-FU enhances sensitivity of cancer cells to radiation therapy.	[97]
First-line metastatic	FOLFOX	Same as above.	[89]
	FOLFIRI	Same as above.	[90]
	CAPOX	Same as above.	[36]
	Bevacizumab + Chemotherapy	Inhibits VEGF, reducing tumor blood supply.	[91]
	Cetuximab + Chemotherapy	Targets EGFR, inhibiting tumor growth.	[98]
	Panitumumab + Chemotherapy	Targets EGFR, blocking downstream signaling.	[99]
Second-line metastatic	FOLFOXIRI	Combines mechanisms of FOLFOX and FOLFIRI for aggressive treatment.	[58]
	Trifluridine/Tipiracil (TAS-102)	Antimetabolite disrupting DNA synthesis.	[77]
	Regorafenib	Multikinase inhibitor targeting tumor growth and angiogenesis.	[91,94]
Recurrent	Re-treatment with FOLFOX/ FOLFIRI	Same as above.	[89,90]
	Targeted therapy	May include agents like Cetuximab or Bevacizumab based on prior treatments.	[91,98]
Unresectable advanced metastatic	FOLFOX + Bevacizumab	Inhibits VEGF; combined with FOLFOX for enhanced efficacy.	[91]
	FOLFIRI + Bevacizumab	Same as above.	[91]
	Regorafenib	Same as above.	[94]

REFERENCES

1. D. Provenzale et al., "NCCN guidelines insights: Colorectal cancer screening, version 2.2020," *Journal of the National Comprehensive Cancer Network*, vol. 18, no. 10, pp. 1312–1320, Oct. 2020, doi: 10.6004/JNCCN.2020.0048.
2. P. Wetwittayakhlang et al., "Stable incidence and risk factors of colorectal cancer in ulcerative colitis: A population-based cohort between 1977–2020," *Clinical Gastroenterology and Hepatology*, vol. 22, no. 1, pp. 191–193.e3, Jan. 2024, doi: 10.1016/j.cgh.2023.03.022.

3. C. C. Murphy and T. A. Zaki, "Changing epidemiology of colorectal cancer—Birth cohort effects and emerging risk factors," *Nature Reviews Gastroenterology & Hepatology*, vol. 21, no. 1, pp. 25–34, Sep. 2023, doi: 10.1038/s41575-023-00841-9.
4. L. Yang, J. Yang, A. Kleppe, H. E. Danielsen, and D. J. Kerr, "Personalizing adjuvant therapy for patients with colorectal cancer," *Nature Reviews Clinical Oncology*, vol. 21, no. 1, pp. 67–79, Nov. 2023, doi: 10.1038/s41571-023-00834-2.
5. P. García-Alfonso et al., "Oral drugs in the treatment of metastatic colorectal cancer," *Therapeutic Advances in Medical Oncology*, vol. 13, Apr. 2021, doi: 10.1177/17588359211009001.
6. K. I. Fujita, Y. Kubota, H. Ishida, and Y. Sasaki, "Irinotecan, a key chemotherapeutic drug for metastatic colorectal cancer," *World Journal of Gastroenterology*, vol. 21, no. 43, p. 12234, Nov. 2015, doi: 10.3748/WJG.V21.I43.12234.
7. H. C. Hsu et al., "Mutations of KRAS/NRAS/BRAF predict cetuximab resistance in metastatic colorectal cancer patients," *Oncotarget*, vol. 7, no. 16, p. 22257, Apr. 2016, doi: 10.18632/ONCOTARGET.8076.
8. M. Hewish and D. Cunningham, "First-line treatment of advanced colorectal cancer," *The Lancet*, vol. 377, no. 9783, pp. 2060–2062, Jun. 2011, doi: 10.1016/S0140–6736(11)60788-5.
9. S. Vodenkova, T. Buchler, K. Cervena, V. Veskrnova, P. Vodicka, and V. Vymetalkova, "5-fluorouracil and other fluoropyrimidines in colorectal cancer: Past, present and future," *Pharmacology & Therapeutics*, vol. 206, p. 107447, Feb. 2020, doi: 10.1016/J.PHARMTHERA.2019.107447.
10. B. R. Hirsch and S. Yousuf Zafar, "Capecitabine in the management of colorectal cancer," *Cancer Management and Research*, vol. 3, no. 1, pp. 79–89, 2011, doi: 10.2147/CMR.S11250.
11. W. Ling, J. Fan, Y. Ma, Y. Ma, and H. Wang, "Capecitabine-based chemotherapy for metastatic colorectal cancer," *Journal of Cancer Research and Clinical Oncology*, vol. 137, no. 6, pp. 927–938, Jun. 2011, doi: 10.1007/S00432-010-0954-0.
12. A. Avallone et al., "Effect of bevacizumab in combination with standard oxaliplatin-based regimens in patients with metastatic colorectal cancer: A randomized clinical trial," *JAMA Network Open*, vol. 4, no. 7, pp. e2118475–e2118475, Jul. 2021, doi: 10.1001/JAMANETWORKOPEN.2021.18475.
13. H. K. Sanoff et al., "Comparative effectiveness of oxaliplatin vs non–oxaliplatin-containing adjuvant chemotherapy for stage III colon cancer," *JNCI: Journal of the National Cancer Institute*, vol. 104, no. 3, pp. 211–227, Feb. 2012, doi: 10.1093/JNCI/DJR524.
14. B. Mohelnikova-Duchonova, B. Melichar, and P. Soucek, "FOLFOX/FOLFIRI pharmacogenetics: The call for a personalized approach in colorectal cancer therapy," *World Journal of Gastroenterology :* WJG, vol. 20, no. 30, p. 10316, Aug. 2014, doi: 10.3748/WJG.V20.I30.10316.
15. S. Degirmencioglu, O. Tanrıverdi, A. G. Demiray, H. Senol, G. G. Dogu, and A. Yaren, "Retrospective comparison of efficacy and safety of CAPOX and FOLFOX regimens as adjuvant treatment in patients with stage III colon cancer," *Journal of International Medical Research*, vol. 47, no. 6, pp. 2507–2515, Jun. 2019, doi: 10.1177/0300060519848258.
16. F. Leal, F. P. Ferreira, and A. D. Sasse, "FOLFOXIRI regimen for metastatic colorectal cancer: A systematic review and meta-analysis," *Clinical Colorectal Cancer*, vol. 16, no. 4, pp. 405–409.e2, Dec. 2017, doi: 10.1016/J.CLCC.2017.03.012.
17. Y. H. Xie, Y. X. Chen, and J. Y. Fang, "Comprehensive review of targeted therapy for colorectal cancer," *Signal Transduction and Targeted Therapy*, vol. 5, no. 1, pp. 1–30, Mar. 2020, doi: 10.1038/s41392-020-0116-z.
18. T. E. A. Botrel, L. G. de O. Clark, L. Paladini, and O. A. C. Clark, "Efficacy and safety of bevacizumab plus chemotherapy compared to chemotherapy alone in previously untreated advanced or metastatic colorectal cancer: A systematic review and meta-analysis," *BMC Cancer*, vol. 16, no. 1, pp. 1–19, Aug. 2016, doi: 10.1186/S12885-016-2734-Y.
19. X. X. Li, L. Liang, L. Y. Huang, and S. J. Cai, "Standard chemotherapy with cetuximab for treatment of colorectal cancer," *World Journal of Gastroenterology :* WJG, vol. 21, no. 22, p. 7022, Jun. 2015, doi: 10.3748/WJG.V21.I22.7022.
20. D. Gornick, A. Kadakuntla, A. Trovato, R. Stetzer, and M. Tadros, "Practical considerations for colorectal cancer screening in older adults," *World Journal of Gastrointestinal Oncology*, vol. 14, no. 6, p. 1086, Jun. 2022, doi: 10.4251/WJGO.V14.I6.1086.
21. A. DeGroff, K. Coa, K. G. Morrissey, E. Rohan, and B. Slotman, "Key considerations in designing a patient navigation program for colorectal cancer screening," *Health Promotion Practice*, vol. 15, no. 4, pp. 483–495, Dec. 2013, doi: 10.1177/1524839913513587.
22. Y. Komatsu et al., "[Second-line chemotherapy for colorectal cancer]," *Gan To Kagaku Ryoho*, vol. 36, no. 5, pp. 721–725, May 2009, doi: 10.3862/jcoloproctology.71.380.

23. J. Souglakos et al., "FOLFOXIRI (folinic acid, 5-fluorouracil, oxaliplatin and irinotecan) vs FOLFIRI (folinic acid, 5-fluorouracil and irinotecan) as first-line treatment in metastatic colorectal cancer (MCC): A multicentre randomised phase III trial from the Hellenic Oncology Research Group (HORG)," *British Journal of Cancer*, vol. 94, no. 6, pp. 798–805, Feb. 2006, doi: 10.1038/sj.bjc.6603011.
24. J. Bennouna et al., "Bevacizumab combined with chemotherapy in the second-line treatment of metastatic colorectal cancer: Results from the phase II BEVACOLOR study," *Clinical Colorectal Cancer*, vol. 11, no. 1, pp. 38–44, Mar. 2012, doi: 10.1016/J.CLCC.2011.05.002.
25. M. H. Cohen, J. Gootenberg, P. Keegan, and R. Pazdur, "FDA drug approval summary: Bevacizumab plus FOLFOX4 as second-line treatment of colorectal cancer," *Oncologist*, vol. 12, no. 3, pp. 356–361, Mar. 2007, doi: 10.1634/THEONCOLOGIST.12–3–356.
26. F. Wang et al., "Right- and left-sided colorectal cancers respond differently to cetuximab," *Cancer Communications*, vol. 34, no. 3, pp. 1–10, Dec. 2015, doi: 10.1186/S40880-015-0022-X.
27. J. Kügler, "Panitumumab (Vectibix®): A treatment for metastatic colorectal cancer," In S. Dübel and J. M. Reichert (eds.), *Handbook of Therapeutic Antibodies: Second Edition*, vol. 4–4, pp. 1855–1870, Dec. 2014, Wiley Blackwell, Weinheim, doi: 10.1002/9783527682423.CH66.
28. S. Dhillon, "Regorafenib: A review in metastatic colorectal cancer," *Drugs*, vol. 78, no. 11, pp. 1133–1144, Jul. 2018, doi: 10.1007/S40265-018-0938-Y.
29. B. M. Chan, H. S. Hochster, and H. J. Lenz, "The safety and efficacy of trifluridine–tipiracil for metastatic colorectal cancer: A pharmacy perspective," *American Journal of Health-System Pharmacy*, vol. 76, no. 6, pp. 339–348, Feb. 2019, doi: 10.1093/AJHP/ZXY006.
30. G. Giordano et al., "Evidence-based second-line treatment in RAS wild-type/mutated metastatic colorectal cancer in the precision medicine era," *International Journal of Molecular Sciences*, vol. 22, no. 14, p. 7717, Jul. 2021, doi: 10.3390/IJMS22147717.
31. N. H. Tran et al., "Precision medicine in colorectal cancer: The molecular profile alters treatment strategies," *Therapeutic Advances in Medical Oncology*, vol. 7, no. 5, pp. 252–262, Jul. 2015, doi: 10.1177/1758834015591952.
32. A. J. Breugom et al., "Adjuvant chemotherapy after preoperative (chemo)radiotherapy and surgery for patients with rectal cancer: A systematic review and meta-analysis of individual patient data," *Lancet Oncology*, vol. 16, no. 2, pp. 200–207, Feb. 2015, doi: 10.1016/S1470–2045(14)71199-4.
33. R. Araujo et al., "Comparison between perioperative and postoperative chemotherapy after potentially curative hepatic resection for metastatic colorectal cancer," *Annals of Surgical Oncology*, vol. 20, no. 13, pp. 4312–4321, Jul. 2013, doi: 10.1245/S10434-013-3162–8.
34. S. Zheng and W. Tao, "Neoadjuvant chemotherapy for colon cancer," *Cancers*, vol. 12, no. 9, p. 2368, Sep. 2020, doi: 10.3390/CANCERS12092368.
35. R. Glynne-Jones, J. Grainger, M. Harrison, P. Ostler, and A. Makris, "Neoadjuvant chemotherapy prior to preoperative chemoradiation or radiation in rectal cancer: Should we be more cautious?" *British Journal of Cancer*, vol. 94, no. 3, pp. 363–371, Feb. 2006, doi: 10.1038/sj.bjc.6602960.
36. R. A. Nagourney, S. Evans, P. H. Tran, A. J. Nagourney, and P. H. Sugarbaker, "Colorectal cancer cells from patients treated with FOLFOX or CAPOX are resistant to oxaliplatin," *European Journal of Surgical Oncology*, vol. 47, no. 4, pp. 738–742, Apr. 2021, doi: 10.1016/J.EJSO.2020.09.017.
37. Y. Wang et al., "Short-course radiotherapy combined with CAPOX and Toripalimab for the total neoadjuvant therapy of locally advanced rectal cancer: A randomized, prospective, multicentre, double-arm, phase II trial (TORCH)," *BMC Cancer*, vol. 22, no. 1, pp. 1–8, Dec. 2022, doi: 10.1186/S12885-022-09348-Z.
38. J. Dunst et al., "Phase I trial evaluating the concurrent combination of radiotherapy and capecitabine in rectal cancer," *Journal of Clinical Oncology*, vol. 20, no. 19, pp. 3983–3991, Oct. 2002, doi: 10.1200/JCO.2002.02.049.
39. G. Brandi et al., "Adjuvant chemotherapy for resected colorectal cancer metastases: Literature review and meta-analysis," *World Journal of Gastroenterology*, vol. 22, no. 2, p. 519, Jan. 2016, doi: 10.3748/WJG.V22.I2.519.
40. S. H. Petersen, H. Harling, L. T. Kirkeby, P. Wille-Jørgensen, and S. Mocellin, "Postoperative adjuvant chemotherapy in rectal cancer operated for cure," *Cochrane Database of Systematic Reviews*, no. 3, Mar. 2012, doi: 10.1002/14651858.CD004078.PUB2.
41. E. Van Cutsem et al., "ESMO consensus guidelines for the management of patients with metastatic colorectal cancer," *Annals of Oncology*, vol. 27, no. 8, pp. 1386–1422, Aug. 2016, doi: 10.1093/ANNONC/MDW235.
42. A. E. Shin, F. G. Giancotti, and A. K. Rustgi, "Metastatic colorectal cancer: Mechanisms and emerging therapeutics," *Trends in Pharmacological Sciences*, vol. 44, no. 4, pp. 222–236, Apr. 2023, doi: 10.1016/J.TIPS.2023.01.003/.

43. M. Dijkstra et al., "The role of neoadjuvant chemotherapy in repeat local treatment of recurrent colorectal liver metastases: A systematic review and meta-analysis," *Cancers*, vol. 13, no. 3, p. 378, Jan. 2021, doi: 10.3390/CANCERS13030378.
44. R. P. Jones, L. Y. W. Lee, P. G. Corrie, S. Danson, and D. Vimalachandran, "Individualized cancer vaccines versus surveillance after adjuvant chemotherapy for surgically resected high-risk stage 2 and stage 3 colorectal cancer: Protocol for a randomized trial," *British Journal of Surgery*, vol. 110, no. 12, pp. 1883–1884, Nov. 2023, doi: 10.1093/BJS/ZNAD332.
45. E. E. van Eeghen, S. D. Bakker, A. van Bochove, and R. J. L. F. Loffeld, "High risk stage 2 and stage 3 colon cancer, predictors of recurrence and effect of adjuvant therapy in a nonselected population," *International Scholarly Research Notices*, vol. 2015, no. 1, p. 790186, Jan. 2015, doi: 10.1155/2015/790186.
46. M. Biniaz, A. Moradi, M. G. Basit, A. A. S. Pashaki, A. Dehghan, and K. Mohammadian, "Combination of neoadjuvant and adjuvant chemotherapy with FOLFOX compared with adjuvant chemotherapy in management of locally advanced rectal cancers: a randomized trial of a promising therapeutic approach," *BMC Cancer*, vol. 24, no. 1, pp. 1–11, Dec. 2024, doi: 10.1186/S12885-024-12634-7.
47. J. M. Loree et al., "Survival impact of CAPOX versus FOLFOX in the adjuvant treatment of stage III colon cancer," *Clinical Colorectal Cancer*, vol. 17, no. 2, pp. 156–163, Jun. 2018, doi: 10.1016/J.CLCC.2018.01.010.
48. A. B. Chakravarthy et al., "Intergroup randomized phase III study of postoperative oxaliplatin, 5-fluorouracil, and leucovorin versus oxaliplatin, 5-fluorouracil, leucovorin, and bevacizumab for patients with stage II or III rectal cancer receiving preoperative chemoradiation: A trial of the ECOG-ACRIN Research Group (E5204)," *Oncologist*, vol. 25, no. 5, pp. e798–e807, May 2020, doi: 10.1634/THEONCOLOGIST.2019-0437.
49. Q. Zhang and H. Teng, "Clinical effect of radiotherapy combined with capecitabine after neoadjuvant therapy for rectal cancer," *Journal of Oncology*, vol. 2021, no. 1, p. 9972051, Jan. 2021, doi: 10.1155/2021/9972051.
50. S. Y. Boostrom et al., "Impact of neoadjuvant chemotherapy with FOLFOX/FOLFIRI on disease-free and overall survival of patients with colorectal metastases," *Journal of Gastrointestinal Surgery*, vol. 13, no. 11, pp. 2003–2010, Nov. 2009, doi: 10.1007/S11605-009-1007-3.
51. F. D. Pouya, Y. Rasmi, I. Y. Camci, Y. Tutar, and M. Nemati, "Performance of capecitabine in novel combination therapies in colorectal cancer," *Journal of Chemotherapy*, vol. 33, no. 6, pp. 375–389, 2021, doi: 10.1080/1120009X.2021.1920247.
52. V. K. Morris et al., "Treatment of metastatic colorectal cancer: ASCO guideline," *Journal of Clinical Oncology*, vol. 41, no. 3, pp. 678–700, Jan. 2023, doi: 10.1200/JCO.22.01690.
53. E. Van Cutsem et al., "Safety and efficacy of first-line bevacizumab with FOLFOX, XELOX, FOLFIRI and fluoropyrimidines in metastatic colorectal cancer: The BEAT study," *Annals of Oncology*, vol. 20, no. 11, pp. 1842–1847, Nov. 2009, doi: 10.1093/ANNONC/MDP233.
54. A. Stein and C. Bokemeyer, "How to select the optimal treatment for first line metastatic colorectal cancer," *World Journal of Gastroenterology :* WJG, vol. 20, no. 4, p. 899, 2014, doi: 10.3748/WJG.V20.I4.899.
55. V. Heinemann et al., "FOLFIRI plus cetuximab versus FOLFIRI plus bevacizumab as first-line treatment for patients with metastatic colorectal cancer (FIRE-3): A randomised, open-label, phase 3 trial," *Lancet Oncology*, vol. 15, no. 10, pp. 1065–1075, Sep. 2014, doi: 10.1016/S1470-2045(14)70330-4.
56. G. Zhao, P. Gao, K. H. Yang, J. H. Tian, and B. Ma, "Capecitabine/oxaliplatin as first-line treatment for metastatic colorectal cancer: A meta-analysis," *Colorectal Disease*, vol. 12, no. 7, pp. 615–623, Jul. 2010, doi: 10.1111/J.1463-1318.2009.01879.X.
57. R. P. Marques et al., "Triplet (FOLFOXIRI) versus doublet (FOLFOX or FOLFIRI) backbone chemotherapy as first-line treatment of metastatic colorectal cancer: A systematic review and meta-analysis," *Critical Reviews in Oncology/Hematology*, vol. 118, pp. 54–62, Oct. 2017, doi: 10.1016/J.CRITREVONC.2017.08.006.
58. F. Loupakis et al., "FOLFOXIRI plus bevacizumab as first-line treatment in BRAF mutant metastatic colorectal cancer," *European Journal of Cancer*, vol. 50, no. 1, pp. 57–63, Jan. 2014, doi: 10.1016/J.EJCA.2013.08.024.
59. A. Ottaiano et al., "Folfiri-aflibercept vs. Folfiri-bevacizumab as second line treatment of RAS mutated metastatic colorectal cancer in real practice," *Frontiers in Oncology*, vol. 9, no. AUG, p. 445482, Aug. 2019, doi: 10.3389/FONC.2019.00766.
60. J. R. Hecht et al., "SPIRITT: A randomized, multicenter, phase II study of panitumumab with FOLFIRI and bevacizumab with FOLFIRI as second-line treatment in patients with unresectable wild type KRAS metastatic colorectal cancer," *Clinical Colorectal Cancer*, vol. 14, no. 2, pp. 72–80, Jun. 2015, doi: 10.1016/J.CLCC.2014.12.009.
61. L. C. Park et al., "Bevacizumab as a second- or later-line of treatment for metastatic colorectal cancer," *World Journal of Gastroenterology :* WJG, vol. 18, no. 10, p. 1104, 2012, doi: 10.3748/WJG.V18.I10.1104.

62. A. Lièvre et al., "Bevacizumab plus FOLFIRI or FOLFOX in chemotherapy-refractory patients with metastatic colorectal cancer: A retrospective study," *BMC Cancer*, vol. 9, no. 1, p. 347, Sep. 2009, doi: 10.1186/1471-2407-9-347.
63. J. García-Foncillas et al., "Distinguishing features of cetuximab and panitumumab in colorectal cancer and other solid tumors," *Frontiers in Oncology*, vol. 9, p. 468024, Sep. 2019, doi: 10.3389/FONC.2019.00849.
64. F. Petrelli, K. Borgonovo, M. Cabiddu, M. Ghilardi, and S. Barni, "Cetuximab and panitumumab in KRAS wild-type colorectal cancer: A meta-analysis," *International Journal of Colorectal Disease*, vol. 26, no. 7, pp. 823–833, Jul. 2011, doi: 10.1007/S00384-011-1149-0.
65. B. Schultheis et al., "Regorafenib in combination with FOLFOX or FOLFIRI as first- or second-line treatment of colorectal cancer: Results of a multicenter, phase Ib study," *Annals of Oncology*, vol. 24, no. 6, pp. 1560–1567, Jun. 2013, doi: 10.1093/ANNONC/MDT056.
66. G. W. Prager et al., "Trifluridine–tipiracil and bevacizumab in refractory metastatic colorectal cancer," *New England Journal of Medicine*, vol. 388, no. 18, pp. 1657–1667, May 2023, doi: 10.1056/NEJMOA2214963/SUPPL_FILE/NEJMOA2214963.
67. C. B. Burness and S. T. Duggan, "Trifluridine/tipiracil: A review in metastatic colorectal cancer," *Drugs*, vol. 76, no. 14, pp. 1393–1402, Sep. 2016, doi: 10.1007/S40265-016-0633-9.
68. A. Nappi et al., "Metastatic colorectal cancer: Role of target therapies and future perspectives," *Current Cancer Drug Targets*, vol. 18, no. 5, pp. 421–429, Feb. 2017, doi: 10.2174/1568009617666170209095143.
69. Q. Y. Feng et al., "Anti-EGFR and anti-VEGF agents: Important targeted therapies of colorectal liver metastases," *World Journal of Gastroenterology :* WJG, vol. 20, no. 15, p. 4263, Apr. 2014, doi: 10.3748/WJG.V20.I15.4263.
70. F. Foubert, T. Matysiak-Budnik, and Y. Touchefeu, "Options for metastatic colorectal cancer beyond the second line of treatment," *Digestive and Liver Disease*, vol. 46, no. 2, pp. 105–112, Feb. 2014, doi: 10.1016/J.DLD.2013.07.002.
71. G. Chong and D. Cunningham, "Second-line therapy for advanced colorectal cancer," *Gastrointestinal Cancer Research*, vol. 1, no. 2, p. 57, Mar. 2007, doi: 10.1007/s11888-005-0010-7.
72. R. M. McQuade, V. Stojanovska, J. C. Bornstein, and K. Nurgali, "Colorectal cancer chemotherapy: The evolution of treatment and new approaches," *Current Medicinal Chemistry*, vol. 24, no. 15, pp. 1537–1557, Jun. 2017, doi: 10.2174/0929867324666170111152436.
73. T. André, S. Lonardi, K. Y. M. Wong, H. J. Lenz, F. Gelsomino, M. Aglietta, M. A. Morse, E. Van Cutsem, R. McDermott, A. Hill, M. B. Sawyer, A. Hendlisz, B. Neyns, S. Abdullaev, A. Memaj, M. Lei, M. Dixon, S. Kopetz, and M. J. Overman, "Nivolumab plus low-dose ipilimumab in previously treated patients with microsatellite instability high mismatch repair-deficient metastatic colorectal cancer: 4-year follow up from CheckMate142," *Ann Oncol.*, Oct. 2022, vol. 33, no. 10, pp. 1052–1060. doi: 10.1016/j.annonc.2022.06.008.
74. D. Coupez, P. Hulo, Y. Touchefeu, C. Bossard, and J. Bennouna, "Pembrolizumab for the treatment of colorectal cancer," *Expert Opinion on Biological Therapy*, vol. 20, no. 3, pp. 219–226, Mar. 2020, doi: 10.1080/14712598.2020.1718095.
75. L. A. Diaz et al., "Pembrolizumab therapy for microsatellite instability high (MSI-H) colorectal cancer (CRC) and non-CRC.," *Journal of Clinical Oncology*, vol. 35, no. 15_suppl, pp. 3071–3071, May 2017, doi: 10.1200/JCO.2017.35.15_SUPPL.3071.
76. M. J. Overman et al., "Nivolumab in patients with metastatic DNA mismatch repair-deficient or microsatellite instability-high colorectal cancer (CheckMate 142): An open-label, multicentre, phase 2 study," *Lancet Oncology*, vol. 18, no. 9, pp. 1182–1191, Sep. 2017, doi: 10.1016/S1470-2045(17)30422-9.
77. R. J. Mayer et al., "Randomized trial of TAS-102 for refractory metastatic colorectal cancer," *New England Journal of Medicine*, vol. 372, no. 20, pp. 1909–1919, May 2015, doi: 10.1056/NEJMOA1414325.
78. J.-Y. Douillard et al., " Panitumumab–FOLFOX4 treatment and RAS mutations in colorectal cancer," *New England Journal of Medicine*, vol. 369, no. 11, pp. 1023–1034, Sep. 2013, doi: 10.1056/NEJMOA1305275/SUPPL_FILE/NEJMOA1305275_DISCLOSURES.PDF.
79. S. Kopetz et al., " Encorafenib, binimetinib, and cetuximab in BRAF V600E–mutated colorectal cancer," *New England Journal of Medicine*, vol. 381, no. 17, pp. 1632–1643, Oct. 2019, doi: 10.1056/NEJMOA1908075.
80. T. Reinert et al., "Analysis of circulating tumour DNA to monitor disease burden following colorectal cancer surgery," *Gut*, vol. 65, no. 4, pp. 625–634, Apr. 2016, doi: 10.1136/GUTJNL-2014-308859.
81. L. H. J. Simkens et al., "Maintenance treatment with capecitabine and bevacizumab in metastatic colorectal cancer (CAIRO3): A phase 3 randomised controlled trial of the Dutch Colorectal Cancer Group," *The Lancet*, vol. 385, no. 9980, pp. 1843–1852, May 2015, doi: 10.1016/S0140-6736(14)62004-3.

82. L. Banias, I. Jung, R. Chiciudean, and S. Gurzu, "From Dukes-MAC staging system to molecular classification: Evolving concepts in colorectal cancer," *International Journal of Molecular Sciences*, vol. 23, no. 16, p. 9455, Aug. 2022, doi: 10.3390/IJMS23169455.
83. N. Wolmark et al., "Adjuvant therapy of Dukes' A, B, and C adenocarcinoma of the colon with portal-vein fluorouracil hepatic infusion: Preliminary results of National Surgical Adjuvant Breast and Bowel Project Protocol C-02," *Journal of Clinical Oncology*, vol. 8, no. 9, pp. 1466–1475, 1990, doi: 10.1200/JCO.1990.8.9.1466.
84. C. Twelves et al., "Capecitabine as adjuvant treatment for stage III colon cancer," *New England Journal of Medicine*, vol. 352, no. 26, pp. 2696–2704, Jun. 2005, doi: 10.1056/NEJMOA043116.
85. T. André et al., "Oxaliplatin, fluorouracil, and leucovorin as adjuvant treatment for colon cancer," *New England Journal of Medicine*, vol. 350, no. 23, pp. 2343–2351, Jun. 2004, doi: 10.1056/NEJMOA032709/ASSET/6C1F77C2-DE3C-4FBC-BAC6-09898169CEC1/ASSETS/IMAGES/LARGE/NEJMOA032709_F3.JPG.
86. T. André et al., "Improved overall survival with oxaliplatin, fluorouracil, and leucovorin as adjuvant treatment in stage II or III colon cancer in the MOSAIC trial," *Journal of Clinical Oncology*, vol. 27, no. 19, pp. 3109–3116, Jul. 2009, doi: 10.1200/JCO.2008.20.6771.
87. J. Y. Douillard et al., "Irinotecan combined with fluorouracil compared with fluorouracil alone. As first-line treatment for metastatic colorectal cancer: A multicentre randomised trial," *Lancet*, vol. 355, no. 9209, pp. 1041–1047, Mar. 2000, doi: 10.1016/S0140-6736(00)02034-1.
88. F. A. Sinicrope, "DNA mismatch repair and adjuvant chemotherapy in sporadic colon cancer," *Nature Reviews Clinical Oncology*, vol. 7, no. 3, pp. 174–177, Mar. 2010, doi: 10.1038/nrclinonc.2009.235.
89. A. de Gramont et al., "Leucovorin and fluorouracil with or without oxaliplatin as first-line treatment in advanced colorectal cancer," *Journal of Clinical Oncology*, vol. 18, no. 16, pp. 2938–2947, Aug. 2000, doi: 10.1200/JCO.2000.18.16.2938.
90. L. B. Saltz et al., "Irinotecan plus fluorouracil and leucovorin for metastatic colorectal cancer," *New England Journal of Medicine*, vol. 343, no. 13, pp. 905–914, Sep. 2000, doi: 10.1056/NEJM200009283431302.
91. H. Hurwitz et al., "Bevacizumab plus irinotecan, fluorouracil, and leucovorin for metastatic colorectal cancer," *New England Journal of Medicine*, vol. 350, no. 23, pp. 2335–2342, Jun. 2004, doi: 10.1056/NEJMOA032691.
92. E. Van Cutsem et al., "Cetuximab and chemotherapy as initial treatment for metastatic colorectal cancer," *New England Journal of Medicine*, vol. 360, no. 14, pp. 1408–1417, Apr. 2009, doi: 10.1056/NEJMOA0805019.
93. F. Loupakis et al., "Initial twith FOLFOXIRI and bevacizumab for metastatic colorectal cancer," *New England Journal of Medicine*, vol. 371, no. 17, pp. 1609–1618, Oct. 2014, doi: 10.1056/NEJMOA1403108.
94. A. Grothey et al., "Regorafenib monotherapy for previously treated metastatic colorectal cancer (CORRECT): An international, multicentre, randomised, placebo-controlled, phase 3 trial," *The Lancet*, vol. 381, no. 9863, pp. 303–312, Jan. 2013, doi: 10.1016/S0140-6736(12)61900-X.
95. J. Cassidy et al., "Randomized phase III study of capecitabine plus oxaliplatin compared with fluorouracil/folinic acid plus oxaliplatin as first-line therapy for metastatic colorectal cancer," *Journal of Clinical Oncology*, vol. 26, no. 12, pp. 2006–2012, 2008, doi: 10.1200/JCO.2007.14.9898.
96. A. Nakamura et al., "Enhancement of 5-fluorouracil-induced cytotoxicity by leucovorin in 5-fluorouracil-resistant gastric cancer cells with upregulated expression of thymidylate synthase," *Gastric Cancer*, vol. 17, no. 1, pp. 188–195, Jan. 2014, doi: 10.1007/S10120-013-0249-7.
97. D. Sebag-Montefiore et al., "Preoperative radiotherapy versus selective postoperative chemoradiotherapy in patients with rectal cancer (MRC CR07 and NCIC-CTG C016): A multicentre, randomised trial," *The Lancet*, vol. 373, no. 9666, pp. 811–820, Mar. 2009, doi: 10.1016/S0140-6736(09)60484-0.
98. E. Van Cutsem et al., "Binimetinib, encorafenib, and cetuximab triplet therapy for patients with BRAF V600E–mutant metastatic colorectal cancer: Safety lead-in results from the phase III BEACON Colorectal Cancer Study," *Journal of Clinical Oncology*, vol. 37, no. 17, pp. 1460–1469, 2019, doi: 10.1200/JCO.18.02459.
99. J. Y. Douillard et al., "Randomized, phase III trial of panitumumab with infusional fluorouracil, leucovorin, and oxaliplatin (FOLFOX4) versus FOLFOX4 alone as first-line treatment in patients with previously untreated metastatic colorectal cancer: The PRIME study," *Journal of Clinical Oncology*, vol. 28, no. 31, pp. 4697–4705, Nov. 2010, doi: 10.1200/JCO.2009.27.4860.

7 Colorectal Cancer Treatment Using Radiation Therapy

Advantages and Disadvantages

Kalsoom Saleem[1] *and Nasrullah Jan*[2,3,4]

[1]Faculty of Pharmaceutical Sciences, Riphah International University, Islamabad, Pakistan.

[2]Department of Rheumatology and Clinical Immunology, the First Affiliated Hospital of Xiamen University, School of Medicine, Xiamen University; Xiamen, China.

[3]Xiamen Municipal Clinical Research Center for Immune Diseases; Xiamen, China.

[4]Xiamen Key Laboratory of Rheumatology and Clinical Immunology; Xiamen, China.

7.1 INTRODUCTION

Colorectal cancer (CRC) is one of the most diagnosed malignant cancers worldwide, with 1.9 million new cases in 2020 and more than 930,000 deaths worldwide, making it the third most common cancer and second leading cause of death. According to the American Cancer Society (ACS), in 2024 an estimated 152,810 cases of CRC have been reported in the United States, among which 106,590 cases were reported for colon cancer alone. By May 2024, total deaths by CRC surpassed 50,000 in the United States (Society, 2024). According to the World Health Organization (WHO), the burden of CRC might report an increase of 63% annually (3.2 million new cases) and an increase of 73% in death rate (1.6 million deaths annually) by 2040. The incident rates were highest in New Zealand, Australia, and Europe, with Eastern Europe ranking highest in mortality rates (World Health Organization, 2023). Thus, there is always an urgent need to find safe and effective methods of CRC treatment and diagnosis. Radiotherapy (radiation therapy, RT) has become one of the primary treatment modalities, along with chemotherapy, hormonal therapy, immunotherapy, and surgery. Since the late 1900s, improvements in existing technologies and technology-driven research leading to innovative delivery methods have enhanced the accuracy of RT, improved treatment delivery, and increased survival rates for patients suffering from cancer (Fiorino et al., 2020). Along with the treatment of CRC, it is also used as palliative care to ameliorate cancer-associated symptoms, i.e., relieving pain for patients suffering from bone metastasis, decreasing cancer size that blocks airways (esophagus), or improving patients' ability to breathe or swallow and improving patients' quality of life (Chen et al., 2021). Hence, radiologists have a two-dimensional goal: first, to safely increase the radiation dose for cancer cells, and second, to protect the normal/healthy cells from radiation toxicities. Incorporating these research-driven, cutting-edge technologies has assisted in the achievement of these goals. For instance, an improved long-term overall survival (OS) was predicted by achieving metastatic control using stereotactic body radiotherapy (SBRT) for patients with lung and liver CRC metastasis (Klement et al., 2019).

Due to advancements in RT, it is considered a first-line treatment option in more than 30% of patients in the United States (Wang and Tepper, 2021), and 50% of patients receive RT during the course of cancer treatment. However, the effects of RT are not cancer-specific and may cause

DOI: 10.1201/9781003503842-7

toxicities to surrounding normal/healthy tissues and organs. Distant metastasis, local cancer recurrence (Anuja et al., 2019; Garden et al., 2013), and resistance mechanisms (Mortezaee and Najafi, 2021) are critical concerns in CRC, leading to poor quality of life and increased mortality. Due to this, it is crucial to comprehend the toxicities of RT and implement specialized RT techniques with enhanced efficacy in clinical settings. These techniques may also serve as an alternative treatment for patients who are unable to undergo surgery or chemotherapy.

7.1.1 Role of Radiation Therapy in Treatment of Cancer

The need for adjuvant treatment to supplement chemotherapy and/or surgery roots from the observation that a significant amount of CRC relapses despite being apparently completely removed. The primary role of RT is to prevent further growth of CRC cells that were not removed through surgery (Gao et al., 2019). Radiation given prior to surgery can potentially inhibit the proliferation or growth of CRC, whether it remains localized or has metastasized to nearby areas at the time. Several preclinical studies done in vitro and in vivo have shown promising results. The results obtained from preclinical studies have encouraged clinical trial translations that are in different phases (NCT01539824, NCT03122509, NCT02909114, and NCT02298946). This was tested by Li and colleagues in 2020, where tumor and normal paired tissue and plasma samples were collected from stage I–III CRC patients. Rectal cancer patients with high LincRNA-p21 expression benefited from postoperative adjuvant chemoradiotherapy with longer OS in comparison to the group of patients with high LincRNA-p21 expression with shorter OS (Li et al., 2020). Underlying this, RT induces tumor microenvironment (TME) remodeling by balancing immunostimulatory signals, leading to increased tumor antigen presentation, T-cell activation (Charpentier et al., 2022), natural killer cell activity (Parikh et al., 2021), and synergistic action with chemotherapeutic drugs (Gao et al., 2019).

7.1.2 Current Standard Approach

The treatment of colon cancer patients is most importantly based on the staging of cancer, upon which the strategy of diagnosis and treatment resides. Though surgery is the primary treatment option in CRC, chemotherapy is preferred in adjuvant settings. Adjuvant RT is classified as preoperative RT, intraoperative RT, and postoperative RT. Its main aim is to improve the local control rate and survival rate of CRC patients. For many years, preoperative RT has been widely employed as the standard of care for locally advanced rectal cancer. According to American Society for Radiation Oncology guidelines, the standard fractionated chemoradiation accepted is 5,000–5,400 cGy in 180–200 cGy per fraction, or for short-course RT, 2,500 cGy in 500 cGy per fraction. Subsequently, this approach evolved with increasing interest in total neoadjuvant therapy (TNT), nonoperative management, or selective RT. Considering TNT for rectal cancer, patients with stage II–III rectal cancer, it is highly recommended. A multicenter stratified randomized clinical trial was conducted by Deng and colleagues, which was meant to evaluate the effectiveness of TNT in stage II–III rectal cancer patients (Wo et al., 2021). The trial results showed an improvement in OS (86.9% vs 76.5%) and incidence of DFS (disease-free survival; 87.0% vs 67.9%) with a lower incidence of distant recurrence (12.5% vs 29.4%) (Deng et al., 2020). For patients with stage II rectal cancer with a lower risk of developing local recurrence, conditional RT can be omitted after discussion with a multidisciplinary team. Chances of sphincter preservation can be improved in patients with cT1–2N0 rectal cancer (Wo et al., 2021).

According to the Japanese Society for Cancer of the Colon and Rectum guidelines, preoperative RT aims to improve the anal sphincter preservation rate and resection rate, and is indicated when the cancer stage is diagnosed as invasion depth cT3 or deeper, or cN positive. Intraoperative RT is indicated when surgical dissection or penetration of the surgical dissection plane is unclear [surgical dissection plane positive, RM1; penetration of surgical dissection (RMX)]. Postoperative RT is indicated when cancer stage is pathologically diagnosed after surgery as invasion depth cT3 or deeper, or cN positive RMX or RM1 status unclear (Watanabe et al., 2018).

A clinical trial, RAPIDO (NCT01558921), aimed at determining the systemic relapse of rectal cancer using short-course RT in combination with chemotherapy. The experimental arm patients received RT (5-5 Gy, 8 days) and six cycles of CAPOX chemotherapy (combination of oxaliplatin and capecitabine), followed by a total mesorectal excision. The findings suggested a decreased probability of disease-related treatment failure in comparison to adjuvant chemotherapy or locoregional failure, distant metastasis, treatment-related deaths, and/or cancer recurrence.

7.2 MECHANISM OF RADIOTHERAPY SENSITIZATION AND DNA DAMAGE RESPONSE

RT is considered an essential approach for the treatment of cancer because it not only saves millions of lives of patients but also provides them with an economical alternative. Therefore, access to RT is expected to become globally accessible in the near future. However, it is also accompanied by unavoidable side effects (discussed in the later part of this chapter). The most widely studied mechanism of RT is direct DNA damage, known as radiation-induced DNA damage (RIDD). Therefore, the DNA damage response of cancer cells and the ability of cancer cells to repair damaged DNA are crucial in determining the outcomes of cancer cells. Once the ionizing radiations (IR) deposit high or low energy onto DNA molecules at the cancer site, a fraction of DNA will sustain two or more damages within either single or double strands of DNA. There is growing evidence that IR induces complex DNA double-strand breaks (DSBs), extremely vital for cellular homeostasis. This represents a threat to genomic integrity. When at least two DNA damage sites are created within one strand, this is called clustered damage sites (CDS), from which DSBs can be generated (Nickoloff et al., 2020). Additionally, IR also compromises base and sugar damage that is thermally responsive (Figure 7.1) (Pilié et al., 2019). Xu et al. provided real-time observation of CDS in irradiated DNA, where two to four damages were produced in a single plasmid (Xu et al., 2019). Previously, the lethality of cancer cells was reported after the introduction of DSBs (Schipler et al., 2016). The damage caused by irradiation forces cancer cells to evolve and develop repair or compensatory mechanisms to maintain genomic integrity (Ferreira and Dutreix, 2019) via several repair pathways, which may result in either RT failure or the development of RT resistance (Shukla et al., 2020) (discussed in Section 7.5).

When a signal is received, various biochemical reactions are initiated in response to DNA damage. This leaves us with key questions: how is the DNA damage signal sensed and recognized, and how are the signaling biochemical reactions initiated or triggered (Maremonti et al., 2020). Moreover, DNA damage sensors also have the capability to recruit RIDD proteins to the damaged site and signal transducers to transmit the signals. Sensors and transducers typically coexist, with transducers often relying on kinase activity to initiate biochemical reactions (Huang and Zhou, 2020). Rad24p is considered the primary responder after DNA damage triggers cell cycle arrest (Shinohara et al., 2019). The Mec3p complex activates a series of biochemical reactions to produce downstream kinases (Liu et al., 2021), such as homologous recombination (HR) in response to CDS (Meyer et al., 2020) and sensor proteins like γH2AX, MDC51, 3BP1, NBS1, BRCA1/2, Nbs1/hMre11/hRad50 complex, and Ku (Ku70/80 dimer) (Huang and Zhou, 2020).

Another widely studied mechanism involves indirectly promoting the deposition of low or high energy onto molecules in the vicinity of cancer DNA. This results in the production of highly reactive oxygen species (ROS), including O_2^- species (superoxide anion), OH^- species (hydroxyl radicals), and H_2O_2 species (hydrogen peroxide) (Liu et al., 2021). These species are toxic to cancer cells and nearby healthy tissues. Additionally, ROS production in mitochondria is also triggered, which alters membrane integrity and further initiates ROS production (McCann et al., 2021). Furthermore, RT also induces disruption in the electron transport chain, causing intracellular redox imbalance (Nilsson and Liu, 2020). This chain of events leads to oxidative stress when ROS reacts with biological molecules (proteins, DNA, and lipids), resulting in protein misfolding, DSBs, CDS, and

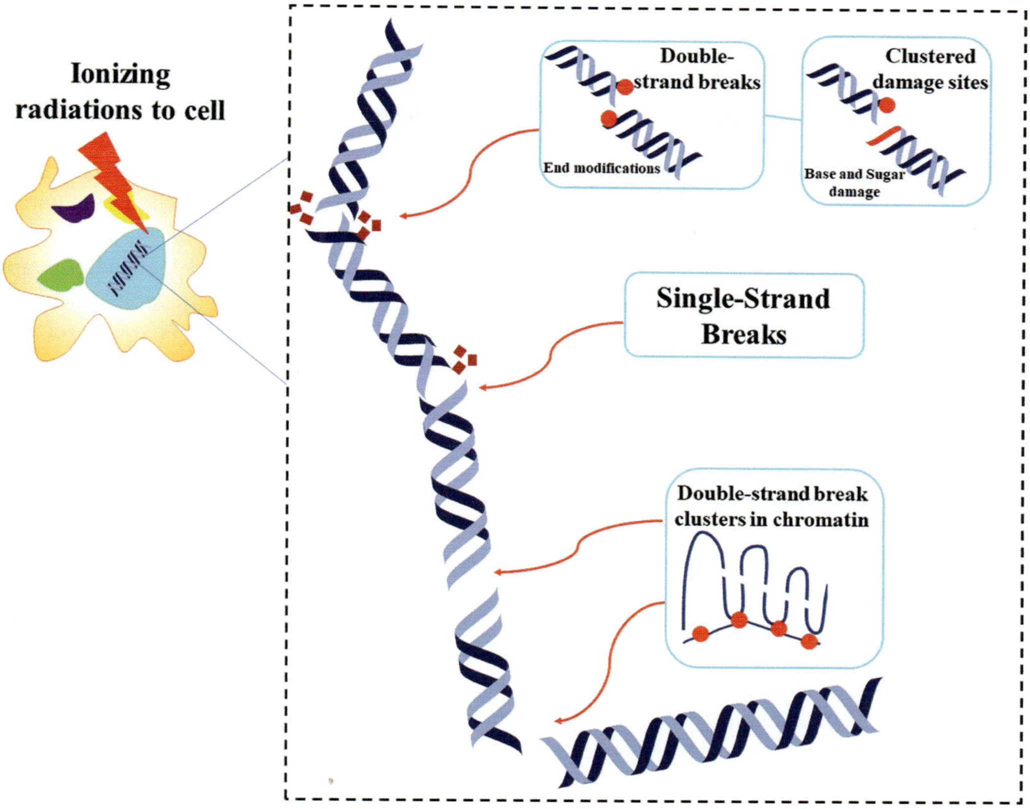

FIGURE 7.1 Radiation-induced DNA damage (RIDD). The major RIDD is induced both by high- and low-energy ionizing radiations such as single-strand breaks (SSBs), double-strand breaks (DSBs), sugar and base pair alterations, and covalent crosslinking. Clustered damage sites (CDS) are induced after low-energy radiation, and CDS increase with an increase in irradiations. DSBs located in the vicinity of chromatin destabilize them.

lipid peroxidation (Ye et al., 2020). As a feedback mechanism, p53 activates redox regulation and promotes detoxification of ROS by upregulating reducing molecules such as NADPH (nicotinamide adenine dinucleotide phosphate) and glutathione (GTH), leading to resistance to RT.

7.3 TECHNIQUES IN RT FOR CRC

7.3.1 Contact Therapy

Previously, contact therapy/Papillon delivered a large dose of radiation in endocavitary (10–40 Gy per fraction) per week or biweekly (a total dose of 45–100 Gy). Due to such high doses, it was claimed that this therapy could achieve optimum CRC control and rectal preservation among patients with early-stage CRC (Gérard et al., 2008; Hull et al., 1994; Kovalic, 1988). It utilizes low-energy X-rays (usually 50 keV) directed toward the tumor surface and penetrates mainly a few millimeters in depth of tumor tissue. Mostly suitable for exophytic tumors (tending to grow outward from the point of origin) rather than deeply infiltrating tumors; hence, it treats only small-volume tumor tissue (<5 cm^3). Due to its access to only a small volume of tumor tissue, it is administered at high radiation doses (≈30 Gy) but within the biological dose (100 Gy) that can be safely delivered in different fractions. When the X-ray beam is targeted directly to the tumor site (5 mm margin), the tumor decreases in layers, i.e., few millimeters every 2 weeks, a fraction during which the tumor suppresses, following which the second treatment is delivered. This also causes minimal damage to normal tissues surrounding the tumor cells that recover during the 2-week interval after the first dose. Following the second treatment dose, a third fraction is given, at the end of which very small tumor tissues can be seen using MRI. However, high-dose-rate endorectal brachytherapy (HDREBT), a variant of contact X-ray radiotherapy, is only also possible in cases where a good response (60% tumor size reduction) to external beam chemoradiotherapy is observed. Minimal side effects (ulceration, bleeding) are observed in patients who usually recover on their own within 3–6 months or up to 1–2 years if the patient is on anticoagulants (Sun Myint et al., 2019). This technique has been reconsidered after the introduction of the Philip RT 50 (contact radiotherapy machine) as a curative and conservative treatment option in early-stage CRC inoperable patients. Advancement in this therapy was reported recently; a photoplethysmography sensor was developed for contact therapy. The investigation showed that the sensor is efficacious in obtaining signals through contact- to non-contact separation of 5 mm along with a variation in intensity between normal and hypoxic cells/tissues (Chatterjee et al., 2020). Due to these advancements, several clinical trials were performed to assess its effectiveness. For instance, a retrospective clinical analysis of the prospective cohort was performed on selected T2 and T3 patients using contact X-ray therapy. The primary objectives were to increase the clinical complete response (cCR) while decreasing local recurrence rates. Here the patients were given contact X-ray radiotherapy (90 Gy for 3 fractions for 4 weeks) in combination with capecitabine. Out of 74 patients, 95% responded with cCR, and the 3-year local recurrence rate was 10% and OS was 88%. Two patients underwent radical proctectomy while 96% were cured with organ preservation (Gérard et al., 2019). The OPERA trial, a multicenter open-label phase 3 clinical trial (NCT02505750), was conducted at 17 cancer centers. Patients with cT2, cT3a, and/or cT3b operable rectal adenocarcinoma received chemoradiotherapy and 45 Gy contact X-ray radiotherapy in 25 fractions. Assessing the primary outcome (3 years OS without radical surgery), it was concluded that contact X-ray radiotherapy significantly improved the 3-year organ preservation rate, especially in patients with tumor sizes smaller than 3 cm, compared to chemoradiotherapy with a boost of external beam radiotherapy (EBRT). This approach was considered the most suitable for avoiding surgery or for organ preservation (Gerard et al., 2023), leading to a more favorable response (Benezery et al., 2020). With similar results, the ACCORD 12 clinical phase III trial conducted on 534 T2 and T3 patients concluded that the organ preservation strategy using contact X-ray radiotherapy yielded a 5-year cancer-specific survival rate comparable to the results obtained from RP-TME (radical proctectomy) (Gerard et al., 2022).

7.3.2 High-Dose-Rate Endorectal Brachytherapy

HDREBT uses high-energy γ-rays or electron beams to irradiate the surface of tumors, initially employed for palliative care. The MAASTRO applicator, designed by Bellezzo and colleagues, is used to enhance contact with tumor tissue. It is a multichannel lateral shielded and angle-edged device (Bellezzo et al., 2020). Similarly, some recently designed applicators also incorporate dose shielding that further reduces the risks to organs and enhances the response to radiotherapy (Devic et al., 2019). Due to these technological advancements, HDREBT now covers large tumor volumes as well as locally advanced CRCs (LACRCs) (Chiang et al., 2020; Garant et al., 2019). Image-guided HDREBT (12–18 Gy) was performed in combination with EBRT (10–13 Gy × 3) for T1-T3 rectal cancer patients. 67% of patients demonstrated cCR, with one patient showing residual tumor. Hence, it was concluded that primary EBRT in combination with HDREBT is feasible, well tolerated, and highly individualized for patients who are not suitable for surgery or refuse to undergo surgery (Fleischmann et al., 2022). Similar results were obtained in clinical phase I and II studies with HDERBT in combination with EBR, which was found to be safe and effective in achieving high tumor suppression with minimal or no side effects (Vuong et al., 2022).

Surprisingly, this treatment can also be applied to locally advanced rectal carcinoma (LARC). A widely used treatment strategy for operable LARC patients is a combination of HDERBT and

EBR. Previously reported literature has shown that dose escalation with HDERBT resulted in higher cCR rates, was well tolerated, and led to better tumor suppression compared to neo-chemotherapy (Saeedian et al., 2023). Similarly, a clinical phase study was conducted for patients who refused surgery or were ineligible for surgery. The study involved EBR with a dose of 40 Gy delivered in 16 fractions, followed by HDERBT with a dose of 30 Gy delivered in 3 fractions applied to the residual tumor tissue. An 86% cCR, 12.8% re-growth proportion, and 72.8% local control rate at 2 years were observed. It was concluded that this technology will likely contribute to challenging nonoperative or organ preservation paradigms for CRC patients (Garant et al., 2019) with excellent long-term quality of life assessment (Dizdarevic et al., 2020) and no significant increase in toxicity or complications in surgery (Saeedian et al., 2023).

7.3.3 Radiation Therapy for Oligometastasis (Liver, Lungs, Spine) or Locally Advanced CRC

7.3.3.1 Intraoperative Radiotherapy

Intraoperative radiotherapy (IORT) is another unique radiotherapy technique that aims to improve treatment outcomes, such as OS and DFS, for LACRC patients. The radiation is applied precisely to the tumor site, minimizing exposure of normal tissues or organs at risk (OAR) by displacing them far away from the irradiation source and shielding them. Additionally, dose escalation for re-irradiation in the recurrent tumors is also possible with it, in comparison to EBRT, where further dose escalation or irradiation is not possible. It is usually delivered in combination with EBRT (IOEBRT), HDERBT (HDR-IORT), or orthovoltage X-rays. Here, IOERT delivers radiation to a depth of 1 cm or more over several minutes, while HDR-IOT delivers radiation to all surfaces for a longer period of time (Tam and Wu, 2019). Recently, the effectiveness of IORT was investigated in patients with locally advanced colon cancer (LACC) treated with low-kV X-rays. Along with clinicopathologic characteristics, OS and DFS were assessed as primary outcomes. The 3-year DFS rate was 82.8%, and OS was 82.1%. At the 20.5-month follow-up, no in-field failures occurred, but locoregional and distant failures were observed in 5% of patients. From the results, Ma et al. concluded that patients with LACC who undergo low-kV X-ray IORT can achieve significant locoregional control, DFS, OS, and distant cancer control without an increase in short- or long-term side effects. This approach would be most effective for patients with pT3 LACC (Ma et al., 2020).

RT given in combination with capecitabine/5-fluorouracil showed cCR and tumor suppression rates of 18% and 81%, respectively. The 3-year DFS was 68.6%, and OS was 82.1%. Grade 3–4 side effects (myelosuppression) were developed in 17 out of 100 patients. Surgery-related complications developed in 7% of patients. Here, it was concluded that chemoradiotherapy was feasible and safe for selected LACC patients with acceptable toxicities (Yuan et al., 2021). Additionally, the addition of IORT to treatment options for LARC patients leads to a lower rate of locoregional recurrence (Fahy et al., 2021). Moreover, it can also be used as an immediate adjuvant treatment when access to EBR is limited. Potemin and colleagues reported that 4-year DFS was 59.4% in the IORT group (Potemin et al., 2020). A meta-analysis also reported enhanced treatment outcomes for CRC patients when treated with IORT (Mirnezami et al., 2013; Tam and Wu, 2019). Similarly, Hawkins and colleagues also reported a 62% superior 5-year OS compared to no neoadjuvant radiotherapy (45.7%) in patients with stage IV colon cancer (Hawkins et al., 2019). Above all, manageable toxicity has been observed in patients. Most commonly in LARC, this includes weakness or chronic pain from peripheral nerve involvement, with neuropathy as secondary to it (Wang and Tepper, 2021).

7.3.3.2 Stereotactic Body Radiotherapy

SBRT is a recently introduced RT technique for pelvic recurrence of CRC, most effectively used for inoperable patients. Patients undergo computed tomography (CT) simulation in a supine position. High-resolution CT images offer better targeting of larger tumor volumes and significantly reduce the risk of OARs compared to EBRT. This also allows for the administration of higher doses of radiation (5–16 Gy) to the tumor site with maximum reduction in mechanical errors (Tam and Wu, 2019). Hence, when planning for SBRT, both CT and MRI images may be obtained for tumor localization. One to three fractions may be suitable for tumors located away from the central liver or OAR, whereas five fractions are preferred otherwise. For tumors in close proximity to or overlapping OAR, an adequate balance between local tumor control and cure and late toxicities must be considered carefully (Lukovic and Dawson, 2023). Modern dose or individualized dose escalation SBRT regimens can achieve 1-year local tumor control rates of 77%–100% for lung metastasis of CRC, 90%–100% for liver metastasis, and 81%–95% for spine metastasis. Toxicities typically experienced by patients are grade 3 or higher, with lower incidence rates below 10% in the lungs, below 5% in the liver, and between 2% and 8% in the spine (including neurological or non-neurological toxicities) (Wild and Yamada, 2017).

Implications for dose escalation and individualized treatment selection can be made with SBRT with the benefit of sparing healthy tissues. This was supported by previous literature where patients with lung metastasis of CRC were associated with significantly lower local tumor control in comparison to non-CRC lung metastasis patients [confidence interval 95% (CI), 1.93–4.45] but higher OS (CI 95%, 0.45–0.82) with no reported mortalities (Cao et al., 2019; Choi et al., 2020). A cohort clinical trial conducted on 100 CRC lung metastasis patients showed an 85% 1-year DFS rate and a 70% 2-year DFS rate in the surgical group, compared to an 87% 1-year DFS rate and a 71% 2-year DFS rate in the SBRT group. The study reported no grade 3 or higher toxicities (Garcia-Exposito et al., 2023). For liver metastasis of CRC, more promising results were obtained compared to radiofrequency ablation. SBRT has been shown to provide 90% local tumor control, compared to 72% from radiofrequency ablation (Yu et al., 2021). This technique has provided a greater opportunity for liver metastasis in CRC patients, as only 20% have clinically resectable liver metastases. SBRT, therefore, has potential application in patients unfit for surgery or those seeking to avoid surgery (Sanuki et al., 2022). A retrospective study conducted on patients with spine and non-spine metastases of CRC investigated local failure as the primary outcome when prescribed with 24 Gy in 2 fractions and 35 Gy in 5 fractions. With a local failure rate of 44%, it was concluded that SBRT is not suitable for bone metastasis (Ito et al., 2020), which contradicted the results of a meta-analysis reported by Guninski and colleagues. They showed high efficacy of SBRT with an overall pain response rate of 83% and cCR of 36%, along with minimal toxicities (9% vertebral fracture) (Guninski et al., 2024). Although large comparative clinical phase studies are lacking, SBRT is regarded as an excellent treatment option for the local treatment of CRC metastases, offering significant outcomes and a low toxicity profile. It is particularly suitable for inoperable patients with co-morbidities, chemotherapy complications, or those who refuse surgery.

7.3.3.3 Radioembolization

When the metastasis is confined to the liver, radioembolization (RE) is a suitable treatment option in addition to chemotherapy. Transarterial RE with yttrium-90 (Y-90) glass microspheres (TARE) delivers micro-sized beta-emitting particles through the blood supply of the liver parenchyma (arteries that feed liver tumors), where these resin/glass-embedded microspheres disperse into the tumor tissue. It then decreases the blood supply to tumor tissues at the microvascular level, minimizing side effects with a selective strategy approach (Tam and Wu, 2019; Türk et al., 2019). When applied to liver metastasis patients of CRC, the patients with left colon cancer have a significantly higher OS compared to right colon cancer patients. Resin-based microspheres showed higher OS and liver progression-free survival compared to glass-based microspheres (Türk et al., 2019). Another randomized open-label multicenter clinical phase II study showed an outcome response of 34% for the TARE group compared to 21.1% for the chemotherapy group. Unfortunately, grade 3 toxicities were frequently reported in the TARE group among 428 patients. Though it can be added as a second-line treatment option for liver metastasis of CRC, subset analysis would be necessary to identify the ideal patient population that would benefit from it (Mulcahy et al., 2021).

7.4 ADVANTAGES OF RT

RT technology has been crucial in the oncology field, improving patient outcomes. For instance, in oligometastatic disease, where the cancer state is in between the early and metastasized stages, the cancer has spread beyond the site of origin, but the metastatic potential is so small representing only its initial stages. The surgical resection in response to oligometastases was the only local treatment until 2003–2007 (Schanne et al., 2019). It is only now, with the help of RT (or SBRT), that nonsurgical local eradication of oligometastases is possible with sufficient safety and efficacy. With technological advancements in RT, there is improved accuracy, a better therapeutic ratio, low toxicity risks, and simultaneous delivery of high doses for local cancer control along with high safety and efficacy in radioresistant patients (Guckenberger et al., 2016; Klement et al., 2019). Not only this, but stereotactic ablative radiotherapy has also been considered the first local therapy for patients with lung oligometastases from CRC in a cohort study (Filippi et al., 2015). It has also solved the treatment question for surgically inoperable, unwilling patients or those not amenable to other alternative treatments. In a phase II clinical trial for SBRT for patients with inoperable liver metastases from CRC, it was found that a local control of 91% was achieved in those patients in 24 months with OS of 29.2 ± 3.7 months. Besides this, no patients experienced any toxicity of >grade 3 or RT-induced liver damage (Scorsetti et al., 2015).

Resectable rectal cancer with clinical T3/T4 or N1/N2 stage was also investigated for acute toxicity, pathological complete response, local recurrence-free survival, and disease-free survival. With dose escalation up to 46 Gy, the treatment was well tolerated, with only 4% of patients reporting grade 3/4 toxicity. A trend toward pathological complete response was obtained with higher doses, along with statistically significant increases in local recurrence-free survival and disease-free survival (Wiltshire et al., 2006). Similarly, improved OS was also obtained in patients with LACC with the use of neoadjuvant RT (Hawkins et al., 2019). As a dose escalation study of HDREBT, it was found to be safe, effective, and well tolerated with a higher cancer regression rate in rectal cancer patients (Vuong et al., 2022). Additionally, one RT option can be used if the other is inaccessible; for instance, IORT can be used as an immediate adjuvant treatment for rectal cancer patients when limited access is present for EBRT (Potemin et al., 2020). With dose escalation, treatment options have widened for CRC patients; for instance, preclinical studies on cell lines and animal models have shown a different response for cancerous cells and healthy cells when compared to conventional doses normally delivered via RT (classically between 0.01 and 0.1 Gys^{-1}), which can increase 400-fold with flash RT (delivering ultra-high-dose range beams) (Bourhis et al., 2019).

7.5 DISADVANTAGES OF RT

7.5.1 RESISTANCE PATHWAYS AND RIDD REPAIR

RT kills cancer cells either directly by causing DNA damage or indirectly by generating ROS. This leads to genomic instability, cell cycle arrest, alteration of checkpoints, apoptosis, and ultimately the death of post-mitotic cancer cells. As a feedback mechanism to these changes, cancer cells evolve DNA damage repair mechanisms. The most studied primary pathways are HR, nonhomologous end joining (NHEJ, which includes PARP proteins), and alternative end joining (Figure 7.2). Though the detailed mechanism of repair remains unclear and requires further investigation, in the HR pathway, DSBs are resected from the 5′–3′ direction on one strand of the DSB, creating a 3′-OH free terminal (Li et al., 2021). Simply put, the HR pathway requires a template for a thorough and accurate restoration. Particularly, CDS can be repaired through HR. Preclinical studies on RAD54B, an effector of the HR pathway, have been conducted by McAndrew et al. They demonstrated that poly(ADP-ribose) polymerase selectively targeted and killed cells deficient in RAD54B. A rise in γ-H2AX (DSBs marker) as well as cleavage in caspase-3 (apoptosis marker) was also observed (McAndrew et al., 2016). Furthermore, CRLX101 (camptothecin nanoparticle drug conjugate) was

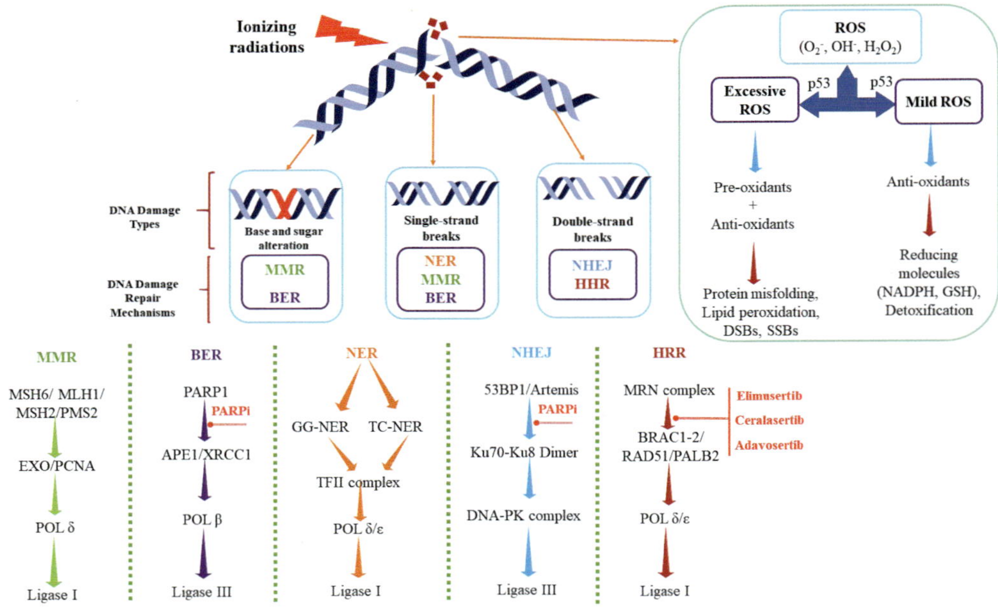

FIGURE 7.2 Different mechanisms in response to DNA damage caused by ionizing radiation, as well as common inhibitors and blockers. Here, DNA damage has been classified as single-strand breaks (SSBs), double-strand breaks (DSBs), sugar, and base pair alterations. DNA insults initiate the DNA damage response/repair pathway (DDR), which involves the activation of multifarious signaling pathways based on DNA insult. MMR and BER signaling pathways are activated in response to base and sugar pair alterations and SSBs; the NER signaling pathway is activated in response to SSBs, while NHEJ and HRR signaling pathways are activated in response to DSBs. The blockers/inhibitors presented are currently under investigation for CRC treatment. Additionally, radiation-induced reactive oxygen species (ROS) response is also seen in ionizing radiations, where ROS levels increase by radiation-induced mitochondrial damage. In the presence of elevated ROS levels, the p53 pathway may significantly ameliorate oxidative stress. *MMR*, mismatch repair; *BER*, base excision repair; *NHEJ*, nonhomologous end joining; *NER*, nucleotide excision repair; *HRR*, homologous recombination repair; *PARPi*, PARP inhibitors; *CRC*, colorectal cancer.

developed to assess its efficacy in enhancing chemoradiotherapy in CRC patients. It was found that the DNA repair mechanism in CRC cells was inhibited, as indicated by the increase in the γ-H2AX marker (Tian et al., 2017).

In the NHEJ pathway, Ku first binds to DNA ends in a ring fashion, encircling the DNA double helix. It serves as a landing pad for recruiting DNA-dependent protein kinase, which is necessary for DNA damage repair. A marker of the NHEJ pathway, MEK1/2 (mitogen-activated protein kinase 1/2 or MAPK), is inhibited by GSK212, leading to delayed DNA damage repair and a persistent increase in γH2Ax levels, indicating sustained DSB damage (Huang and Zhou, 2020).

Activation of cell cycle checkpoints is another mechanism that governs DNA damage repair. It allows cell components to be replicated and transferred to the next generations of cancer cells. A large number of proteins, namely, cyclin-dependent kinases (CDKs), CDK inhibitors (CKIs), Ink4 family members (p15–16, p18–19), Cip/Kip family members (p21, p27, p57), CDC25, p53 family proteins, and MDM2, are activated in response to initiate the repair process. Targeting these proteins in cancer cells is another strategy to overcome radiation resistance. p53, a transcription factor, plays a vital role in cell cycle regulation, primarily by monitoring the G1 and G2 phase checkpoints (Liu et al., 2019), and contributes to resistance (Deng et al., 2019). CDK5 also activates signaling to confer radiation resistance by upregulating TAZ (Zeng et al., 2020). Xie et al. investigated the effect of the CKI palbociclib and found that it amplified the radiosensitivity of cancer cells and even suppressed DNA repair (Xie et al., 2019). MDMX2 inhibitors have also paved the way for radiosensitization in preclinical studies involving various types of cancers such as lung, gastric, and colon cancer and glioblastoma (Miles et al., 2021; Stewart-Ornstein et al., 2021; Yi et al., 2018). Radiosensitization has been shown to improve after the administration of graphene-functionalized silver nanoprisms (Habiba et al., 2019). Besides these pathways, several other pathways are involved in radiation resistance (Catalano et al., 2022; Huang and Zhou, 2020; Kim et al., 2019; Liu et al., 2021). The nuclear factor-κβ pathway regulates several genes that are responsible for producing radiation resistance.

7.5.2 Toxicities

Radiation-induced injuries to normal cells are summarized in Figure 7.3, where RT directly or indirectly damages the DNA of healthy cells by initiating a cascade of reactions. These side effects can be acute, subacute, or chronic. Acute toxicities typically occur within 1–2 weeks of RT, often manifesting as inflammatory reactions, while chronic toxicities usually develop years after RT, characterized by fibrosis, vascular injury, or organ failure resulting from the impact on dividing cells. Residual DNA damage may rarely lead to carcinogenesis (Wang and Tepper, 2021).

Skin, breast, and connective tissues are where **radiation dermatitis** is predominant. Radiation dermatitis is commonly observed in RT of head and neck cancer (Bonomo et al., 2019), vulvar cancer (Teng et al., 2022), breast cancer (Xie et al., 2021a), and melanoma/skin cancer (Ashack et al., 2020). These effects are due to cutaneous epithelial depopulation including symptoms such as erythema or desquamation. In a study conducted previously, hypofractionated RT was found to reduce the risk of developing radiation dermatitis compared to conventional fractionated RT. Sequential bolus and boost significantly associated with an increased risk of radiation dermatitis, while no association was found with chemotherapy, hormonal therapy, trastuzumab therapy, or nodal radiation (Xie et al., 2021b). For instance, in breast cancer patients, the risk is highest in postmastectomy

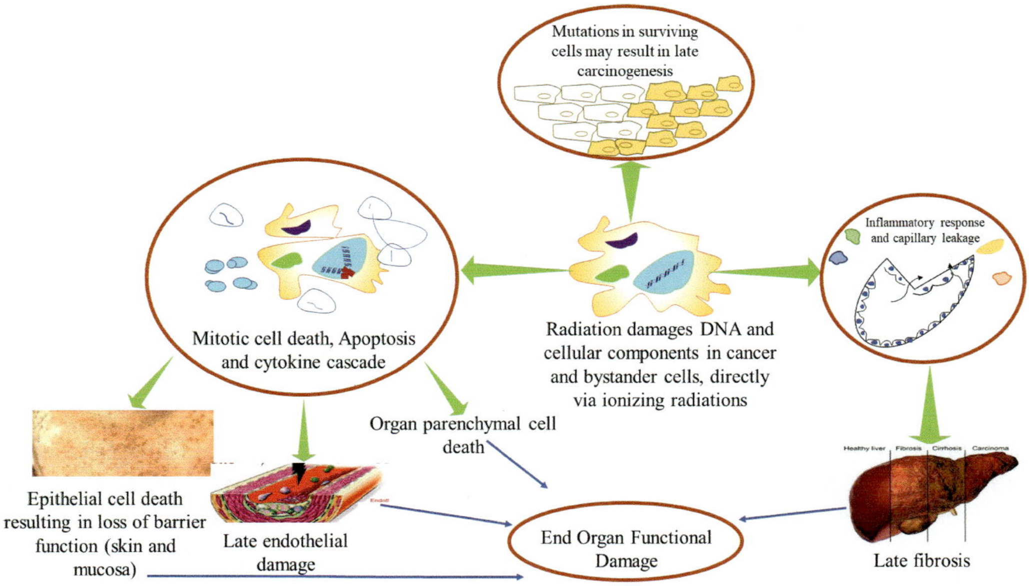

FIGURE 7.3 The pathophysiology of radiation-induced injury to healthy/normal cells. Ionizing radiations cause RIDD, activating a cascade of events leading to toxicity. Commonly, acute effects include inflammation or epithelial cell death. Late effects commonly include fibrosis, vascular injury, carcinogenesis, and organ parenchymal cell death, which may result in end-organ functional damage. *RIDD*, radiation-induced DNA damage.

patients, where the skin receives higher radiation doses (20%–40% grade 2–3 desquamation). Groin areas of skin, including the perineum and inframammary folds (defining the shape and structure of the chest in women), are prone to radiation dermatitis. Previously the management was done by topical steroid-based creams (mometasone) or avoiding radiations, which were found to decrease desquamation by one-third in randomized clinical trials (Ho et al., 2018).

Recently, many advancements have taken their place. In 2022, a clinical phase II randomized double-blind placebo-controlled trial was conducted on radiation dermatitis in breast cancer patients. The intervention group patients were given boron-based gel (3% sodium pentaborate pentahydrate gel) 15 minutes prior to radiation, while the placebo arm received a placebo gel (gel containing no chemical substance). Primary outcomes were comparison of dermatitis, desquamation (dry and moist), and erythema between both arms. The results presented that after 14 days of applications, the intervention group had much less toxicity compared to the placebo group (dermatitis: 9.9% vs 98.7%; dry desquamation: 3.9% vs 50%; moist desquamation: 0.6% vs 18.4%; erythema: 12.2% vs 96.1%) (Sahin et al., 2022). Similarly, another randomized controlled clinical trial was conducted on 171 breast cancer patients provided with curcumin gel, HPR Plus™, and a placebo. HPR Plus™ is a medical device approved by the FDA for radiation dermatitis. From the first day until 1 week post-RT, patients were advised to apply topical agents. Although most differences weren't observed in interventional groups, prophylactic treatment with curcumin gel proved effective in minimizing the radiation dermatitis and associated pain (Ryan Wolf et al., 2020). Another interesting, rather easy management is the use of barrier films (polyurethane dressings remaining on the skin for the duration of RT) and mepitel film (Behroozian et al., 2023), which prove to be excellent tools for the prevention of such toxicities (Dejonckheere et al., 2023).

Soft tissue fibrosis and changes in breast appearance can develop because of high-dose RT or combination treatments involving surgery, chemoradiotherapy, and other modalities. This may also limit movement and cause tissue hardening. Either through acute inflammation or injury, fibroblast proliferation, vascular insufficiency, and increased extracellular matrix deposition occur (Wang and Tepper, 2021). Fibrosis may severely affect breast cosmesis, and although 80%–85% of women who undergo lumpectomy plus RT report good results, they are at higher risk of breast cosmesis issues. (Naoum et al., 2020). RT also presents a significant risk of failure over time for tissue expanders or permanent implants in breast cancer patients (Dicuonzo et al., 2020; Reinders et al., 2020). Vascular insufficiency and fibrosis may also result in impaired wound healing, fat necrosis, graft/implant failure, and infection in autologous graft reconstruction (Kang et al., 2019). With the local administration of vitamin E and pentoxifylline, a lower risk of fibrosis and improved tissue healing can be achieved (Abbas et al., 2022; Wang and Tepper, 2021).

Lymphedema can arise for any peripheral cancer site that requires nodal management (axilla in breast cancer, groin areas in vulvar cancer, and both for melanoma) and occurs due to impaired lymphatic drainage (fluid retention and tissue swelling) from fibrosis or nodal removal. Swelling, pain, and limited activity in areas that underwent nodal removal are symptoms that may develop acutely or chronically (Allam et al., 2020). Risk ranges from 5% with or without RT to 20%–30% after lymph node removal followed by RT. After the removal of nodes, currently, lymphedema is irreversible; hence, early detection is important to avoid delayed symptoms. Initial management includes physical therapy and compression. Besides this, more rigorous management and complete decongestive therapy involve manual drainage of retention fluids, compression, skincare, and exercise. Even more rigorous techniques, such as lymphatic bypass and lymph node transfer, have provided promising results (Wang and Tepper, 2021). Drug therapy has also been included in the management of lymphedema and is currently in clinical trials (NCT03783715). The potential of ketoprofen in decreasing lymphedema was investigated in this clinical trial that showed reduced skin thickness, decreased plasma granulocyte CSF expression, and improved histopathology (Nassif et al., 2020). Later on, the study was terminated due to COVID-19 and the unavailability of ketoprofen (Bethesda, 2023).

Another unavoidable acute effect of RT on **bones** is hematological suppression and inflammation. Among chronic changes, it includes dysregulation of mesenchymal stem cell function, reduced osteoblasts on the bone surface, an increase in bone marrow adiposity, and osteocyte apoptosis. Moreover, damage caused by RT in bones remains difficult to treat, especially in old-aged patients, due to the extent of damage caused. For instance, in gynecological cancers, various location fractures were observed, with 83% of fractures occurring in the sacral area, 10% in the sacrum and pubis area, and 3.5% in the sacrum, acetabulum, and iliac crest. Combining RT with chemotherapeutic drugs (i.e., cisplatin) may also contribute to bone damage (Chandra et al., 2019). A study by Jacob and colleagues demonstrated the long-term effects of RT on pelvic bones in patients with chemoradiotherapy for rectal cancer. From a total of 513 patients, 432 patients underwent pelvic imaging and were diagnosed with pelvic insufficiency fractures, representing 10.7% of the reassessed patients (Rijpma-Jacobs et al., 2023). Similarly, in pediatric patients, four out of five developed late skeletal effects due to impaired bone growth (grade 3–4 toxicities). Hence, when delivering radiation to pediatric patients, four practical approaches must be considered: (i) target the entire vertebral body while minimizing RT dose to femoral heads; (ii) reduce radiation dose, treat with bladder as much as possible to minimize urinary system-related damages; (iii) minimize damaging effects on the reproductive system, shield the scrotum and reduce dose to ovaries; and (iv) contour vagina and vulva as OARs (Blitzer and Bradley, 2021).

Following RT in or near brain and spinal neoplasms/cancers, **CNS** late effects are observed both in pediatric patients and adult patients. Patients receiving cranial RT are frequently associated with headaches, nausea, vomiting, and generalized edema. The worst IQ and psychological health have been associated with pediatric patients following RT, while Alzheimer's disease and vascular dementia may be associated with adult patients (Wang and Tepper, 2021). A microglial inflammatory response with an abnormal increase in cytokines causes negative effects on neurogenesis and cognition (Pazzaglia et al., 2020). More concern has been shown for low-dose RT to CNS lymphoma, as high doses of 30 Gy have shown irreversible or permanent neurotoxicity (Manickam Gurusamy et al., 2023). *Radiation necrosis* has also been reported in patients administered focal RT (even after sparing the cranial area). Here, a series of reactions occur in CNS edema and/or inflammation leading to increased intracranial pressure, glial cell injury along with demyelination of neurons, and vascular endothelial damage leading to hypoxia that results in over-expression of vascular endothelial growth factor (VEGF) and abnormal angiogenesis of all leaky vasculature (Ali et al., 2019). Treatment mainly includes steroids with bevacizumab (VEGF inhibitor), followed by surgery if cancer progression is suspected (Liao et al., 2021). RT of pituitary adenomas or skull cancers may lead to endocrinopathies (toxicity of endocrine glands), and these effects are correlated with dose. Growth hormone deficiency plus prolactin and thyroid deficiency appear at <30 Gy, while ACTH deficiency appears at higher doses: >50 Gy (Wang and Tepper, 2021; Xu et al., 2020).

RT-induced **cardiotoxicity** is another major toxicity encountered; ischemic heart disease and myocardial infarction are among the most recognized ones, even after decades of treatment. Although "tangential" radiation fields are currently in practice to minimize these toxicities (Karpf et al., 2019), adjacent tissues may receive incidental radiation doses that are directly associated with cardiac toxicity (Alcorn et al., 2020). Common management techniques include deep inspiratory breath hold (separating heart from breast in breast cancer patients), allowing minimal cardiac risk (Simonetto et al., 2019; Stowe et al., 2022). Besides these, inflammatory pericarditis, valvular diseases, arrhythmias, and several types of cardiomyopathies may also occur (Wang and Tepper, 2021).

The **gastrointestinal tract** is another victim of RT-induced toxicities. *Acute esophagitis* commonly occurs after lung and esophageal cancer RT treatment. Ensuring adequate liquid and solid intake is critical, whereas solutions containing viscous lidocaine (as an analgesic), antacids, and analgesics may also be given for its management (Thawani and Vyas, 2021). For patients receiving focal RT in hepatocellular carcinoma and liver metastasis of cancer and with baseline function of the liver, RT may lead to *liver function* decline, increasing ascites, within weeks or months, which may worsen to encephalopathy and even death. It may be minimized by proper patient selection (excluding patients with cirrhosis or baseline function of the liver) and proper treatment selection (i.e., SBRT, IMRT, etc.) (Miften et al., 2021). *Stomach and small bowel*: here, RT induces mild gastritis, ulcer, bleeding, nausea, epithelial depletion, inflammation, abscess production, diarrhea, and impaired gastric mobility. Management usually includes antacids, proton pump inhibitors, avoidance of irritants, anti-emetics, brief radiation breaks, avoiding intake of solid seeds, skins, leafy greens, etc. However, fatigue, alopecia, and cytopenia are also observed, which may be linked with the systemic effects of RT (Wang and Tepper, 2021).

Other toxicities may include direct urothelial damage, prostate inflammation, nocturia, urgency, and hematuria in **urethral and bladder toxicities** (Chorbińska et al., 2021; Zuppone et al., 2020), which may arise either due to direct urothelial damage and/or inflammation, edema, or hyperemia in the vicinity. Most preferably, α1-blockers (tamsulosin) are given to alleviate the symptoms that persist over months (Helissey et al., 2021). Although kidneys are radiosensitive at low radiation doses of 10–20 Gy, **kidney toxicities** arise when whole-body irradiation is done for bone marrow transplantation and pediatric malignancies of the abdomen and are even more prominent in patients with baseline kidney functions. These effects arise due to capillary endothelial damage that leads to progressive glomerulosclerosis, radiation nephropathy, and tubulointerstitial fibrosis (Figure 7.4) (Klaus et al., 2021). Similarly, **ureter toxicities** also arise from EBRT followed by surgery, which most commonly leads to hydronephrosis. Proper ureteral shielding from radiation helps in minimizing risk while maximizing cancer control (Martin et al., 2022). In patients of reproductive age, **reproductive toxicities** arise from chemoradiotherapy for pelvis-associated cancers. This may cause adverse effects on reproduction, hormones, and fertility (teratogenesis) and is highly not recommended in pregnant patients. Oocytes depletion and follicles damage due to RT also lead to accelerated depletion in progesterone and estrogen even at low doses (2–5 Gy). Even though uterine irradiation decreases uterus size, fibrotic changes and vascular functionality result in reduced chances of pregnancy, higher chances of pregnancy complications, preterm birth, and low-birth-weight infants (Griffiths et al., 2019). Cryoprotection/cryopreservation of the fetus/embryo and transportation of ovaries from the irradiation site are highly recommended if the patient

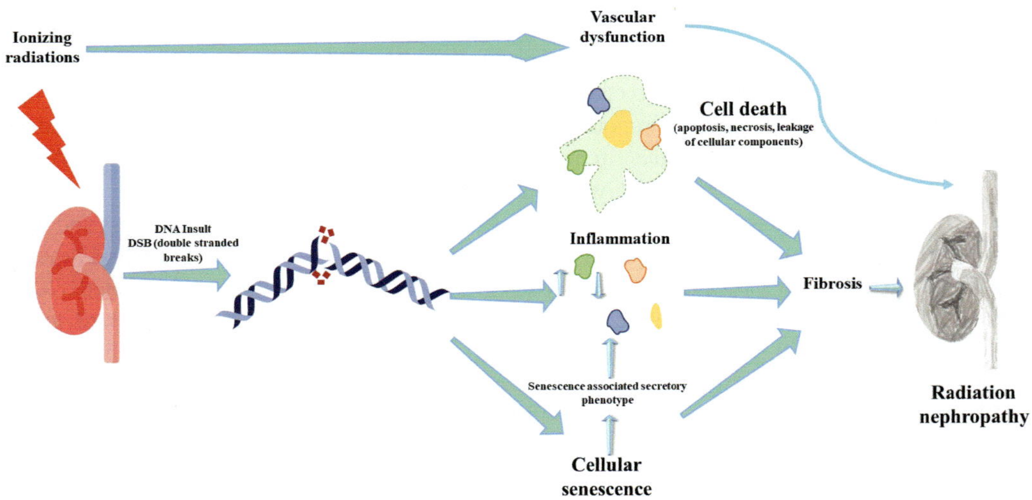

FIGURE 7.4 Pathophysiology of radiation-induced kidney toxicities. Along with RIDD, activation of renin-angiotensin-aldosterone (RASS) and vascular dysfunction also occur that collectively contribute to radiation nephropathy. DSBs initiate cellular senescence, inflammation, and cell death. With conflicting clinical data available, both cell death and inflammation may enter into a vicious cycle. Senescence-associated phenotype also contributes to inflammation. All these events combine to cause loss of nephron mass and fibrosis (hallmark of radiation nephropathy). *RIDD*, radiation-induced DNA damage; *DSBs*, double-strand breaks.

has to undergo RT (McGettigan et al., 2023; Shah et al., 2019). **Hematological toxicities** such as thrombocytopenia, lymphopenia, leukopenia and other immunosuppressive as well as immunostimulatory properties are also linked with RT (Terrones-Campos et al., 2021). Pelvic bone marrow stem cells are highly radiosensitive, and patients undergoing pelvic nodal RT most commonly end up with lymphopenia, which may sometimes lead to a dismal prognosis of cancer. In such cases, bone marrow-sparing techniques must be applied to minimize hematological toxicities while considering bone marrow as OAR (Iorio et al., 2021).

Carcinogenesis is another terrible adverse effect of RT, as the sublethal damage that accompanies chromosomal abnormalities or mutations may predispose to or lead to cancer. Therapeutic radiations and diagnostic radiations both have carcinogenic potential. For instance, a 5 fold-15 fold increase in incidence of adult and early-onset breast cancer has been found in patients who underwent RT for benign breast cancers (Wang and Tepper, 2021). In comparison, a sixfold increase in subsequent malignant cancers has been found in childhood cancer survivors, who are prone to the development of second or later cancers. The incidence of CRC may also increase by 70% with each 10 Gy increase in dose and volume (Bright et al., 2019). Improvement in RT delivery may minimize the incidence of RT-induced carcinogenesis. Proton therapy must be used instead, which works with minimum exposure of the body to irradiations (Meijer et al., 2020; Simone et al., 2020).

7.6 SUMMARY

Various radiotherapy techniques are available for CRC patients at different stages of cancer. These techniques have the potential to achieve organ preservation and improve the quality of life of CRC patients. Outcomes from clinical studies have established improved pain relief and local control of CRC. While some clinical trials have been conducted in the past to assess the balance between the benefits and risks of radiotherapy techniques, including disease control and toxicity rates, more clinical studies are needed to thoroughly evaluate these techniques as the standard of care. Despite advancements in radioresistance and RIDD repair in response to radiotherapy, it remains a conventional cancer treatment. In this chapter, we have detailed the mechanism through which cancer rewires intracellular functions to enhance resistance to radiotherapy. Hence, ongoing research is focusing on developing therapeutics that target radioresistance by creating radiosensitizers with the goal of enhancing radiotherapy for CRC patients.

REFERENCES

Abbas, D. B., Lavin, C. V., Fahy, E. J., Griffin, M., Guardino, N. J., Nazerali, R. S., Nguyen, D. H., Momeni, A., Longaker, M. T., and Wan, D. C. (2022). Fat grafts augmented with vitamin e improve volume retention and radiation-induced fibrosis. *Aesthetic Surgery Journal, 42*(8), 946–955.

Alcorn, S. R., Sloan, L., McNutt, T. R., Stinson, S. F., Asrari, F., Croog, V. J., Floreza, B., Weaver, A., and Wright, J. L. (2020). Acute toxicity outcomes and dosimetric implications from incidental irradiation of adjacent tissues in tangent field hypofractionated breast radiotherapy. *Reports of Practical Oncology and Radiotherapy, 25*(3), 345–350.

Ali, F. S., Arevalo, O., Zorofchian, S., Patrizz, A., Riascos, R., Tandon, N., Blanco, A., Ballester, L. Y., and Esquenazi, Y. (2019). Cerebral radiation necrosis: incidence, pathogenesis, diagnostic challenges, and future opportunities. *Current Oncology Reports, 21*(1), 1–17.

Allam, O., Park, K. E., Chandler, L., Mozaffari, M. A., Ahmad, M., Lu, X., and Alperovich, M. (2020). The impact of radiation on lymphedema: a review of the literature. *Gland Surgery, 9*(2), 596.

Anuja, K., Chowdhury, A. R., Saha, A., Roy, S., Rath, A. K., Kar, M., and Banerjee, B. (2019). Radiation-induced DNA damage response and resistance in colorectal cancer stem-like cells. *International Journal of Radiation Biology, 95*(6), 667–679.

Ashack, K. A., Kuritza, V., Visconti, M. J., and Ashack, L. (2020). Dermatologic sequelae associated with radiation therapy. *American Journal of Clinical Dermatology, 21*(1), 541–555.

Behroozian, T., Milton, L., Karam, I., Zhang, L., Ding, K., Lou, J., Gallant, F., Rakovitch, E., Tran, W., and Soliman, H. (2023). Mepitel film for the prevention of acute radiation dermatitis in breast cancer: a randomized multicenter open-label phase III trial. *Journal of Clinical Oncology, 41*(6), 1250–1264.

Bellezzo, M., Fonseca, G. P., Voncken, R., Verrijssen, A.-S., Van Beveren, C., Roelofs, E., Yoriyaz, H., Reniers, B., Van Limbergen, E. J., and Berbée, M. (2020). Advanced design, simulation, and dosimetry of a novel rectal applicator for contact brachytherapy with a conventional HDR 192Ir source. *Brachytherapy, 19*(4), 544–553.

Benezery, K., Montagne, L., Evesque, L., Schiappa, R., Hannoun-Levi, J.-M., Francois, E., Thamphya, B., and Gerard, J.-P. (2020). Clinical response assessment after contact X-Ray brachytherapy and chemoradiotherapy for organ preservation in rectal cancer T2-T3 M0: The time/dose factor influence. *Clinical and Translational Radiation Oncology, 24*(1), 92–98.

Bethesda. (2023). *Characterization of Treatment Responses in Lymphedema*. Retrieved from https://clinicaltrials.gov/study/NCT03783715

Blitzer, G. C., and Bradley, K. A. (2021). Four ways to decrease late toxicity from pelvic radiation therapy in children and young adults. *Practical Radiation Oncology, 11*(6), 434–440.

Bonomo, P., Desideri, I., Loi, M., Ciccone, L. P., Lo Russo, M., Becherini, C., Greto, D., Simontacchi, G., Pimpinelli, N., and Livi, L. (2019). Management of severe bio-radiation dermatitis induced by radiotherapy and cetuximab in patients with head and neck cancer: emphasizing the role of calcium alginate dressings. *Supportive Care in Cancer, 27*(1), 2957–2967.

Bourhis, J., Montay-Gruel, P., Jorge, P. G., Bailat, C., Petit, B., Ollivier, J., Jeanneret-Sozzi, W., Ozsahin, M., Bochud, F., and Moeckli, R. (2019). Clinical translation of FLASH radiotherapy: why and how? *Radiotherapy Oncology, 139*, 11–17.

Bright, C. J., Reulen, R. C., Winter, D. L., Stark, D. P., McCabe, M. G., Edgar, A. B., Frobisher, C., and Hawkins, M. M. (2019). Risk of subsequent primary neoplasms in survivors of adolescent and young adult cancer (Teenage and Young Adult Cancer Survivor Study): a population-based, cohort study. *The Lancet Oncology, 20*(4), 531–545.

Cao, C., Wang, D., Tian, D. H., Wilson-Smith, A., Huang, J., and Rimner, A. (2019). A systematic review and meta-analysis of stereotactic body radiation therapy for colorectal pulmonary metastases. *Journal of Thoracic Disease, 11*(12), 5187–5198.

Catalano, F., Borea, R., Puglisi, S., Boutros, A., Gandini, A., Cremante, M., Martelli, V., Sciallero, S., and Puccini, A. (2022). Targeting the DNA damage response pathway as a novel therapeutic strategy in colorectal cancer. *Cancers, 14*(6), 1388.

Chandra, A., Park, S. S., and Pignolo, R. J. (2019). Potential role of senescence in radiation-induced damage of the aged skeleton. *Bone, 120*, 423–431.

Charpentier, M., Spada, S., Van Nest, S. J., and Demaria, S. (2022). Radiation therapy-induced remodeling of the tumor immune microenvironment. *Seminars in Cancer Biology, 86*(1), 737–747.

Chatterjee, S., Patel, Z., Thaha, M. A., and Kyriacou, P. A. (2020). In silico and in vivo investigations using an endocavitary photoplethysmography sensor for tissue viability monitoring. *Journal of Biomedical Optics, 25*(2), 027001–027001.

Chen, J. J., Rawal, B., Krishnan, M. S., Hertan, L. M., Shi, D. D., Roldan, C. S., Huynh, M. A., Spektor, A., and Balboni, T. A. (2021). Patterns of specialty palliative care utilization among patients receiving palliative radiation therapy. *Journal of Pain and Symptom Management, 62*(2), 242–251.

Chiang, C. L., Lee, S. F., Lee, V. W. Y., Yeung, C. S. Y., Cheung, S. K. T., Poon, M. C. M., Foo, D. C. C., Lo, S. H., Lam, K. O., Lee, F. A. S., Wong, F. C. S., and Chan, M. K. H. (2020). Toxicity outcome of endorectal brachytherapy boost in medically inoperable patients. *Strahlentherapie und Onkologie, 196*(11), 993–997.

Choi, H. S., Jeong, B. K., Kang, K. M., Jeong, H., Song, J. H., Ha, I. B., and Kwon, O.-Y. (2020). Tumor control and overall survival after stereotactic body radiotherapy for pulmonary oligometastases from colorectal cancer: a meta-analysis. *Cancer Research and Treatment, 52*(4), 1188–1198.

Chorbińska, J., Krajewski, W., and Zdrojowy, R. (2021). Urological complications after radiation therapy—nothing ventured, nothing gained: a narrative review. *Translational Cancer Research, 10*(2), 1096.

Dejonckheere, C. S., Dejonckheere, E., Layer, J. P., Layer, K., Sarria, G. R., Koch, D., Abramian, A., Kaiser, C., Lindner, K., Bachmann, A., Anzböck, T., Röhner, F., Schmeel, F. C., and Schmeel, L. C. (2023). Barrier films for the prevention of acute radiation dermatitis in breast cancer: a systematic review and meta-analysis of randomised controlled trials. *The Breast, 71*, 31–41.

Deng, X., Liu, P., Jiang, D., Wei, M., Wang, X., Yang, X., Zhang, Y., Wu, B., Liu, Y., and Qiu, M. (2020). Neoadjuvant radiotherapy versus surgery alone for stage II/III mid-low rectal cancer with or without high-risk factors: a prospective multicenter stratified randomized trial. *Annals of Surgery, 272*(6), 1060–1069.

Deng, Y.-R., Chen, X.-J., Chen, W., Wu, L.-F., Jiang, H.-P., Lin, D., Wang, L.-J., Wang, W., and Guo, S.-Q. (2019). Sp1 contributes to radioresistance of cervical cancer through targeting G2/M cell cycle checkpoint CDK1. *Cancer Management Research, 11*(1),5835–5844.

Devic, S., Bekerat, H., Garant, A., and Vuong, T. (2019). Optimization of HDRBT boost dose delivery for patients with rectal cancer. *Brachytherapy, 18*(4), 559–563.

Dicuonzo, S., Leonardi, M. C., Radice, D., Morra, A., Gerardi, M. A., Rojas, D. P., Surgo, A., Dell'Acqua, V., Luraschi, R., Cattani, F., Rietjens, M., De Lorenzi, F., Veronesi, P., Galimberti, V., Marvaso, G., Fodor, C., Orecchia, R., and Jereczek-Fossa, B. A. (2020). Long-term results and reconstruction failure in patients receiving postmastectomy radiation therapy with a temporary expander or permanent implant in place. *Plastic and Reconstructive Surgery, 145*(2), 317–327.

Dizdarevic, E., Hansen, T. F., Pløen, J., Jensen, L. H., Lindebjerg, J., Rafaelsen, S., Jakobsen, A., and Appelt, A. (2020). Long-term patient-reported outcomes after high-dose chemoradiation therapy for nonsurgical management of distal rectal cancer. *International Journal of Radiation Oncology* Biology* Physics, 106*(3), 556–563.

Fahy, M. R., Kelly, M. E., Power Foley, M., Nugent, T. S., Shields, C. J., and Winter, D. C. (2021). The role of intraoperative radiotherapy in advanced rectal cancer: a meta-analysis. *Colorectal Disease, 23*(8), 1998–2006.

Ferreira, S., and Dutreix, M. (2019). DNA repair inhibitors to enhance radiotherapy: progresses and limitations. *Cancer/Radiothérapie, 23*(8), 883–890.

Filippi, A. R., Badellino, S., Ceccarelli, M., Guarneri, A., Franco, P., Monagheddu, C., Spadi, R., Ragona, R., Racca, P., and Ricardi, U. (2015). Stereotactic ablative radiation therapy as first local therapy for lung oligometastases from colorectal cancer: a single-institution cohort study. *International Journal of Radiation Oncology*Biology*Physics, 91*(3), 524–529.

Fiorino, C., Guckenberger, M., Schwarz, M., van der Heide, U. A., and Heijmen, B. (2020). Technology-driven research for radiotherapy innovation. *Molecular Oncology, 14*(7), 1500–1513.

Fleischmann, M., Diefenhardt, M., Trommel, M., Scherf, C., Ramm, U., Chatzikonstantinou, G., Fokas, E., Rödel, C., and Tselis, N. (2022). Image-guided high-dose-rate brachytherapy for rectal cancer: technical note and first clinical experience on an organ-preserving approach. *Strahlentherapie und Onkologie, 198*(7), 654–662.

Gao, H., Zhang, X., Ding, Y., Qiu, R., Hong, Y., and Chen, W. (2019). Synergistic suppression effect on tumor growth of colorectal cancer by combining radiotherapy with a TRAIL-armed oncolytic adenovirus. *Technology in Cancer Research & Treatment, 18*(1), 1533033819853290.

Garant, A., Magnan, S., Devic, S., Martin, A.-G., Boutros, M., Vasilevsky, C.-A., Ferland, S., Bujold, A., DesGroseilliers, S., Sebajang, H., Richard, C., and Vuong, T. (2019). Image guided adaptive endorectal brachytherapy in the nonoperative management of patients with rectal cancer. *International Journal of Radiation Oncology*Biology*Physics, 105*(5), 1005–1011.

Garcia-Exposito, N., Ramos, R., Navarro-Perez, V., Molina, K., Arnaiz, M. D., Padrones, S., Ruffinelli, J. C., Santos, C., Guedea, F., and Navarro-Martin, A. (2023). Stereotactic body radiotherapy versus surgery for lung metastases from colorectal cancer: single-institution results. *Cancers, 15*(4), 1195.

Garden, A. S., Dong, L., Morrison, W. H., Stugis, E. M., Glisson, B. S., Frank, S. J., Beadle, B. M., Gunn, G. B., Schwartz, D. L., Kies, M. S., Weber, R. S., Ang, K. K., and Rosenthal, D. I. (2013). Patterns of disease recurrence following treatment of oropharyngeal cancer with intensity modulated radiation therapy. *International Journal of Radiation Oncology*Biology*Physics, 85*(4), 941–947.

Gérard, J.-P., Barbet, N., Gal, J., Dejean, C., Evesque, L., Doyen, J., Coquard, R., Gugenheim, J., Benizri, E., Schiappa, R., Baudin, G., Benezery, K., and François, E. (2019). Planned organ preservation for early T2–3 rectal adenocarcinoma: a French, multicentre study. *European Journal of Cancer, 108*(1), 1–16.

Gerard, J.-P., Barbet, N., Schiappa, R., Magné, N., Martel, I., Mineur, L., Deberne, M., Zilli, T., Dhadda, A., and Myint, A. S. (2023). Neoadjuvant chemoradiotherapy with radiation dose escalation with contact x-ray brachytherapy boost or external beam radiotherapy boost for organ preservation in early cT2–cT3 rectal adenocarcinoma (OPERA): a phase 3, randomised controlled trial. *The Lancet Gastroenterology & Hepatology, 8*(4), 356–367.

Gerard, J.-P., Montagne, L., Thamphya, B., Doyen, J., Schiappa, R., Benezery, K., Gourgou, S., Dejean, C., and Hannoun-Levi, J.-M. (2022). Propensity score analysis of radical proctectomy versus organ preservation using contact X-ray brachytherapy for rectal cancer. *Clinical and Translational Radiation Oncology, 33*(1), 70–76.

Gérard, J.-P., Ortholan, C., Benezery, K., Ginot, A., Hannoun-Levi, J.-M., Chamorey, E., Benchimol, D., and François, E. (2008). Contact X-ray therapy for rectal cancer: experience in centre antoine-lacassagne, Nice, 2002–2006. *International Journal of Radiation Oncology*Biology*Physics, 72*(3), 665–670.

Griffiths, M. J., Winship, A. L., and Hutt, K. J. (2019). Do cancer therapies damage the uterus and compromise fertility? *Human Reproduction Update, 26*(2), 161–173.

Guckenberger, M., Klement, R. J., Allgäuer, M., Andratschke, N., Blanck, O., Boda-Heggemann, J., Dieckmann, K., Duma, M., Ernst, I., and Ganswindt, U. (2016). Local tumor control probability modeling of primary and secondary lung tumors in stereotactic body radiotherapy. *Radiotherapy Oncology, 118*(3), 485–491.

Guninski, R. S., Cuccia, F., Alongi, F., Andratschke, N., Belka, C., Bellut, D., Dahele, M., Josipovic, M., Kroese, T. E., Mancosu, P., Minniti, G., Niyazi, M., Ricardi, U., Munck af Rosenschold, P., Sahgal, A., Tsang, Y., Verbakel, W. F. A. R., and Guckenberger, M. (2024). Efficacy and safety of SBRT for spine metastases: a systematic review and meta-analysis for preparation of an ESTRO practice guideline. *Radiotherapy and Oncology, 190*(1), 109969.

Habiba, K., Aziz, K., Sanders, K., Santiago, C. M., Mahadevan, L. S. K., Makarov, V., Weiner, B. R., Morell, G., and Krishnan, S. (2019). Enhancing colorectal cancer radiation therapy efficacy using silver nanoprisms decorated with graphene as radiosensitizers. *Scientific Reports, 9*(1), 17120.

Hawkins, A. T., Ford, M. M., Geiger, T. M., Hopkins, M. B., Kachnic, L. A., Muldoon, R. L., and Glasgow, S. C. (2019). Neoadjuvant radiation for clinical T4 colon cancer: a potential improvement to overall survival. *Surgery, 165*(2), 469–475.

Helissey, C., Cavallero, S., Brossard, C., Dusaud, M., Chargari, C., and François, S. (2021). Chronic inflammation and radiation-induced cystitis: molecular background and therapeutic perspectives. *Cells, 10*(1), 21.

Ho, A. Y., Olm-Shipman, M., Zhang, Z., Siu, C. T., Wilgucki, M., Phung, A., Arnold, B. B., Porinchak, M., Lacouture, M., and McCormick, B. (2018). A randomized trial of mometasone furoate 0.1% to reduce high-grade acute radiation dermatitis in breast cancer patients receiving postmastectomy radiation. *International Journal of Radiation Oncology* Biology* Physics, 101*(2), 325–333.

Huang, R.-X., and Zhou, P.-K. (2020). DNA damage response signaling pathways and targets for radiotherapy sensitization in cancer. *Signal Transduction and Targeted Therapy, 5*(1), 60.

Hull, T. L., Lavery, I. C., and Saxton, J. P. (1994). Endocavitary irradiation. *Diseases of the Colon and Rectum, 37*(12), 1266–1270.

Iorio, G. C., Spieler, B. O., Ricardi, U., and Dal Pra, A. (2021). The impact of pelvic nodal radiotherapy on hematologic toxicity: a systematic review with focus on leukopenia, lymphopenia and future perspectives in prostate cancer treatment. *Critical Reviews in Oncology/Hematology, 168*, 103497.

Ito, K., Yamaguchi, T., Ogawa, H., Nakajima, Y., and Karasawa, K. (2020). Stereotactic body radiotherapy for bone metastases in patients with colorectal cancer. *Japanese Journal of Clinical Oncology, 50*(12), 1442–1446.

Kang, D., Luan, J., Fu, S., Mu, D., Liu, C., Xin, M., and Xu, B. (2019). Factors of fat necrosis after breast reconstruction with autologous fat graft. *Chinese Journal of Medical Aesthetics* Cosmetology, *6(1)*, 21–24.

Karpf, D., Sakka, M., Metzger, M., and Grabenbauer, G. G. (2019). Left breast irradiation with tangential intensity modulated radiotherapy (t-IMRT) versus tangential volumetric modulated arc therapy (t-VMAT): trade-offs between secondary cancer induction risk and optimal target coverage. *Radiation Oncology, 14*(1), 1–11.

Kim, W., Lee, S., Seo, D., Kim, D., Kim, K., Kim, E., Kang, J., Seong, K. M., Youn, H., and Youn, B. (2019). Cellular stress responses in radiotherapy. *Cells, 8*(9), 1105.

Klaus, R., Niyazi, M., and Lange-Sperandio, B. (2021). Radiation-induced kidney toxicity: molecular and cellular pathogenesis. *Radiation Oncology, 16*(1), 43.

Klement, R. J., Abbasi-Senger, N., Adebahr, S., Alheid, H., Allgaeuer, M., Becker, G., Blanck, O., Boda-Heggemann, J., Brunner, T., Duma, M., Eble, M. J., Ernst, I., Gerum, S., Habermehl, D., Hass, P., Henkenberens, C., Hildebrandt, G., Imhoff, D., Kahl, H., Klass, N. D., Krempien, R., Lewitzki, V., Lohaus, F., Ostheimer, C., Papachristofilou, A., Petersen, C., Rieber, J., Schneider, T., Schrade, E., Semrau, R., Wachter, S., Wittig, A., Guckenberger, M., and Andratschke, N. (2019). The impact of local control on overall survival after stereotactic body radiotherapy for liver and lung metastases from colorectal cancer: a combined analysis of 388 patients with 500 metastases. *BMC Cancer, 19*(1), 173.

Kovalic, J. J. (1988). Endocavitary irradiation for rectal cancer and villous adenomas. *International Journal of Radiation Oncology*Biology*Physics, 14*(2), 261–264.

Li, L.-y., Guan, Y.-d., Chen, X.-s., Yang, J.-m., and Cheng, Y. (2021). DNA repair pathways in cancer therapy and resistance. *Frontiers in Pharmacology, 11*, 629266.

Li, Y., Castellano, J. J., Moreno, I., Martínez-Rodenas, F., Hernandez, R., Canals, J., Diaz, T., Han, B., Muñoz, C., Biete, A., Monzo, M., and Navarro, A. (2020). LincRNA-p21 levels relates to survival and post-operative radiotherapy benefit in rectal cancer patients. *Life, 10*(9), 172.

Liao, G., Khan, M., Zhao, Z., Arooj, S., Yan, M., and Li, X. (2021). Bevacizumab treatment of radiation-induced brain necrosis: a systematic review. *Frontiers in Oncology, 11*, 593449.

Liu, C., Nie, J., Wang, R., and Mao, W. (2019). The cell cycle G2/M block is an indicator of cellular radiosensitivity. *Dose-Response, 17*(4), 1559325819891008.

Liu, Y.-P., Zheng, C.-C., Huang, Y.-N., He, M.-L., Xu, W. W., and Li, B. (2021). Molecular mechanisms of chemo- and radiotherapy resistance and the potential implications for cancer treatment. *MedComm, 2*(3), 315–340.

Lukovic, J., and Dawson, L. A. (2023). Stereotactic body radiation therapy for colorectal cancer liver metastases. *Journal of Gastrointestinal Oncology, 1*(1), 1–9.

Ma, L., Qiang, J., Yin, H., Lin, L., Jiao, Y., Ma, C., Li, X., Dong, L., Cui, J., Wei, D., Sharma, A. M., Schwartz, D. L., Gu, W., and Chen, H. (2020). Low-kilovolt x-ray intraoperative radiotherapy for pT3 locally advanced colon cancer: a single-institution retrospective analysis. *World Journal of Surgical Oncology, 18*(1), 132.

Manickam Gurusamy, V., Raveendran Divakar, S., Halsnad Chandramouli, S., Kunheri, B., Hussain Al-Abdulla, H., Shaikh, G., Chaudary Apsani, R., Riyaz Poolakundan, M., Caparrotti, P., Wafiq Hammoud, R., and Al-Hammadi, N. (2023). The role of radiotherapy in newly diagnosed primary CNS lymphoma: a descriptive review and a pragmatic approach to clinical practice. *Clinical and Translational Radiation Oncology, 39*, 100559.

Maremonti, E., Eide, D. M., Rossbach, L. M., Lind, O. C., Salbu, B., and Brede, D. A. (2020). In vivo assessment of reactive oxygen species production and oxidative stress effects induced by chronic exposure to gamma radiation in Caenorhabditis elegans. *Free Radical Biology and Medicine, 152*(1), 583–596.

Martin, J. M., Richardson, M., Siva, S., Cardoso, M., Handmer, M., and Sidhom, M. (2022). Mechanisms, mitigation, and management of urinary toxicity from prostate radiotherapy. *The Lancet Oncology, 23*(12), e534–e543.

McAndrew, E. N., Lepage, C. C., and McManus, K. J. (2016). The synthetic lethal killing of RAD54B-deficient colorectal cancer cells by PARP1 inhibition is enhanced with SOD1 inhibition. *Oncotarget, 7*(52), 87417.

McCann, E., O'Sullivan, J., and Marcone, S. (2021). Targeting cancer-cell mitochondria and metabolism to improve radiotherapy response. *Translational Oncology, 14*(1), 100905.

McGettigan, M., Thomas, K., and Kamath, A. (2023). Cancer in pregnancy: treatment effects. *Abdominal Radiology, 48*(5), 1774–1783.

Meijer, T. W. H., Scandurra, D., and Langendijk, J. A. (2020). Reduced radiation-induced toxicity by using proton therapy for the treatment of oropharyngeal cancer. *The British Journal of Radiology, 93*(1107), 20190955.

Meyer, F., Becker, S., Classen, S., Parplys, A. C., Mansour, W. Y., Riepen, B., Timm, S., Ruebe, C., Jasin, M., Wikman, H., Petersen, C., Rothkamm, K., and Borgmann, K. (2020). Prevention of DNA replication stress by CHK1 leads to chemoresistance despite a DNA repair defect in homologous recombination in breast cancer. *Cells, 9*(1), 238.

Miften, M., Vinogradskiy, Y., Moiseenko, V., Grimm, J., Yorke, E., Jackson, A., Tomé, W. A., Ten Haken, R. K., Ohri, N., and Romero, A. M. (2021). Radiation dose-volume effects for liver SBRT. *International Journal of Radiation Oncology* Biology* Physics, 110*(1), 196–205.

Miles, X., Vandevoorde, C., Hunter, A., and Bolcaen, J. (2021). MDM2/X inhibitors as radiosensitizers for glioblastoma targeted therapy. *Frontiers in Oncology, 11*, 703442.

Mirnezami, R., Chang, G. J., Das, P., Chandrakumaran, K., Tekkis, P., Darzi, A., and Mirnezami, A. H. (2013). Intraoperative radiotherapy in colorectal cancer: systematic review and meta-analysis of techniques, long-term outcomes, and complications. *Surgical Oncology, 22*(1), 22–35.

Mortezaee, K., and Najafi, M. (2021). Immune system in cancer radiotherapy: resistance mechanisms and therapy perspectives. *Critical Reviews in Oncology/Hematology, 157*(1), 103180.

Mulcahy, M. F., Mahvash, A., Pracht, M., Montazeri, A. H., Bandula, S., Martin, R. C. G., 2nd, Herrmann, K., Brown, E., Zuckerman, D., Wilson, G., Kim, T. Y., Weaver, A., Ross, P., Harris, W. P., Graham, J., Mills, J., Yubero Esteban, A., Johnson, M. S., Sofocleous, C. T., Padia, S. A., Lewandowski, R. J., Garin, E., Sinclair, P., and Salem, R. (2021). Radioembolization with chemotherapy for colorectal liver metastases: a randomized, open-label, international, multicenter, phase III trial. *Journal of Clinical Oncology, 39*(35), 3897–3907.

Naoum, G. E., Salama, L., Niemierko, A., Vieira, B. L., Belkacemi, Y., Colwell, A. S., Winograd, J., Smith, B., Ho, A., and Taghian, A. G. (2020). Single stage direct-to-implant breast reconstruction has lower complication rates than tissue expander and implant and comparable rates to autologous reconstruction in patients receiving postmastectomy radiation. *International Journal of Radiation Oncology* Biology* Physics, 106*(3), 514–524.

Nassif, T. M., Brunelle, C. L., Gillespie, T. C., Bernstein, M. C., Bucci, L. K., Naoum, G. E., and Taghian, A. G. (2020). Breast cancer-related lymphedema: a review of risk factors, radiation therapy contribution, and management strategies. *Current Breast Cancer Reports, 12*(4), 305–316.

Nickoloff, J. A., Sharma, N., and Taylor, L. (2020). Clustered DNA double-strand breaks: biological effects and relevance to cancer radiotherapy. *Genes, 11*(1), 99.

Nilsson, R., and Liu, N.-A. (2020). Nuclear DNA damages generated by reactive oxygen molecules (ROS) under oxidative stress and their relevance to human cancers, including ionizing radiation-induced neoplasia part I: physical, chemical and molecular biology aspects. *Radiation Medicine Protection, 1*(3), 140–152.

World Health Organization. (2023). *Colorectal Cancer*. Retrieved from https://www.who.int/news-room/fact-sheets/detail/colorectal-cancer?gad_source=1&gclid=Cj0KCQiAz8GuBhCxARIsAOpzk8wAH9GFRex6TSkf2SWEg51LU_IVUYrMa_V-M4tA9Ifrk856rNvTUo0aAjiREALw_wcB

Parikh, A. R., Szabolcs, A., Allen, J. N., Clark, J. W., Wo, J. Y., Raabe, M., Thel, H., Hoyos, D., Mehta, A., Arshad, S., Lieb, D. J., Drapek, L. C., Blaszkowsky, L. S., Giantonio, B. J., Weekes, C. D., Zhu, A. X., Goyal, L., Nipp, R. D., Dubois, J. S., Van Seventer, E. E., Foreman, B. E., Matlack, L. E., Ly, L., Meurer, J. A., Hacohen, N., Ryan, D. P., Yeap, B. Y., Corcoran, R. B., Greenbaum, B. D., Ting, D. T., and Hong, T. S. (2021). Radiation therapy enhances immunotherapy response in microsatellite stable colorectal and pancreatic adenocarcinoma in a phase II trial. *Nature Cancer, 2*(11), 1124–1135.

Pazzaglia, S., Briganti, G., Mancuso, M., and Saran, A. (2020). Neurocognitive decline following radiotherapy: mechanisms and therapeutic implications. *Cancers, 12*(1), 146.

Pilié, P. G., Tang, C., Mills, G. B., and Yap, T. A. (2019). State-of-the-art strategies for targeting the DNA damage response in cancer. *Nature Reviews Clinical oncology, 16*(2), 81–104.

Potemin, S., Kübler, J., Uvarov, I., Wenz, F., and Giordano, F. (2020). Intraoperative radiotherapy as an immediate adjuvant treatment of rectal cancer due to limited access to external-beam radiotherapy. *Radiation Oncology, 15*(1), 11.

Reinders, F., Young-Afat, D., Batenburg, M., Bruekers, S., van Amerongen, E., Macare van Maurik, J., Braakenburg, A., Zonnevylle, E., Hoefkens, M., and Teunis, T. (2020). Higher reconstruction failure and less patient-reported satisfaction after post mastectomy radiotherapy with immediate implant-based breast reconstruction compared to immediate autologous breast reconstruction. *Breast Cancer, 27*, 435–444.

Rijpma-Jacobs, L., van der Vlies, E., Meek, D. B., Bollen, T. L., Siersema, P. D., Weusten, B. L. A. M., Intven, M., van Lelyveld, N., and Los, M. (2023). Pelvic insufficiency fractures and pelvic bone metastases after neoadjuvant (chemo)radiotherapy for rectal cancer. *Acta Oncologica, 62*(10), 1295–1300.

Ryan Wolf, J., Gewandter, J. S., Bautista, J., Heckler, C. E., Strasser, J., Dyk, P., Anderson, T., Gross, H., Speer, T., Dolohanty, L., Bylund, K., Pentland, A. P., and Morrow, G. R. (2020). Utility of topical agents for radiation dermatitis and pain: a randomized clinical trial. *Supportive Care in Cancer, 28*(7), 3303–3311.

Saeedian, A., Lashkari, M., Ghalehtaki, R., Taherioun, M., Razmkhah, M., Kazemian, A., and Aghili, M. (2023). Complication and response assessment of high-dose-rate endorectal brachytherapy boost in neo-adjuvant chemoradiotherapy of locally advanced rectal cancer with long-term outcomes. *Journal of Contemporary Brachytherapy, 15*(2), 117–122.

Sahin, F., Pirouzpanah, M. B., Bijanpour, H., Mohammadzadeh, M., Eghdam Zamiri, R., Ghasemi Jangjoo, A., Nasiri, B., Saboori, H., Doğan, A., Demirci, S., Ayşan, E., Çağrı Büke, A., Naseri, A. R., Shakouri, S. K., Aghamohammadi, D., Alizade-Harakiyan, M., and Seyed Nejad, F. (2022). The preventive effects of boron-based gel on radiation dermatitis in patients being treated for breast cancer: a phase III randomized, double-blind, placebo-controlled clinical trial. *Oncology Research and Treatment, 45*(4), 197–204.

Sanuki, N., Takeda, A., Tsurugai, Y., and Eriguchi, T. (2022). Role of stereotactic body radiotherapy in multidisciplinary management of liver metastases in patients with colorectal cancer. *Japanese Journal of Radiology, 40*(10), 1009–1016.

Schanne, D. H., Heitmann, J., Guckenberger, M., and Andratschke, N. H. (2019). Evolution of treatment strategies for oligometastatic NSCLC patients–a systematic review of the literature. *Cancer Treatment Reviews, 80*, 101892.

Schipler, A., Mladenova, V., Soni, A., Nikolov, V., Saha, J., Mladenov, E., and Iliakis, G. (2016). Chromosome thripsis by DNA double strand break clusters causes enhanced cell lethality, chromosomal translocations and 53BP1-recruitment. *Nucleic Acids Research, 44*(16), 7673–7690.

Scorsetti, M., Comito, T., Tozzi, A., Navarria, P., Fogliata, A., Clerici, E., Mancosu, P., Reggiori, G., Rimassa, L., Torzilli, G., Tomatis, S., Santoro, A., and Cozzi, L. (2015). Final results of a phase II trial for stereotactic body radiation therapy for patients with inoperable liver metastases from colorectal cancer. *Journal of Cancer Research and Clinical Oncology, 141*(3), 543–553.

Shah, N. M., Scott, D. M., Kandagatla, P., Moravek, M. B., Cobain, E. F., Burness, M. L., and Jeruss, J. S. (2019). Young women with breast cancer: fertility preservation options and management of pregnancy-associated breast cancer. *Annals of Surgical Oncology, 26*(5), 1214–1224.

Shinohara, M., Bishop, D. K., and Shinohara, A. (2019). Distinct functions in regulation of meiotic crossovers for DNA damage response clamp loader Rad24(Rad17) and Mec1(ATR) kinase. *Genetics, 213*(4), 1255–1269.

Shukla, V., Halabelian, L., Balagere, S., Samaniego-Castruita, D., Feldman, D. E., Arrowsmith, C. H., Rao, A., and Aravind, L. (2020). HMCES functions in the alternative end-joining pathway of the DNA DSB repair during class switch recombination in B cells. *Molecular Cell, 77*(2), 384–394. e384.

Simone, C. B., Plastaras, J. P., Jabbour, S. K., Lee, A., Lee, N. Y., Choi, J. I., Frank, S. J., Chang, J. Y., and Bradley, J. (2020). Proton reirradiation: expert recommendations for reducing toxicities and offering new chances of cure in patients with challenging recurrence malignancies. *Seminars in Radiation Oncology, 30*(3), 253–261.

Simonetto, C., Eidemüller, M., Gaasch, A., Pazos, M., Schönecker, S., Reitz, D., Kääb, S., Braun, M., Harbeck, N., Niyazi, M., Belka, C., and Corradini, S. (2019). Does deep inspiration breath-hold prolong life? Individual risk estimates of ischaemic heart disease after breast cancer radiotherapy. *Radiotherapy and Oncology, 131*(1), 202–207.

Society, A. C. (2024). *Explore Cancer Statistics*. Retrieved from https://cancerstatisticscenter.cancer.org/?_gl=1*u5ta1n*_ga*MTYyMjM3MzA4OS4xNjcwNDI5MDYz*_ga_12CJLLFFQT*MTY5MDI5NTUxMy4yOTQuMS4xNjkwMjk2MTA3LjAuMC4w&_ga=2.188866859.1869760132.1690197259-1622373089.1670429063#!/cancer-site/Colorectum

Stewart-Ornstein, J., Iwamoto, Y., Miller, M. A., Prytyskach, M. A., Ferretti, S., Holzer, P., Kallen, J., Furet, P., Jambhekar, A., and Forrester, W. C. (2021). p53 dynamics vary between tissues and are linked with radiation sensitivity. *Nature Communications, 12*(1), 898.

Stowe, H. B., Andruska, N. D., Reynoso, F., Thomas, M., and Bergom, C. (2022). Heart sparing radiotherapy techniques in breast cancer: a focus on deep inspiration breath hold. Breast Cancer: Targets Therapy, *14*, 175–186.

Sun Myint, A., Stewart, A., Mills, J., Sripadam, R., Whitmarsh, K., Roy, R., Franklin, A., and Dhadda, A. (2019). Treatment: the role of contact X-ray brachytherapy (Papillon) in the management of early rectal cancer. *Colorectal Disease, 21*(1), 45–52.

Tam, S. Y., and Wu, V. W. C. (2019). A review on the special radiotherapy techniques of colorectal cancer. *Frontiers in Oncology, 9*(1).

Teng, X., Zhang, X., Zhi, X., Chen, Y., Xu, D., Meng, A., and Zhu, Y. (2022). Risk factors of dermatitis during radiation for vulvar carcinoma. *Precision Medical Sciences, 11*(3), 106–110.

Terrones-Campos, C., Ledergerber, B., Vogelius, I. R., Helleberg, M., Specht, L., and Lundgren, J. (2021). Hematological toxicity in patients with solid malignant tumors treated with radiation – temporal analysis, dose response and impact on survival. *Radiotherapy and Oncology, 158*, 175–183.

Thawani, N., and Vyas, S. (2021). Mucosal and Esophageal Toxicities of Radiation Therapy. *Handbook of Cancer Treatment-Related Symptoms and Toxicities E-Book*, (pp. 291), Elsevier.

Tian, X., Nguyen, M., Foote, H. P., Caster, J. M., Roche, K. C., Peters, C. G., Wu, P., Jayaraman, L., Garmey, E. G., Tepper, J. E., Eliasof, S., and Wang, A. Z. (2017). CRLX101, a nanoparticle–drug conjugate containing camptothecin, improves rectal cancer chemoradiotherapy by inhibiting DNA repair and HIF1α. *Cancer Research, 77*(1), 112–122.

Türk, G., Eldem, G., Kılıçkap, S., Bozkurt, F. M., Salancı, B. V., Çil, B. E., Peynircioğlu, B., Yalçın, Ş., and Balkancı, F. (2019). Outcomes of radioembolization in patients with chemorefractory colorectal cancer liver metastasis: a single-center experience. *Journal of Gastrointestinal Cancer, 50*(2), 236–243.

Vuong, T., Garant, A., Vendrely, V., Martin, A. G., and Devic, S. (2022). Clinical applications of high dose rate endorectal brachytherapy for patients with rectal cancer. *Cancer/Radiothérapie, 26*(6), 879–883.

Wang, K., and Tepper, J. E. (2021). Radiation therapy-associated toxicity: etiology, management, and prevention. *CA: A Cancer Journal for Clinicians, 71*(5), 437–454.

Watanabe, T., Muro, K., Ajioka, Y., Hashiguchi, Y., Ito, Y., Saito, Y., Hamaguchi, T., Ishida, H., Ishiguro, M., Ishihara, S., Kanemitsu, Y., Kawano, H., Kinugasa, Y., Kokudo, N., Murofushi, K., Nakajima, T., Oka, S., Sakai, Y., Tsuji, A., Uehara, K., Ueno, H., Yamazaki, K., Yoshida, M., Yoshino, T., Boku, N., Fujimori, T., Itabashi, M., Koinuma, N., Morita, T., Nishimura, G., Sakata, Y., Shimada, Y., Takahashi, K., Tanaka, S., Tsuruta, O., Yamaguchi, T., Yamaguchi, N., Tanaka, T., Kotake, K., and Sugihara, K. (2018). Japanese Society for Cancer of the Colon and Rectum (JSCCR) guidelines 2016 for the treatment of colorectal cancer. *International Journal of Clinical Oncology, 23*(1), 1–34.

Wild, A. T., and Yamada, Y. (2017). Treatment options in oligometastatic disease: stereotactic body radiation therapy - focus on colorectal cancer. *Visceral Medicine, 33*(1), 54–61.

Wiltshire, K. L., Ward, I. G., Swallow, C., Oza, A. M., Cummings, B., Pond, G. R., Catton, P., Kim, J., Ringash, J., Wong, C. S., Wong, R., Siu, L. L., Moore, M., and Brierley, J. (2006). Preoperative radiation with concurrent chemotherapy for resectable rectal cancer: effect of dose escalation on pathologic complete response, local recurrence-free survival, disease-free survival, and overall survival. *International Journal of Radiation Oncology*Biology*Physics, 64*(3), 709–716.

Wo, J. Y., Anker, C. J., Ashman, J. B., Bhadkamkar, N. A., Bradfield, L., Chang, D. T., Dorth, J., Garcia-Aguilar, J., Goff, D., Jacqmin, D., Kelly, P., Newman, N. B., Olsen, J., Raldow, A. C., Ruiz-Garcia, E., Stitzenberg, K. B., Thomas, C. R., Wu, Q. J., and Das, P. (2021). Radiation therapy for rectal cancer: executive summary of an ASTRO clinical practice guideline. *Practical Radiation Oncology, 11*(1), 13–25.

Xie, X., Zheng, W., Chen, T., Lin, W., Liao, Z., Liu, J., and Ding, Y. (2019). CDK4/6 inhibitor palbociclib amplifies the radiosensitivity to nasopharyngeal carcinoma cells via mediating apoptosis and suppressing DNA damage repair. *OncoTargets Therapy, 12*, 11107.

Xie, Y., Wang, Q., Hu, T., Chen, R., Wang, J., Chang, H., and Cheng, J. (2021a). Risk factors related to acute radiation dermatitis in breast cancer patients after radiotherapy: a systematic review and meta-analysis. *Frontiers in Oncology, 11*, 738851.

Xie, Y., Wang, Q., Hu, T., Chen, R., Wang, J., Chang, H., and Cheng, J. (2021b). Risk factors related to acute radiation dermatitis in breast cancer patients after radiotherapy: a systematic review and meta-analysis. *Frontiers in Oncology, 11*(1).

Xu, X., Nakano, T., Tsuda, M., Kanamoto, R., Hirayama, R., Uzawa, A., and Ide, H. (2019). Direct observation of damage clustering in irradiated DNA with atomic force microscopy. *Nucleic Acids Research, 48*(3), e18–e18.

Xu, Y., Sun, Y., Zhou, K., Xie, C., Li, T., Wang, Y., Zhang, Y., Rodriguez, J., Zhang, X., and Shao, R. (2020). Cranial irradiation alters neuroinflammation and neural proliferation in the pituitary gland and induces late-onset hormone deficiency. *Journal of Cellular Molecular Medicine, 24*(24), 14571–14582.

Ye, L. F., Chaudhary, K. R., Zandkarimi, F., Harken, A. D., Kinslow, C. J., Upadhyayula, P. S., Dovas, A., Higgins, D. M., Tan, H., and Zhang, Y. (2020). Radiation-induced lipid peroxidation triggers ferroptosis and synergizes with ferroptosis inducers. *ACS Chemical Biology, 15*(2), 469–484.

Yi, H., Yan, X., Luo, Q., Yuan, L., Li, B., Pan, W., Zhang, L., Chen, H., Wang, J., and Zhang, Y. (2018). A novel small molecule inhibitor of MDM2-p53 (APG-115) enhances radiosensitivity of gastric adenocarcinoma. *Journal of Experimental Clinical Cancer Research, 37*(1), 1–10.

Yu, J., Kim, D. H., Lee, J., Shin, Y. M., Kim, J. H., Yoon, S. M., Jung, J., Kim, J. C., Yu, C. S., Lim, S.-B., Park, I. J., Kim, T. W., Hong, Y. S., Kim, S. Y., Kim, J. E., Park, J.-h., and Kim, S. Y. (2021). Radiofrequency ablation versus stereotactic body radiation therapy in the treatment of colorectal cancer liver metastases. *Cancer Research and Treatment, 54*(3), 850–859.

Yuan, Y., Xiao, W.-W., Xie, W.-H., Cai, P.-Q., Wang, Q.-X., Chang, H., Chen, B.-Q., Zhou, W.-H., Zeng, Z.-F., Wu, X.-J., Liu, Q., Li, L.-R., Zhang, R., and Gao, Y.-H. (2021). Neoadjuvant chemoradiotherapy for patients with unresectable radically locally advanced colon cancer: a potential improvement to overall survival and decrease to multivisceral resection. *BMC Cancer, 21*(1), 179.

Zeng, Y., Liu, Q., Wang, Y., Tian, C., Yang, Q., Zhao, Y., Liu, L., Wu, G., and Xu, S. (2020). CDK5 activates hippo signaling to confer resistance to radiation therapy via upregulating TAZ in lung cancer. *International Journal of Radiation Oncology* Biology* Physics, 108*(3), 758–769.

Zuppone, S., Bresolin, A., Spinelli, A. E., Fallara, G., Lucianò, R., Scarfò, F., Benigni, F., Di Muzio, N., Fiorino, C., and Briganti, A. (2020). Pre-clinical research on bladder toxicity after radiotherapy for pelvic cancers: state-of-the art and challenges. *Frontiers in Oncology, 10*, 527121.

8 Colon Cancer Treatment Using Immunotherapy
Advantages and Disadvantages

Syed Arman Rabbani[1], *Mohamed El-Tanani*[1], *Yahia El-Tanani*[2], *Alaa A. Aljabali*[3], *Mohammad Ahmed Khan*[4], and *Shrestha Sharma*[5]

[1]RAK College of Pharmacy, RAK Medical and Health Sciences University, Ras Al Khaimah, United Arab Emirates

[2]Royal Cornwall Hospital Trust, NHS, United Kingdom

[3]Department of Pharmaceutics and Pharmaceutical Technology, Pharmacy, Yarmouk University, Irbid, Jordan

[4]School of Pharmaceutical Education and Research, Jamia Hamdard, New Delhi, India

[5]Amity Institute of Pharmacy, Amity University, Haryana, India

8.1 INTRODUCTION

Colon cancer, or colorectal cancer (CRC), particularly referred to as adenocarcinoma cancer, is the third most frequent cancer all over the world and is observed to be increasing at a raised trend in developing countries [1]. This cancer develops from glandular epithelial cells of the large intestine. It starts as a genetic or epigenetic change that enables these cells to divide and form benign adenomas that may progress to carcinoma and metastasize with time [2]. CRC is the third most prevalent form of malignancy globally, constituting 10%–11% of total cases reported by GLOBOCAN 2022, and ranks second in mortality among all cancer types [3,4].

CRC is more common in developed countries, particularly Hungary, with the highest incidence among males, and Norway among females. The recent rise in Japan and South Korea requires new approaches to screening and prevention. In 2022, approximately 1,918,143 new CRC cases (including anal cancers) and 904,091 deaths were estimated, which are nearly 10% of total cancer incidence and mortality. Increased staging of CRC reduces survival rates [3].

The overall 5-year survival rate for rectal cancer usually decreases as the stage increases. As a general rule, the rate for Stage I is about 88%. It reduces to about 50% of Stage IIB and further declines to about 81% of Stage IIA. The following is the overall survival (OS) of patients who underwent surgery for Stage III: survival rate of Stage IIIA=83%; survival rate of Stage IIIB=72%; and survival rate of Stage IIIC=58%. However, in patients diagnosed with Stage IV rectal cancer, the OS is very low at 13%. The above finding highlights the role of cancer staging in determining the survival of the patient and the need to have an early diagnosis as a key to improving the survival rates [3,5].

Conventional therapy, such as surgery, and adjuvant therapies, including chemotherapy and radiation therapy, remain the cornerstone of CRC therapy. Surgery is useful only in the initial stage of the disease, but it becomes practically ineffective once metastases occur [6,7]. A common treatment for cancer is chemotherapy, which adversely affects healthy body cells, causing side effects, and its effectiveness declines as the body develops resistance to the drugs. This treatment is very useful in rectal cancer; however, when used for colon cancer, it is less efficient since it is difficult to aim at the

DOI: 10.1201/9781003503842-8

colon without damaging nearby tissues. Although targeted therapies are aimed at bringing about the desired end result, they only last for some time because of resistance.[[8,9].

Immunotherapy as a form of cancer treatment is relatively new, but it represents an essential step forward because it helps to improve the immunological ability to detect and destroy cancer cells [10]. Different from conventional treatments, immunotherapy works on the genetic features of cancer cells that enable them to escape the immune system, for example, protein directories such as programmed death-1 (PD-1), programmed death-ligand 1 (PD-L1), and cytotoxic T-lymphocyte-associated protein 4 (CTLA-4). Figure 8.1, depicting a CRC patient receiving various types of immunotherapy, provides a visual summary of these approaches. However, it has its advantages and is mainly indicated for target patients with specific features such as microsatellite instability-high (MSI-H) or deficient mismatch repair (dMMR). The treatment's immune-related side effects arise from the enhanced stimulation of the immune system [11–13].

The objective of this chapter is to briefly summarize the current knowledge on immunotherapy in CRC, to delineate its mode of action, and to evaluate its potential to overcome the drawbacks of conventional therapies. It will also cover biomarkers and combination therapies, such as cancer vaccines and adoptive cell transfer, while examining the challenges in expanding immunotherapy

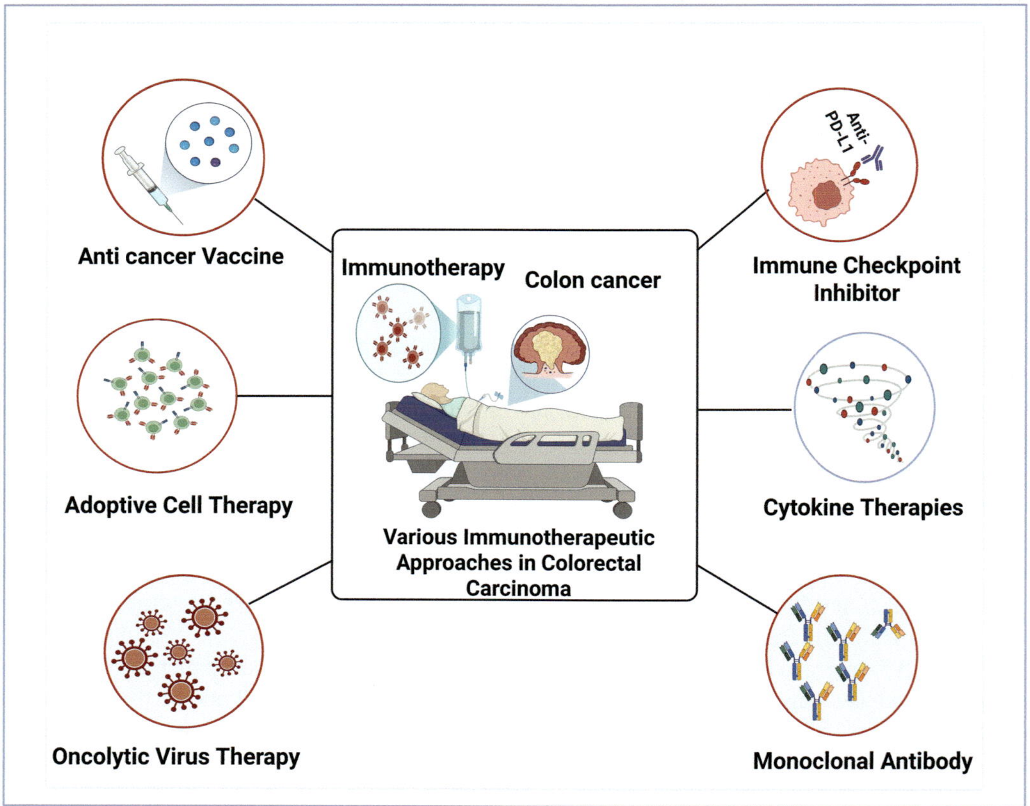

FIGURE 8.1 Immunotherapeutic approaches used in treating colorectal cancer. Various immunotherapeutic approaches are currently being used in the clinic to suppress colon cancer growth and proliferation. These include adoptive cell therapy (engineered immune cells to boost immune response), oncolytic virus (genetically modified virus to target cancer cells), immune checkpoint inhibitors (anti-PD-1/PD-L1 or CTLA-4 inhibitors activating T-cells), anti-cancer vaccines (non-infectious component to stimulate T-cell activation and development of memory immune cells), cytokine therapy (pro-inflammatory cytokines increase immune activation), and monoclonal antibodies (specifically bind to cancer cells). These agents represent comprehensive immunomodulatory approaches in combating colorectal cancer.

to a wider patient population. Based on the present review of current literature and future trends, the focus is made to equip the readers, who are clinicians and researchers, with a valuable reference for applying immunotherapy to CRC.

8.2 MECHANISMS OF IMMUNOTHERAPY IN COLON CANCER

8.2.1 Immune Checkpoint Inhibitors PD-1/PD-L1 Pathway Blockade

T-cells express the PD-1 receptor, a co-inhibitory receptor that plays a crucial role in downregulating the T-cell response and the immune response in general [11]. PD-1 binds specifically to its ligands, PD-L1 and PD-L2, both of which are often overexpressed both in the tumor and in specific immunocompetent cells within the tumor stroma. This interaction acts as a suppressive signal to limit T-cell activation, proliferation, and function, which facilitates the immune escape of tumors [11,12].

Therefore, upon T-cell activation, PD-1 is upregulated during constant antigen stimulation. PD-L1 is constitutively expressed in many tumor cell types as well as in tumor-infiltrating macrophages, whereas PD-L2 is expressed in dendritic cells (DCs), monocytes, and activated T-cells [14]. The displayed mRNA and protein levels of these ligands compared to normal tissue controls induce an immunosuppressive phenotype in the tumor microenvironment (TME). It has been demonstrated that when PD-1 engulfs PD-L1/PD-L2, it suppresses intracellular signal transduction, which indirectly results in suppression of TCR signal generation with decreased cytokine production of already established cytokines such as interferon γ (IFN-γ) and tumor necrosis factor α (TNF-α). This interaction prevents T-cell proliferation and leads to T-cell exhaustion, described by blunted function and continual expression of negative receptors [14,15].

To overcome this immunosuppression, new therapies like PD-1 inhibitors, including pembrolizumab and nivolumab, have been developed. These monoclonal antibodies block PD-1; thus, it is unable to bind to PD-L1 and PD-L2 proteins. This blockade stops the inhibitory signals that suppress the action of T-cells, reviving the cells' ability to target and kill cancer cells [15,16]. The suppression of PD-1 results in strengthened T-cell activity due to stimulation of proliferation, enhancement of cytotoxic activity, and production of effector cytokines. Memory T-cells can migrate to tumor tissue and mount potent anti-tumor immune responses upon reactivation [12,15].

In addition, restoration of T-cell function enhances pro-inflammatory cytokine secretion, leading to enhanced immune response and increased recruitment of additional immune cells to the tumor site [17]. Activated T-cells have a better capacity to recognize tumor-associated antigens—a contribution that is important for sparing normal tissues and achieving maximum therapeutic efficacy by targeting cancer cells [15,18].

8.2.2 CTLA-4 Inhibition

Cytotoxic T-lymphocyte-associated protein 4 (CTLA-4) has been identified as an immunological character that plays a crucial function in controlling T-cell signaling and thus immune response [19]. The drawbacks of CTLA-4 are that it counter-competes with another costimulatory receptor known as CD28, which works on the same ligands, the CD80 and CD86 on the APCs. In contrast, when CTLA-4 is signaled, it transmits signals that essentially suppress T-cell activation and proliferation, thus limiting the immune response against tumor cells. This mechanism is specific for TGF-ß and is especially important in the tumor environment, where immune control evasion is a frequent event in cancer cells [19,20].

This inhibitory interaction is aimed at being blocked by CTLA-4 inhibitors like Ipilimumab by capturing CTLA-4 to preclude its action. This blockade helps to increase T-cell activation and, consequently, the number of T-cells that are good enough to defeat the pathogen [21]. Therefore, the formation of effector T-cells, which have the potential to evoke powerful anti-tumor reactions, is enhanced effectively. Reactivated T-cells have a higher efficiency of responsive and effective destruction of cancer cells, enhancing the chances of immunotherapy [21,22]. This mechanism shows that CTLA-4 has two functions: to maintain immune homeostasis and for possible intervention to develop immune checkpoints to enhance anti-tumor immunity, pointing to the future management of different cancers. CTLA-4 inhibitors have also been shown to have clinical utility in various trials, and current usage of CTLA-4 blockage in cancer treatment supports the necessity of CTLA-4 inhibition in cancer immunotherapy [22,23]. Figure 8.2, illustrating the interaction between T-cell receptors (TCRs) and peptide–major histocompatibility complex (MHC) complexes, along with the engagement of immune checkpoints, provides a visual understanding of how checkpoint inhibitors reawaken T-cells to target and destroy cancer cells.

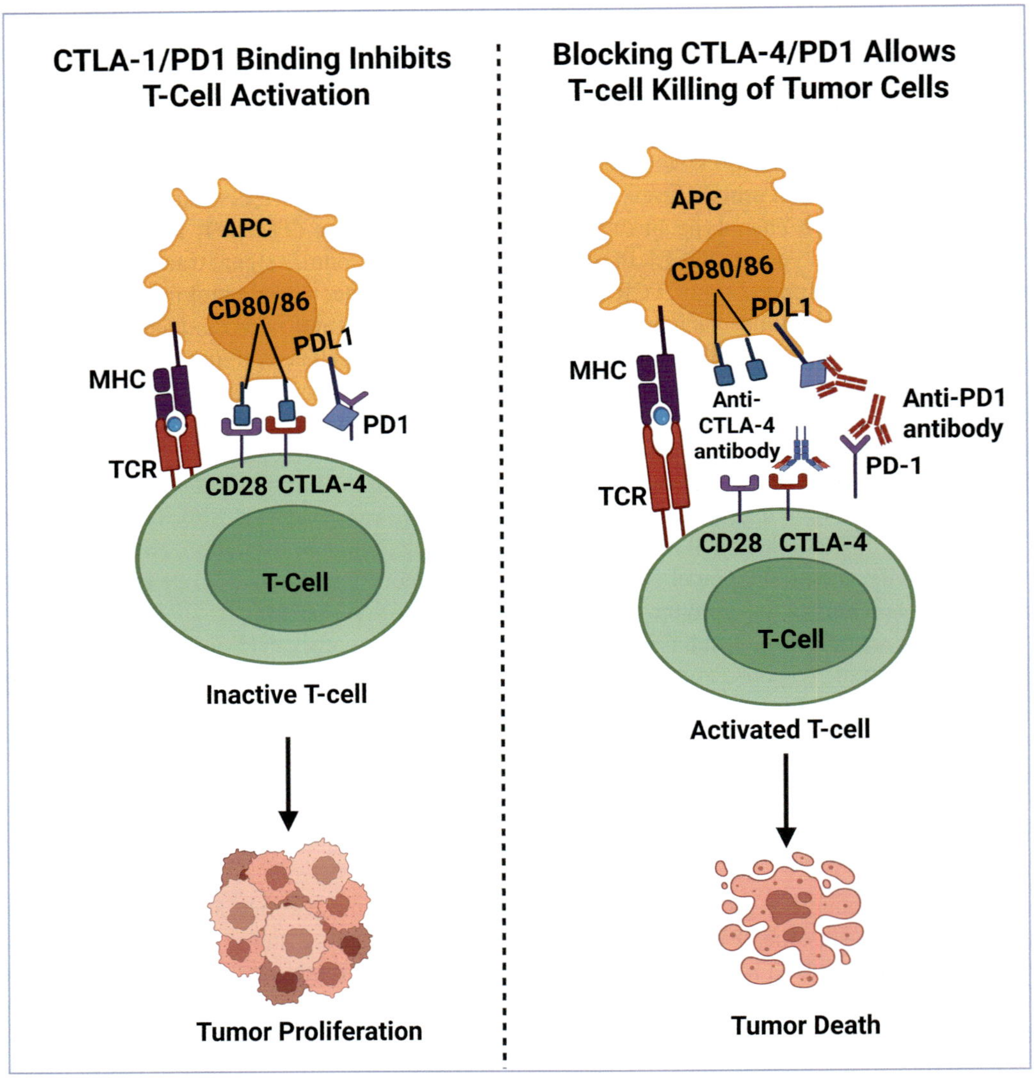

FIGURE 8.2 Immune checkpoint inhibitors as a therapeutic approach in colorectal cancer. Checkpoint proteins, such as CD80/CD86 and PD-L1 on antigen-presenting cells (APCs) and CTLA-4 and PD-1 on T-cells, help keep the body's immune responses in check. When the T-cell receptor (TCR) binds to antigen and major histocompatibility complex (MHC) proteins on the APC and CD28 and PD-1 bind to CD80/CD86 and PD-L1 on the APC, respectively, the T-cell can be activated. However, the binding of CD80/CD86 to CTLA-4 keeps the T-cells in an inactive state so they are not able to kill tumor cells in the body (left panel). Blocking the binding of CD80/CD86 to CTLA-4 and PD-L1 to PD-1 with immune checkpoint inhibitors (anti-CTLA-4 antibody/anti-PD-1 antibody) allows the T-cells to be active and to kill tumor cells (right panel).

8.2.3 Adoptive Cell Therapies

8.2.3.1 T-Cell Receptor Therapy

Adoptive cell therapy (ACT) represents an emerging novel immunotherapeutic approach, utilizing recent advances in specifically targeting and eradicating cancer cells using T-cells. However, one of the most commonly described forms of ACT is TCR therapy, in which T-cells are genetically modified by TCRs specific to tumors. T-cells are first temporarily harvested from the patient's blood before being activated by cytokines such as interleukin 2 (IL-2) and by antigen-presenting cells (APCs) that display costimulatory molecules such as CD80/CD86. Such activation results in increased expression of activation markers (e.g., CD25) and leads to a signaling cascade triggering the phosphoinositide 3-kinase/Akt pathway and stimulating cell survival and proliferation [23,24].

After that, the specific TCR genes are transduced into these activated T-cells by employing viral vectors since these vectors rely on the specific cell's transcription to generate fully functional TCR proteins. The TCR complex is made of TCRα and β chains, which are transduced genes and rearranged to create a receptor that can accurately identify tumor-associated antigens. Subsequently, the reformed T-cells display these engineered TCRs on their surface and are capable of binding with the peptide-MHC complex on cancer cells [24,25].

When the modified T-cells are reinfused into the patient, they recognize and mechanically bind to specific antigens presented on the surface of MHC molecules of tumor cells. Interaction of the TCR with its ligand on the APC activates a series of intracellular signaling events, including activation of the mitogen-activated protein kinase (MAPK) and nuclear factor kappa B pathways, which are initiated primarily by the interacting TCR/CD3 complex. These pathways promote effector cytokine production, principally IFN-γ and TNF-α, which augment T-cell proliferation and cytotoxicity [25,26].

In addition, TCR binding to tumor antigens leads to the release of perforin and granzymes, cytotoxic molecules that destroy target malignancy cells by inducing apoptosis. TCR-engineered T-cells can exert a potent anti-tumor effect precisely via this molecular mechanism, as it significantly enhances the specificity and activity of a tumoricidal T-cell immune response while minimizing damage to normal tissue [26,27].

8.2.4 Chimeric Antigen Receptor T-Cell Therapy

Chimeric antigen receptor T-cell (CAR-T) therapy is a novel strategy that belongs to adoptive cell immunotherapy through the engineering of autologous T-cells to express designed synthetic chimeric antigen receptors (CARs). These CARs are designed to bind selectively with particular antibodies present on the surface of cancer cells, thereby increasing the chances of T-cells to work efficiently against tumors [28]. It starts with the collection of T-cells from the patient (most often conducted by leukapheresis), followed by T-cell activation by cytokines such as IL-2, IL-7, or IL-15, as well as APCs that act as stimulators of T-cell expansion and performance [29].

After activation, T-cells undergo genetic modification via viral vectors (usually lentiviral or retroviral) to insert the CAR gene. The gene expresses an extracellular single-chain variable fragment (scFv) specific for TAA, a transmembrane domain that tethers the CAR within the T-cell membrane, and intracellular domains, including the CD3ζ chain and costimulatory domains such as CD28 or 4-1BB. Once transduced, the CAR-T cells are expanded in culture under strict quality control, assuring requisite CAR expression and functional cell capabilities before reinfusion into the patient [29,30].

Instead, the CAR-T cells are injected after treatment and migrate to the TME. These engineered T-cells bind to TAAs on cancer cells utilizing the scFv region, bypassing the requirement of MHC presentation, which is critical for targeting tumors that may downregulate MHC to avoid immune detection. This binding activates intracellular signaling through the CD3ζ chain and costimulatory domains, initiating several key events: The expression of these potentially protective functions included increased production of pro-inflammatory cytokines (IL-2, IFN-γ, and TNF-α), enhanced

T-cell proliferation, and potent cytotoxic activity. The released cytolytic granules are activated CAR-T cells releasing perforin and granzyme that ultimately result in the apoptotic (programmed cell death) death of target cancer cells [31,32].

In addition, CAR-T therapy engenders the generation of long-lived memory T-cells found in the blood, which remain in the body and continue to express protection against tumor recurrence. This comprehensive mechanism illustrates the potential for CAR-T therapy to create robust and sustained anti-tumor immune responses [32].

8.2.5 Cancer Vaccines

Cancer vaccines are new treatment strategies intended to induce strong immune anti-tumor responses against tumor-associated antigens (TAAs). These vaccines 'present' these antigens—either the free antigens themselves, if they are stripped from the tumor cell surface, or the antigens themselves, exhibited on the outermost layer of the tumor cell—to the immune system so that they can be targeted at, and hence destroyed by, the immune system. Squamous cell carcinoma antigens recognized by T-cells 3, survivin, carcinoembryonic antigen (CEA), melanoma-associated antigen, insulin-like growth factor II mRNA-binding protein 3, and Wilms tumor 1 protein have been identified as several TAAs in CRC. The release of these antigens activates local immune cells, leading to increased infiltration of T-cells and DCs in the TME [33,34].

Vaccination strategies can be classified based on the molecular composition of the antigens into three primary categories: cell-based vaccines, vector-based vaccines, and molecular-based vaccines. Molecular-based vaccines include DNA and mRNA vaccines, as well as peptide vaccines. In clinical trials, mRNA and DNA vaccines have shown significant anti-tumor efficacy and favorable safety profiles in contrast to traditional peptide vaccines [35]. Cell-based vaccines, using native or genetically modified tumor cells in conjunction with activated DCs, have not proved very efficacious in some trials but are expected to improve the effectiveness of chemotherapy and immunotherapy when used together. Live-attenuated viruses, yeasts, or bacterial vectors are the basis of vector-based vaccines. Despite having novel therapeutic avenues, oncolytic viral vectors have proven problematic because of immunogenicity concerns. However, the strategies utilizing live-attenuated bacterial and yeast vectors are under investigation as promising therapeutic options against CRC, providing novel approaches to strengthening the immune response to cancer [36].

Cancer vaccines have a major therapeutic application in CRC by promoting immune recognition and targeting TAAs. Several TAAs, including CEA, Wilms tumor 1 protein, and survivin, are potential candidates to be used in the formulation of vaccines [36]. Some previous studies have demonstrated that vaccines can provoke particular T-cell reactions; this results in tumor regression and increased survival rates in CRC patients [37]. A detailed overview of nanovaccine types is shown in Figure 8.3, illustrating various subclasses, including lipid-based, polymeric, inorganic, and biologically derived nanovaccines, each designed to deliver specific antigenic material effectively.

It has also been found that the efficacy of cancer vaccines can be increased if combined with other treatment regimens, including immune checkpoint inhibitors as well as chemotherapy. For instance, the use of vaccines together with agents such as pembrolizumab or nivolumab that prevent PD-1 helps to superimpose T-cell activation and maintain immune action against tumors. Additionally, an ongoing study is being done on the more targeted use of vaccines designed for each individual based on the antigenic constitution of the patient's tumor [37,38].

8.2.6 Modulation of the Tumor Microenvironment

8.2.6.1 Role of the Microenvironment in Immune Evasion

The TME is a multifaceted structural representation of the ground substrate in which it is lodged and with which it has an intricate relationship that enhances its immune escape mechanism, making it capable of growing and metastasizing. The TME includes tumor cells, cancer-associated fibroblasts, endothelial cells, and immune cells that together contribute to immunosuppression [39,40].

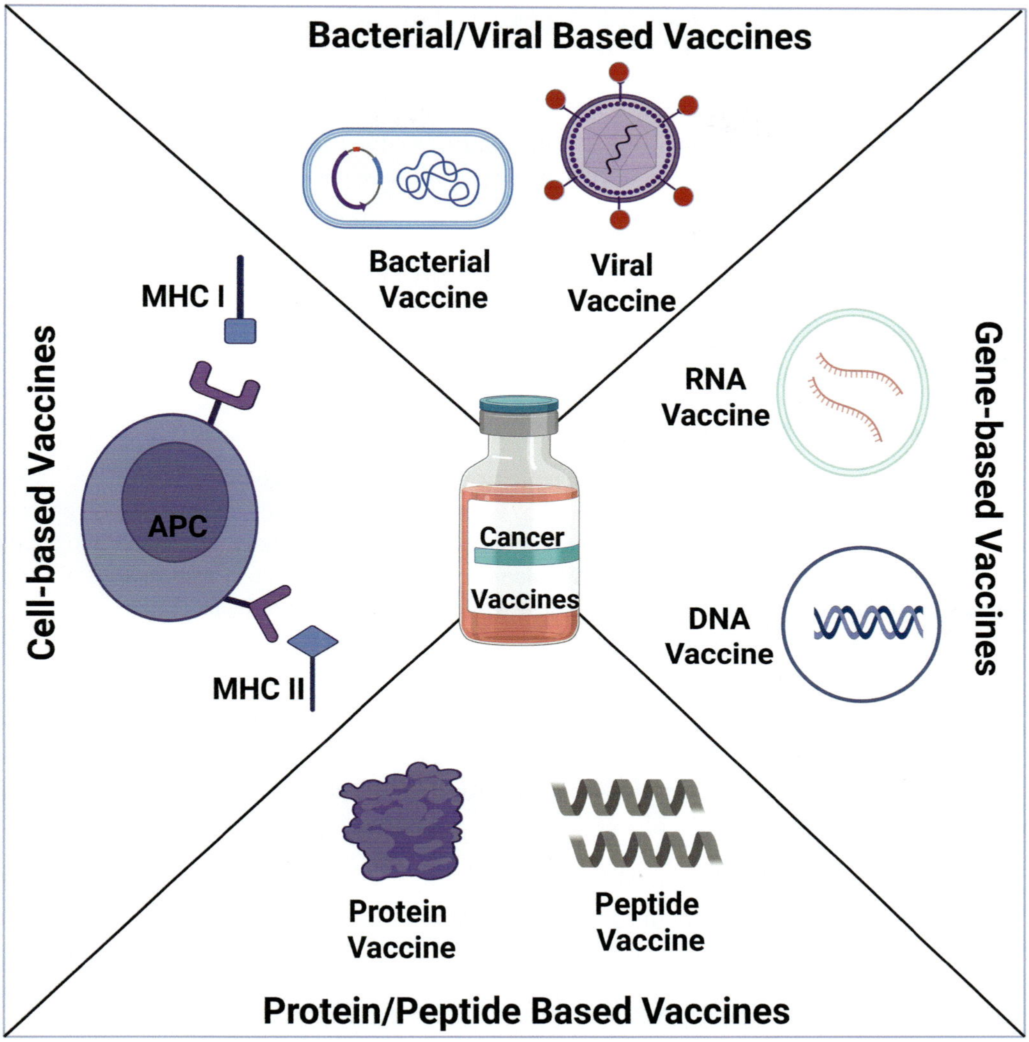

FIGURE 8.3 Vaccines used in cancer therapy. Cancer vaccines are among the most promising strategies in immunotherapy for cancer treatment. They work by activating or reactivating the immune response to tumors through the administration of antigens recognized as foreign by the immune system. Four types of vaccines have been developed to impede the growth of colorectal cancer cells. Whole cell-based vaccines (patient's own cancer cells are injected as a vaccine, triggering both innate and adaptive immune responses), bacterial/viral-based vaccines (modified bacteria or virus are used as vaccines, triggering both innate and adaptive immunity), gene-based or nucleic acid-based vaccines (DNA plasmids or conventional non-replicating mRNA encoding antigens of interest are used as vaccine, triggering both innate and adaptive immunity), and protein/peptide-based vaccines (proteins or peptides bind with the MHC molecule expressed on antigen-presenting cells and activate T-cells). MHC, Major histocompatibility complex.

Immunosuppressive cytokines, such as interleukin 10 (IL-10) and transforming growth factor-beta (TGF-beta), induced by tumor cells, can suppress T-cell activation and promote the development of regulatory T-cells (Tregs) [40]. In addition, effector T-cell responses are further suppressed by Tregs, substantially diminishing the efficacy of anti-tumor immunity. As well, myeloid-derived suppressor cells (MDSCs) are recruited to the TME and secrete factors downregulating T-cell function and promoting tumor progression [40,41].

Tumor metabolic pathways that support tumor survival can also be influenced by the TME. As a simple example, tumor cells may use tryptophan metabolism through indoleamine 2,3-dioxygenase to deplete tryptophan in the local microenvironment to facilitate T-cell proliferation and function. This depletion results in T-cell anergy or apoptosis, further decreasing the anti-tumor immune response [42,43].

Furthermore, the extracellular matrix (ECM) in the TME can literally obstruct T-cell infiltration and shape immune cell behavior. An alteration in ECM composition can create a barrier to the movement of immune cells toward tumor cells while providing signals that favor tumor cell survival and proliferation [44,45].

8.3 STRATEGIES TO ENHANCE IMMUNOGENICITY

To enhance the immunogenicity of cancer therapies, several strategies can be employed, each targeting different aspects of the immune response and the TME.

8.3.1 Combination Therapies

Immunotherapies can improve the immunogenicity of tumors when combined with traditional approaches, such as chemotherapy and radiation. Immune cell death induced by chemotherapeutic agents can promote the release of tumor-associated antigens and damage-associated molecular pattern moieties, which activate DCs and enhance overall T-cell priming. It also increases T-cell recognition of tumors through mutations and de novo expression of neoantigens [15,26].

8.3.2 Use of Adjuvants

Agents that can enhance the immune response to vaccines are called adjuvants. Instead, they stimulate innate immune cells, help mature DCs, and present antigens to T-cells. Toll-like receptor agonists, which stimulate pattern recognition receptor activation, can enhance cytokine production and increase T-cell activity [26,39].

8.3.2.1 Modulation of the Tumor Microenvironment

MDSCs, Tregs, and inhibitory cytokines are all found in the TME, which often suppress the immune system [41–43]. Strategies to regulate the TME might include:

Depleting MDSCs: Targeting MDSCs can ameliorate the suppression of T-cell activity, thereby enhancing T-cell proliferation and effector function [41].

Targeting Tregs: Inhibiting the function of Tregs inside the tumor may restore effector T-cell activity and produce a more powerful anti-tumor immune response [42].

Blocking Immunosuppressive Cytokines: Inhibiting cytokines such as IL-10 and TGF-β, which induce immunosuppression, may improve T-cell responses [43].

8.3.2.2 Oncolytic Virus Therapy

Oncolytic viruses selectively infect and destroy cancer cells while simultaneously inducing a systemic immune response. Figure 8.4 illustrates the detailed mechanism of oncolytic virus therapy. As these viruses propagate within tumors, they can expose tumor antigens to the immune system, enhancing cancer cell recognition and targeting. This dual action can lead to a more significant and sustained anti-tumor immune response [34,35]. To offer a complete review of current advancements in immunotherapy for CRC, Table 8.1 highlights various cell-based and vector-based vaccines. These vaccines represent promising avenues in enhancing immune responses and improving patient survival rates in CRC.

8.3.2.3 Gene Editing Techniques

CRISPR/Cas9 and other advanced gene editing technologies may be used to alter tumor cells to express more immunogenic antigens or to eliminate inhibitory receptors that suppress the immune system. T-cells may be better able to identify tumor cells as a result, increasing the possibility of successful immune-mediated elimination [46].

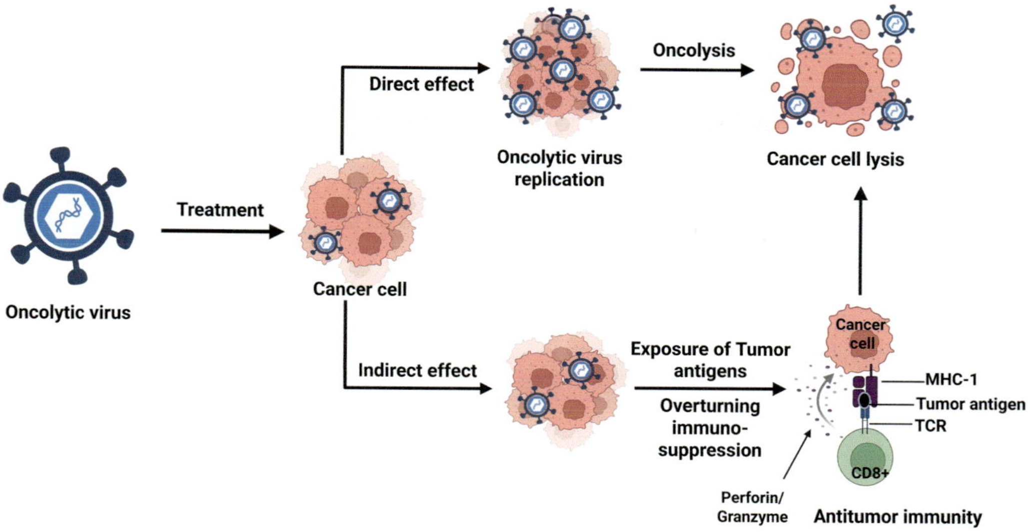

FIGURE 8.4 Oncolytic virus (OV) therapy in colorectal cancer. OVs represent a novel class of cancer immunotherapy agents that preferentially infect and kill cancer cells and promote protective anti-tumor immunity. OVs destroy cancer cells either directly by undergoing replication in tumor cells, causing cancer cell lysis, or indirectly by stimulating host anti-tumor immune response, causing cell death.

8.3.2.4 Personalized Cancer Vaccines

The immune response may be greatly improved by creating customized vaccinations that target certain neoantigens particular to each person's tumor. Researchers may generate customized vaccinations that trigger a strong immune response by analyzing the tumor's genome to find mutations that yield neoantigens [35].

8.3.2.5 Dendritic Cell Vaccination

An essential function of DCs is to bridge the gap between innate and adaptive immunity. The capacity of DCs to deliver these antigens to T-cells may be improved by vaccination tactics that load them with TAAs or entire tumor lysates. This will encourage T-cell activation, proliferation, and memory formation. A stronger anti-tumor response may result from this [46,47].

8.4 BIOMARKERS PREDICTING IMMUNOTHERAPY RESPONSE

8.4.1 Microsatellite Instability-High and Mismatch Repair Deficiency

It is a molecular phenotype resulting from defects in the DNA mismatch repair (MMR) system, which is responsible for fixing base mismatches during DNA replication in humans and most other organisms. When MMR proteins, mainly MLH1, MSH2, MSH6, and PMS2, are deficient, there will be a poor repair process, leading to the formation of errors in DNA sequences known as microsatellites [48,49]. These short repetitive sequences of DNA, a microsatellite, become unstable; they acquire both insertion and deletion mutations in tumor cells, hence the term microsatellite instability, or MSI. The repair process for MSI-H tumors lacks an appropriate mechanism to address these numerous mutations, and that leads to cancer [49].

8.4.1.1 Molecular Mechanism

Initial recognition of DNA mismatches is dependent on the MSH2/MSH6 heterodimer. The system scans newly replicated DNA strands for mismatches of single bases (base-base) or small insertion-deletion loops (indels), such as those characteristic of microsatellites [50]. The repair process starts once MSH2/MSH6 binds to the site after a mismatch has been detected. The MSH2/MSH6

TABLE 8.1
Major Clinical Trials of Cancer Vaccines Targeting Colorectal Cancer: Neoantigen, Biological Vector-Based, and Colorectal Cancer-Specific Vaccines [35]

NCT Number (Status)	Tumor Type/Stage	Intervention Type	Interventions	Adjuvant/Combined Therapy	Primary Objective	Results	Phases (Enrollment)	Completion Date
NCT05141721 (Not yet recruiting)	mCRC	Molecular-based vaccine	GRT-C901, GRT-R902	Fluoropyrimidine, oxaliplatin, bevacizumab, ipilimumab	Safety, efficacy	Pending	Phase II/III (665)	Mar-27
NCT03639714 (Active, not recruiting)	MSS solid tumors (including CRC)	Molecular-based vaccine	GRT-C901, GRT-R902	Nivolumab, Ipilimumab	Safety, immune response	Pending	Phase I/II (214)	Mar-23
NCT03953235 (Recruiting)	MSS solid tumors (including CRC)	Molecular-based vaccine	GRT-C903, GRT-R904	Nivolumab, Ipilimumab	Safety, immune response	Pending	Phase I/II (144)	Dec-23
NCT04912765 (Recruiting)	mCRC, hepatocellular cancer	Dendritic cell vaccine	Neoantigen dendritic cell vaccine	Nivolumab	Safety, immune response	Pending	Phase II (60)	May-25
NCT05243862 (Not yet recruiting)	mCRC	Molecular-based vaccine	PolyPEPI1018	Montanide ISA 51, Atezolizumab	Safety, immune response	Pending	Phase II (28)	Mar-26
NCT05078866 (Not yet recruiting)	Lynch syndrome	Viral vector-based vaccine	GAd-209-FSP, MVA-209-FSP	—	Safety, immune response	Pending	Phase Ib/II (45)	Dec-25
NCT01885702 (Active, not recruiting)	Colorectal cancer (MSI) or lynch syndrome	Dendritic cell vaccine	DC vaccination	—	Safety, immune response	Pending	Phase I/II (25)	Dec-22
NCT05130060 (Recruiting)	mCRC	Molecular-based vaccine	PolyPEPI1018	Montanide ISA 51, TAS-102	Safety, immune response	Pending	Phase I (15)	May-24
NCT04117087 (Recruiting)	MSS mCRC, pancreatic cancer	Peptide-based vaccine	KRAS peptide vaccine	Poly-ICLC, nivolumab, ipilimumab	Safety, immune response	Pending	Phase I (30)	Jun-24
NCT04853017 (Recruiting)	KRAS/NRAS mutated solid tumor (G12D or G12R)	Peptide-based vaccine	ELI-002 2P	Amph-CpG-7909	Safety, immune response	Pending	Phase I (18)	Sep-24
NCT04147078 (Recruiting)	Solid tumors (including CRC)	Dendritic cell vaccine	Neoantigen-primed DC vaccine	—	Safety, immune response	Pending	Phase I (80)	Jun-23
NCT04799431 (Not yet recruiting)	mCRC, metastatic pancreatic cancer	Neoantigen vaccine	Neoantigen vaccine	Poly-ICLC, retifanlimab	Safety, immune response	Pending	Phase I (12)	Feb-26

(Continued)

TABLE 8.1 (*Continued*)
Major Clinical Trials of Cancer Vaccines Targeting Colorectal Cancer: Neoantigen, Biological Vector-Based, and Colorectal Cancer-Specific Vaccines [35]

NCT Number (Status)	Tumor Type/Stage	Intervention Type	Interventions	Adjuvant/Combined Therapy	Primary Objective	Results	Phases (Enrollment)	Completion Date
NCT01147965 (Complete)	Advanced CRC	Viral vector-based vaccine	Ad5 [E1-, E2b-]-CEA (6D)	—	Safety, immune response	Minimal toxicity; OS at 12 months is 48%.	Phase I/II (32)	—
NCT01972737 (Complete)	Stage I/II CRC	Viral vector-based vaccine	Ad5-hGCC-PADRE vaccine	—	Safety, immune response	GUCY2C-specific T-cell responses detected in 40% of patients; minimal adverse events observed.	Phase I (10)	—
NCT00103142 (Complete)	mCRC	Therapeutic autologous DC vaccine	PANVAC (viral vector), DCs	GM-CSF	RFS, immune response	RFS at 2 years was similar; CEA-specific T-cell responses were statistically similar.	Phase II (76)	—
NCT02981524 (Complete)	Advanced pMMR CRC	Cancer cell vaccine	GVAX/ Cy+pembrolizumab	—	ORR, safety, PFS, OS, DOR	Median PFS: 82 days; Median OS: 213 days; toxicities were acceptable.	Phase II (17)	—
NCT00924092 (Complete)	mCRC	Live-attenuated yeast vaccine	GI-6207 (yeast-CEA)	—	Safety, immune response	Minimal toxicity observed; antigen-specific T-cell responses and CEA stabilization detected.	Phase I (22)	—
NCT01519817 (Complete)	Advanced CRC	Live-attenuated yeast vaccine	GI-6301 (yeast brachyury vaccine)	—	Immune response, safety	Brachyury-specific T-cell responses observed; no autoimmunity or serious adverse events.	Phase I (11)	—
NCT01966289 (Complete)	Advanced CRC	Cancer cell vaccine	GM-CSF, cyclophosphamide/ guadecitabine	—	Immune response, safety	No significant increase in CD45RO+ cells; Grade 3–4 toxicities were manageable.	Phase I (18)	—
NCT01348256 (Complete)	mCRC	Dendritic cells vaccine	Ex vivo TNF-a, IL1, IL6, and PGE2	—	Safety, OS, RFS	Median OS: 7.4 months; median tumor progression: 2.4 months.	Phase I/II (20)	—

MSS, microsatellite stable; CRC, colorectal cancer; mCRC, metastatic colorectal cancer; MSI, microsatellite instability; KRAS, Kirsten rat sarcoma viral oncogene; NRAS, neuroblastoma RAS viral oncogene; OS, overall survival; CEA, carcinoembryonic antigen; ORR, objective response rate; PFS, progression-free survival; IL-2, interleukin 2.

complex, once it identifies the mismatch, recruits the MLH1/PMS2 heterodimer to the site of the error [49]. This heterodimer serves as a molecular scaffolding complex that coordinates the excision of the erroneous DNA segment. Other components of the repair machinery interact with the MLH1/PMS2 complex and with each other to cleave the DNA strand near the mismatch. After the heterodimer (MLH1/PMS2) is recruited, an exonuclease enzyme removes the mismatched region by removing the erroneous nucleotides. The gap is then filled with the correct nucleotide sequence by DNA polymerase, guided by the template strand, and sealed by DNA ligase, which severs the DNA and restores genome integrity [50,51].

8.4.1.2 Detection of MSI-H and MMR Deficiency

The detection of MSI-H status in tumors is critical for diagnosis, prognosis, and therapeutic decision-making, particularly in cancers like CRC, endometrial cancer, and others. Several methods are used to identify MSI-H:

Polymerase Chain Reaction (PCR): In PCR-based detection of MSI, a panel of microsatellite markers (usually five) is analyzed to compare microsatellite lengths between tumor tissue and normal tissue. BAT 25, 26, D2S 123, D5S 346, and D17S 250 are the most-used microsatellite markers. If two or more of these markers show instability, then a tumor is said to be MSI-H [51,52].

Next-Generation Sequencing (NGS): As NGS covers more of the tumor genome and detects MSI, the utility of the test is also increased. MSI testing using NGS can evaluate a greater number of microsatellites and other genomic alterations together. The advantage of this technique is that it provides higher sensitivity and can be included in multi-gene panels in cancer diagnostics [53,54].

Immunohistochemistry (IHC): Expression of MMR proteins (MLH1, MSH2, MSH6, and PMS2) is assessed in tumor tissues by IHC. IHC is routinely used in clinical settings because it is simple and cost-effective, and there are definite associations between the loss of expression of one or more of these proteins and dMMR, which is strongly associated with MSI-H [55].

MLH1 Promoter Methylation Testing: Further testing for promoter methylation of the MLH1 gene is warranted in cases where MLH1 is found to be deficient by IHC [56]. In most sporadic MSI-H CRCs, methylation of the MLH1 promoter is a common cause of its inactivation. This allows discrimination of sporadic MSI-H from hereditary Lynch syndrome with MMR germline mutation [56,57].

8.4.1.3 Clinical Significance

In early-stage colorectal cancer, the favorable prognosis associated with MSI-H/dMMR is reflected in better OS and lower metastasis rates than with microsatellite stable (MSS) tumors [53]. As a predictive biomarker, it predicts response to immune checkpoint inhibitors such as pembrolizumab and nivolumab in MSI-H/dMMR tumors and leads to durable responses. However, these tumors are inherently resistant to standard chemotherapy agents, including fluorouracil and oxaliplatin. Detection of dMMR is also important in diagnosing Lynch syndrome, a hereditary cancer syndrome [54]. Similar to MSI-H tumors, MSI tumors also show increased immune cell infiltration, including increased CD8+ T-cells, which contribute to the responsiveness to immunotherapies and may be a potential therapeutic target to improve treatment outcomes [55].

8.4.2 Tumor Mutation Burden

Tumor mutation burden (TMB) is the entire number of mutations that are contained within the tumor's DNA. However, it may be a potential biomarker of how a tumor will respond to immunotherapy, such as immune checkpoint inhibitors. Although tumors with high TMB (TMB-H) often harbor more neoantigens than those with low TMB (TMB-L), these neoantigens are more detectable by the immune system. This enhances the likelihood of a real, live immune response, specifically in tandem with therapies like PD-1/PD-L1 research inhibitors [58]. TMB is quantified by sequencing some portion of the tumor genome using NGS techniques. The number of mutations per molbase (mut/Mb) of the sequenced genome is calculated, with a threshold usually used to define the tumors as TMB-high (e.g., ≥10 mut/Mb) or TMB-low [59,60]. This analysis is performed on different panels, each with different gene coverages, making panel selection important to get consistent TMB values. Analysis of its correlation with an MSI-H status has been shown to have a strong correlation with this status [60,61]. Mutations are more numerous in MSI-H tumors (which arise from MMR defects), and TMB-H is common in all tumors. That raises them as prime targets for immunotherapy—an approach that trains the immune response to get rid of cancer, and in their case that immune response is induced by checkpoint inhibitors like pembrolizumab or nivolumab—because the increased mutation burden means there are more neoantigens made that can cause an immune reaction [62,63].

8.4.3 PD-L1 Expression

The programmed death-ligand 1 (PD-L1) expressed on the surface of the cells of the tumor and of the immune cells has a role in immune evasion and is referred to as PD-L1 expression [11]. When cancer tumors express PD-L1, they can bind to the PD-1 receptor, arresting T-cells. This prevents T-cells from attacking cancer cells, and as a result, prevents the immune system from finding and killing them. The likelihood of response to immunotherapy, namely immune checkpoint inhibitors, is determined in part by the expression level of PD-L1 [12,20].

8.4.4 Assessment Methods

The expression of PD-L1 is often carried out through IHC. This method involves the laboratory staining and standardization of tumor tissue samples to decipher the extent to which PD-L1 is present on the outer membrane of the cells [64]. Several IHC assays are available, each with other antibody and staining scoring, such as tumor proportion score (TPS) and combined positive score (CPS) [65,66]. TPS quantifies the percentage of cancer cells that are stained, while CPS quantifies the abreast with cancer cells and immune cells. Both of these scoring systems have different cutoff values depending on the type of cancer being diagnosed and the therapy under consideration [66].

8.4.5 Prognostic vs. Predictive Value

Prognostic Value: Recently, several studies have reported that PD-L1 expression is not suitable to represent the prognostic factor in all types of cancer. Its relevance to the understanding of the natural history of cancer without intervention is not well understood. Predictive Value: PD-L1 is a prognostic biomarker used to determine response to immunotherapy, especially immune checkpoint inhibitors such as pembrolizumab and nivolumab. It has also been seen that if the immune protein PD-L1 is high, then these therapies are more likely to be effective. However, it is not ideal in this sense since patients with low or no PD-L1 can equally benefit from immunotherapy, especially when administered with other treatments or other biomarkers like TMB [64,67]. Prognostic and predictive biomarkers are particularly important in the diagnosis and treatment of CRC. Prognostic biomarkers give insights into the progression and prognosis of the disease while, on the other hand, giving insights into the response to a particular treatment. Table 8.2 shows the tissue-derived and blood-derived biomarkers outlined according to whether they have prognostic or predictive implications in CRC.

8.4.6 Emerging Biomarker

For the best result of immunotherapy in CRC, the investigation of biomarkers that are potentially linked to response to treatment must be done. The significant biomarkers are categorized into PD-L1 expression, immune responses' history, tumor mutations, and microbiome. TMB calculated

TABLE 8.2
Prognostic and Predictive Biomarkers in Colorectal Cancer (CRC) [67,106]

Biomarker Type	Tissue-Derived	Blood-Derived
Prognostic biomarkers	MUC2 (Mucin 2), SATB2 (Special AT-rich sequence-binding protein 2), CK20/CDX2 (Cytokeratin 20/Caudal type homeobox 2)	Preoperative CEA (Carcinoembryonic Antigen)
Prognostic biomarkers	VEGF (Vascular Endothelial Growth Factor), IMP3 (Insulin-like Growth Factor II mRNA-binding protein 3), TNIK (Traf2 and Nck-interacting kinase), KRAS (Kirsten rat sarcoma viral oncogene), NRAS (Neuroblastoma RAS viral oncogene), P53 (Tumor protein 53), BRAF (B-Raf proto-oncogene), miRNAs (MicroRNAs), MSI (Microsatellite Instability)	Postoperative CEA (Carcinoembryonic Antigen), CA19-9 (Carbohydrate Antigen 19-9), CTC (Circulating Tumor Cells)
Predictive biomarkers	KRAS/NRAS (Kirsten/Neuroblastoma RAS viral oncogene), BRAF (B-Raf proto-oncogene), PI3K (Phosphoinositide 3-kinase)	MSI (Microsatellite Instability)
Predictive biomarkers	KRAS/NRAS (Kirsten/Neuroblastoma RAS viral oncogene), BRAF (B-Raf proto-oncogene), PI3K (Phosphoinositide 3-kinase)	MSI (Microsatellite Instability), CD133 (Prominin-1)

the total number of somatic mutations in the tumor genome; the higher TMB suggests higher immunogenicity and better responses to ICIs. Characteristics similar to those related to MSI-H cancer also exist for MSS cancer, although the values are lower [68]. Preclinical models by REGONIVO and CCTG CO26 showed enhanced progression-free survival (PFS) and OS in patients with raised TMB. Notably, in MSS CRC patients, higher TMB was associated with improved response to PD-L1 and CTLA-4 inhibitors, making TMB a potential biomarker [68,69].

The exploratory analysis that followed the REGONIVO trial showed that in patients with high TMB, the median PFS reached 12.5 months, while in the low TMB patients, the PFS was 7.9 months. Likewise, in the CCTG CO26 trial, which utilized ctDNA to calculate plasma TMB, researchers found that 28 mutations per megabase determined patients who might benefit from combination therapy with durvalumab and tremelimumab. These observations draw attention to the possibility that TMB may be a definitive marker that can be used to direct immunotherapy during CRC treatment [69,70].

8.4.7 Microbiome Influences

Recent investigations reveal that gut microbiota can predict immune checkpoint blockade (ICB) outcomes in CRC patients to the best of the authors' knowledge. Current findings indicate that microbiota found in the gut contributes to CRC development through stimulation of inflammation and prevention of immune response. Bacteria like *Bifidobacterium* spp., *Bacteroides fragilis*, and *Akkermansia muciniphila* are associated with ICB outcomes [71,72]. In pMMR/MSS CRC murine models, the changes in microbiome influence immune cytokines in the TME, leading to the impact on response toward PD-1 antibody treatment. Interestingly, the combination of *Bifidobacterium* with anti-PD-L1 treatment augments the CD8+ T-cell activity to increase the effectiveness of the medicine [72].

In the clinic, both *Bifidobacterium longum* and *Enterococcus faecium* were associated with increased anti-PD-1 efficacy in melanoma; *A. muciniphila* was also similar to that in non-small cell lung cancer and renal cancer patients. The Lachnospiraceae family of bacteria was identified to be predominant in the responders among those treated with chemotherapy, bevacizumab, and atezolizumab, whereas Veillonellaceae and *Fusobacterium nucleatum* were reported to have poor prognostic significance in the recent CRC study (atezoTRIBE trial). These data imply a potential to improve immunotherapy effectiveness in CRC by remodeling the gut microbiota [73,74].

8.5 EMERGING THERAPIES AND INVESTIGATIONAL AGENTS

Antibodies currently being researched or used in CRC treatments are as follows:

8.5.1 Cetuximab

Since 2004, cetuximab, a chimeric monoclonal antibody against EGFR, has been authorized for usage in metastatic colorectal cancer (mCRC) [74]. In mCRC patients, it is utilized separately or in combination with the FOLFIRI (chemotherapy regimen). The clinical behavior of cetuximab in increasing OS and PFS in KRAS WT mCRC patients has demonstrated that this agent is an important therapeutic agent in this subtype of CRC.

8.5.2 Panitumumab

A commercially available humanized monoclonal antibody (panitumumab) approved on June 29, 2017, was authorized for the therapy of mCRC patients, especially those with wild-type KRAS and RAS mutations [75–77]. In WT KRAS mCRC patients added to the FOLFOX regimen (chemotherapy), panitumumab has shown improvements in PFS, objective response, and trends in OS.

8.5.3 Nimotuzumab

The humanized monoclonal antibody nimotuzumab against EGFR is currently being tested in a phase II clinical trial (NCT05278728) in conjunction with radiotherapy and capecitabine in mCRC patients [78]. But, to date, this ongoing study has brought no obvious results.

8.5.4 Necitumumab

Necitumumab is another human monoclonal antibody designed to bind to EGFR for mCRC patients with cetuximab failure [79]. It was investigated in a completed phase II trial (NCT00835185) when used in combination with mFOLFOX6 in patients with locally advanced or mCRC. Necitumumab has been proven to be active with reasonable toxicity in these patients.

8.5.5 Bevacizumab

Bevacizumab is a recombinant, humanized monoclonal antibody that targets VEGF [80,81]. It was approved on February 26, 2004, for mCRC. It is most commonly combined with a 5-fluorouracil-containing chemotherapy regimen. Bevacizumab has improved survival in mCRC patients, making bevacizumab a CRC therapy. Drawing from the above evidence, as shown in the chapter, bevacizumab is an important composition in the CRC involving angiogenesis inhibition.

8.5.6 Ramucirumab

Ramucirumab, a human monoclonal antibody directed against VEGFR-2, was approved on April 24, 2015, for the treatment of second-line mCRC in combination with FOLFIRI [82]. The RAISE trial demonstrated that this combination benefits patients with mCRC who had progressed following first-line therapy.

8.5.7 Nivolumab

Nivolumab is an agent that targets PD-1 human monoclonal antibodies that are non-approved for MSI-H and/or dMMR mCRC, beginning on August 1, 2017 [83]. It offers long-lasting outcomes and disease control in pretreated MSI-H/dMMR mCRC patients, and it is a valuable immunotherapy option in such a setting.

8.5.8 Ipilimumab

The CTLA-4-blocking human monoclonal antibody Ipilimumab received FDA approval on July 11, 2018, in combination with low-dose nivolumab for MSI-H/dMMR mCRC [84]. This combination has been found to act for a long time and is therefore deemed to be a reliable treatment for these patients.

8.5.9 Cemiplimab

Cemiplimab is an anti-PD-1 human monoclonal antibody studied in a phase 2 trial (NCT04157985) for a range of solid tumors, including mCRC [85]. They are evaluating it with other inhibitors, and trials are on; however, results have not yet been reported.

8.5.10 Atezolizumab

The immunotherapeutic agent atezolizumab, targeting the PD-L1 receptor, is in clinical research trials (NCT05118724) in conjunction with neoantigen therapies for the treatment of EL3 Stage III CRC patients for whom oxaliplatin-based adjuvant therapy is not feasible [86]. Earlier trials failed to achieve the major objectives, but further studies are being pursued.

8.5.11 Avelumab

Much like pembrolizumab, avelumab is another PD-L1 specially designed monoclonal antibody under investigation in combination with regorafenib in patients with mCRC (NCT03475953) [87,88]. The current trial is still in progress, and therefore no one can give out the results yet.

8.5.12 Dostarlimab

Dostarlimab is a humanized monoclonal antibody that targets PD-1 [89]. It is currently under clinical research for locally advanced rectal cancer (LARC) in the dMMR cohort (NCT04165772). First, the available data show that tumors in this group are highly responsive to anti-PD-1 treatment, but more extended follow-up is needed.

8.5.13 Pembrolizumab

On June 29, 2020, a PD-1-targeting humanized monoclonal antibody, pembrolizumab, was approved as a first-line treatment for unresectable metastatic MSI-H or dMMR CRC [85]. As a result, it is a very effective treatment for this subset of CRC patients, as it reduced disease progression risk by 40% in the KEYNOTE-177 [90] trial.

8.5.14 Relatimab

As a LAG-3 and human monoclonal antibody, relatlimab (NCT05328908) is now being studied in conjunction with nivolumab in advanced MSS CRC [89]. This ongoing trial is to study the efficacy of a combination of LAG-3 and PD-1 inhibitors in the treatment of CRC.

8.6 DURABLE RESPONSES AND SURVIVAL BENEFITS

8.6.1 Tumor-Infiltrating Lymphocytes

The AtezoTRIBE trial, which included 171 patients with Stage IV CRC receiving chemotherapy, bevacizumab, and either atezolizumab or a placebo, demonstrated that patients exhibiting low tumor-infiltrating lymphocytes (TILs) had diminished benefits from the addition of atezolizumab. In Stage IV CRC patients receiving chemotherapy and immunotherapy, a positive correlation was observed between regulatory T-cell (FoxP3+) tumor infiltration and tumor response [91].

8.6.2 Immune Score

The participants had greater intratumoral CD3+ and CD8+ densities than non-responders in a trial of 12 Stage IV CRC patients with dMMR treated with pembrolizumab. Immunotherapy response in patients with pMMR was shown to be positively correlated with CD3+ and CD8+ T-cell concentrations in neoadjuvant ICB treatment research, including localized-stage rectal cancer (dMMR and pMMR). Higher IS was also linked to stronger tumor response in pMMR patients, according to the phase II NICHE study, which examined early-stage colon cancer treated with nivolumab and ipilimumab. Patients with poor TIL infiltration benefited less from atezolizumab addition, according to the AtezoTRIBE study, which included 171 Stage IV CRC patients treated with chemotherapy, bevacizumab, atezolizumab, or placebo. Additionally, a favorable connection between tumor response and regulatory T-cell (FoxP3+) tumor infiltration was seen in individuals with Stage IV CRC receiving immunotherapy and chemotherapy [92–94].

8.6.3 PD-L1 Expression

The results obtained in the KEYNOTE-016 trial in multiple tumor types treated with pembrolizumab revealed that co-expression is more favorable for improving the response to ICB [95].

8.7 GENOMIC ALTERATIONS

8.7.1 MSI-H/dMMR

In the KEYNOTE-177 trial, pertuzumab versus chemotherapy was compared in patients with Stage IV CRC and MSI-H/dMMR; there were better PFS and objective response rate (ORR) rates with ICB. Likewise, CheckMate-142 and CheckMate-8HW trials of nivolumab and ipilimumab in Stage IV CRC patients with MSI-H/dMMR showed the same. In early-stage dMMR colon cancer, the NICHE and NICHE-2 trials with nivolumab/ipilimumab reported a post-treatment pathological response rate of 99%. Also in the above Cercek et al. dMMR rectal cancer study done at a localized level, dostarlimab was given concomitantly as neoadjuvant therapy and also produced a clinical complete response in most patients [92,96,97].

8.7.2 TMB (Tumor Mutational Burden)

In the Stage IV CRC patients, Fabrizio et al. identified a pMMR/MSS population with 3% of patients showing high TMB, who are good candidates for ICB. However, Rousseau et al. stated that high TMB was not an efficient biomarker for ICB in Stage IV CRC patients [98].

8.7.3 POLE Mutation

Some articles reported about Stage IV CRC patients that had good responses; more responses were noted in the case of V411L and F367S mutations of POLE [99,100].

8.7.4 BRAF Mutation

In the KEYNOTE-177 trial, it was established that 25% of Stage IV MSI-H/dMMR CRC patients presented BRAF V600E mutation. These patients had better PFS; the study found the HR=0.48 in relation to the wild-type patients. In the CheckMate-142 trial, the objective response in BRAF V600E-mutant and wild-type populations was comparable, 55% [49,96].

8.7.5 Gut Microbiota

The AtezoTRIBE trial recruited 171 Stage IV CRC patients receiving chemotherapy, bevacizumab, and atezolizumab or a placebo and took fecal samples for WGS. In pMMR patients, the relative abundance of immunogenic SGBs—Lachnospiraceae from the responder cohort—was identified while evaluating the prognosis of Veillonellaceae and pathobionts [101,102].

8.8 PATIENT QUALITY OF LIFE IMPROVEMENTS

In an attempt to enhance QoL in patients receiving immunotherapy, there has been tremendous development in mCRC treatment plans. One of the nutritional evaluation tools used in this evaluation is the health-related quality of life (HRQoL), which depicts an index of patients' well-being. Cohort studies reported more recently prove that immunotherapy and targeted therapy have a better impact on HRQoL compared to chemotherapy. Immunotherapy, especially with agents like pembrolizumab, significantly enhanced the patients' perceived HRQoL, especially those who had MSI-H/dMMR tumors. These outcomes utilized the European Organization for the Research and Treatment of Cancer (EORTC) QLQ-C30 Global Health Status (GHS) and the EuroQoL-five dimension index questionnaire (EQ-5D) [103,104].

For example, it was observed that increased quantity and quality of treatment-related immune-related adverse events (irAEs) with nivolumab and ipilimumab combination improved the HRQoL of patients with MSI-H/dMMR mCRC, mainly regarding physical and emotional functioning in the CheckMate-142 study [83,84]. This is in concord with KEYNOTE-177, which compared pembrolizumab to chemotherapy in the survival rate and also on the reduction in side effects affecting patients' quality of life, including fatigue, nausea, and diarrhea. Realizing improved QoL outlined in our previous blogs, pembrolizumab patients had fewer hospitalizations and fewer treatment-linked inconveniences compared to chemotherapy regimens over time [104,105].

Similarly, the effect of immunotherapy in altering the HRQoL preferentially among patients under monotherapy was noted; compared to combination therapy or NAT, patients on monotherapy immunotherapy reported significantly higher GHS scores. Monotherapies like cetuximab or pembrolizumab were reported to be the most preferred to enhance QoL, though cetuximab demonstrated statistically significant enhancements in both the short-term and long-haul appraisals. For example, cetuximab demonstrated an improvement in GHS/QoL with a mean difference (MD) of 24.11 (95% CI: 4.34 and 47.64), while the pembrolizumab regimen showed an MD of 9.8 (95% CI: 2.59–17.09). However, combination therapies drew more negative responses due to more side effects experienced by patients besides having to travel frequently to the hospital [103]–[106].

8.9 TAILORING TREATMENTS BASED ON BIOMARKERS

New developments in molecular targeting for CRC have advanced toward a biomarker-based approach to increasing therapeutic outcomes. Personalized medicine aims at the fact that different treatments depend on the features of the patient, especially genetic and epigenetic features that determine the reaction to a specific medication. MSI-H, KRAS, NRAS, BRAF, and PIK3CA mutations are vital biomarkers that define the direction of management of the CRC patient. For example, tumors with MSI-H/dMMR respond to immune checkpoint inhibitors like pembrolizumab, but

KRAS/NRAS mutations are related to the absence of cetuximab and panitumumab absorption. The findings based on these molecular markers will help to avoid negative outcomes and enhance the effectiveness of the treatment in general [107,108].

8.9.1 Opportunities for Individualized Therapy Plans

However, the improved awareness of the hereditary basis of CRC makes particular therapy strategies a much more viable concept. By using both gene expression profiles and tumor mutation data, the clinicians can then proceed to treat patients with therapies that will inhibit specific molecular abnormalities related to cancer development. Specifically, Wnt signaling, RAS-MAPK, and TGF-β signaling are most commonly affected in CRC and thus considered major candidates for individualized therapy. Furthermore, immuno-oncology agents that enhance the patient's immune system to target tumor cells—like PD-1/PD-L1—will be personalized according to TMB or immune profile. This helps in bringing up the right therapy for the patient since the tumor type in each patient varies genetically and biologically, hence enhancing the cure and the quality of life [107–108].

8.10 FUTURE DEVELOPMENTS IN COLORECTAL CANCER THERAPEUTICS

The theragnostic field of CRC is rapidly growing as several approaches continue to be developed for better patient care. Despite chemotherapy and surgery still being strong tropes today, the future of CRC treatment is now aimed at a more precise approach. The rise of molecular profiling, immunotherapy, and precision medicine is the basis of new approaches to building therapies that target the genetics and molecular markers of each tumor. This trend not only guarantees an increase in effectiveness but also plans to minimize the adverse effects, providing more personalized treatment of CRC. Future opportunities for radar include the prospects of future technology and novel drug approaches based on breakthrough technologies that may transform CRC treatments. Below is the list of current clinical trials in CRC that have the potential to lead to better therapeutics, herein illustrating the emergent advancements made with regard to cancer therapeutics.

8.10.1 Sunitinib (TKI)

Sunitinib belongs to the phylogenetic family of receptor tyrosine kinase inhibitors (TKIs) or VEGF/R TKIs specifically active against VEGFR1–3. It has been investigated in phase II for CRC in combination with FOLFIRI for the treatment of patients with mCRC. The trial decreed that the addition of sunitinib to FOLFIRI did not produce any profound improvement in the effectiveness of the treatment for mCRC, which limits the utility of the targeting agent when added to conventional chemotherapy regimens [109].

8.10.2 Axitinib (TKI)

Another TKI, axitinib, specific to VEGFR1–3, also underwent a phase II trial (NCT00460603) either in combination with bevacizumab or FOLFOX-6 for mCRC first-line treatment. This investigation demonstrated that neither axitinib, a single agent, nor axitinib in combination with bevacizumab can extend ORR, PFS, or OS compared with bevacizumab alone in mCRC treatment [110].

8.10.3 Sorafenib (Kinase Inhibitor)

Sorafenib, prepared as an inhibitor of VEGFR, was investigated as well in a phase II trial in combination with cetuximab for the treatment of mCRC harboring KRAS mutation (NCT00326495). The study offered no factual answers and showed that sorafenib is not very beneficial for this kind of patient [111].

8.10.4 Regorafenib (Multikinase Inhibitor)

Regorafenib is a multi-targeted receptor TKI that acts on VEGFR 1–3, TIE2, KIT, RET, RAF, PDGFR-B, and FGFR. These conditions include advanced colorectal cancer and mCRC, for which Vectibix remains a targeted therapy to this date since its FDA approval for use on September 27, 2012 [112].

8.10.5 Encorafenib (Kinase Inhibitor)

The BRAF-V600E kinase inhibitor, encorafenib, which also targets wild-type BRAF and CRAF, was granted FDA approval on 8th April 2020 specifically for the treatment of mCRC aligned with BRAF-V600E mutation. They say there is progress in targeted therapies for this exact mutation in CRC [113].

8.10.6 Simtuzumab (Monoclonal Antibody)

LOXL2-targeting monoclonal antibody simtuzumab was tested in a phase II trial (NCT01479465) with FOLFIRI enrolled as second-line treatment for metastatic CRC with KRAS mutation. The inclusion of simtuzumab into the regimen did not result in improved clinical outcomes, providing evidence of moderate utility of the treatment for these patients [114].

8.10.7 Lenvatinib (TKI)

Lenvatinib is a TKI for VEGFR1–3, KIT, RET, and PDGFR-alpha receptors. Presently, there is an open phase 3 trial (NCT04776148) where lenvatinib is combined with pembrolizumab in mCRC. The findings of this trial are still pending to explore the impact of combined immunotherapy using this regimen [115].

8.10.8 Tivozanib (TKI)

Tivozanib from the group of inhibitors of VEGFR1–3 was investigated in the phase II trial together with everolimus in patients with treatment-refractory mCRC. The combination was manageable, and 50% of patients attained stable disease, supporting the use of tivozanib in a combination setting for resistant mCRC [116].

8.10.9 Tipifarnib (Farnesyltransferase Inhibitor)

Tipifarnib is an orally administered farnesyltransferase inhibitor studied in a phase II trial (Clinical trial ID: NCT00005833) as a first-line single agent in mCRC. Tipifarnib did not have any therapeutic value for mCRC, and thus the compound does not offer clinical benefit when administered alone [117].

8.10.10 D-1553 (Small Molecule KRasG12C Inhibitor)

D-1553, a KRAS G12C small molecule inhibitor, is set to be tested in a phase I trial (NCT04585035) for mCRC that harbors a KRAS G12C mutation. This trial of an ongoing nature is hoped to determine the safety and efficacy of the same in perhaps providing a breakthrough in the treatment of this hard target mutation [118].

8.10.11 Aflibercept (Recombinant Fusion Protein)

Aflibercept, a humanized recombinant fusion protein that binds to VEGF-A, VEGF-B, and PGF, was assessed in a phase II trial (NCT02181556) as first-line FOLFIRI in combination with mCRC. Despite the absence of achieving the primary endpoint, the combination was comparable to the median PFS and OS to standards of care [119].

8.10.12 Berberine (Alkaloid)

The anti-proliferative effect of berberine, an alkaloid, was investigated in vitro using HCT 116 cell lines of CRC. These outcomes proved that berberine possessed the ability to suppress telomerase activity, shorten telomeres, and arrest the cell cycle and therefore can act as a potential therapeutic for CRC [120].

8.10.13 Piper nigrum Ethanolic Extract (Alkaloid)

Recent research established that the given EEPN has in vivo antioxidative effects and inhibits the growth of colorectal carcinoma cell lines. The study's findings showed a time- and dose-dependent response, enhancing its value as a supplementary treatment for CRC [121].

8.10.14 Fucoidan (Polysaccharide)

Fucoidan, a polysaccharide extracted from brown seaweed, was also treated with vitamin C in vitro for the test on the HCT-116 colon cell viability. The outcomes revealed that it displayed strong antiproliferative and antiangiogenic activity in vitro, which raises its possibilities as a co-treatment in CRC [122].

8.10.15 Curcumin (Polyphenol)

Curcumin, an apoptotic and anti-angiogenic polyphenol, has been examined in the phase I/II trial (NCT01490996) with the FOLFOX regimen for the mCRC. Describing curcumin as safe and tolerable, the study showed better PFS and OS, indicating that it has potential as an adjunct therapy [123].

8.10.16 Gingerol (Polyphenol)

Polyphenol, gingerol having antioxidative and anti-inflammatory properties, was studied in a phase II clinical trial (NCT01344538) for CRC prevention. The authors of these studies proposed that gingerol possessed the ability to decrease cell proliferation and increase the level of apoptosis, which confirmed that gingerol may be protective against CRC [124].

8.10.17 Eicosapentaenoic Acid (Polyunsaturated Fatty Acids)

The eicosapentaenoic acid (EPA) - a polyunsaturated fatty acid with documented inhibitory effects on angiogenic factors—was only further tried in a phase IV trial stemming from marine microalgae, NCT00398333. Further confirmation of the therapeutic effect in CRC will require a better sample size, and thus the study was stopped midway [125].

8.10.18 Epigallocatechin gallate (Polyphenol)

The epigallocatechin gallate (EGCG) that has been shown to induce apoptosis is now in the earlier phase I trial (NCT02891538) in CRC. The study is still ongoing, and the best results are still to be produced to measure the efficacy of EGCG in the treatment of CRC [126].

8.10.19 Polysaccharide-K (Polysaccharide)

Polysaccharide-K (PSK), polysaccharide, was studied for its efficacy in phase III level trials as an adjuvant of tegafur/uracil for CRC. Reduction in the recurrence and mortality rate in Stages I and II CRC patients who received PSK was demonstrated as 43.6% and 40.2%, respectively [127].

8.10.20 Resveratrol (Polyphenol)

In a phase I study of resectable CRC (NCT00920803), resveratrol, a polyphenol, has been tested. The study also showed that resveratrol interferes with the Wnt signaling pathway, reaffirming the importance of the study toward the development of a therapeutic target for CRC [128].

8.10.21 Topotecan (Alkaloid)

Topotecan, an alkaloid tracker with antimitotic action, enters a phase II evaluation trial (EORTC) as CRC therapy. The results showed that topotecan had very low activity in the form of single-agent therapy, implying its restricted role in the treatment of CRC [129,130].

8.10.22 Metformin (Alkaloid)

Metformin, an alkaloid with potent antiproliferative and antimetastatic activity, was tested in conjunction with aspirin in a phase II trial (NCT03047837) for Stages I and II CRC. The findings indicated that the integration could effectively postpone the onset and enhance the outcome, which provided a theoretical basis for metformin application in early-stage CRC [130].

8.10.23 Everolimus (Macrolide)

Everolimus, a macrolide derived from sirolimus presenting antiproliferative and antimetastatic effects, has been tested in association with tivozanib for refractory mCRC in a phase II trial (NCT01058655). According to the findings of this study, the oral combination was proved to be safe and revealed stable disease dynamics in 50% of patients [131].

8.10.24 Andrographolide (Diterpenoid)

Andrographolide, which belongs to the diterpenoid class of compounds, was investigated in vitro for its combination with 5-FU in CRL 1606 CRC cells. The work indicated that andrographolide can improve the therapeutic efficacy of 5-FU by suppressing the c-MET signaling pathway, indicating its possibility as an adjuvant therapy [132].

8.10.25 Silymarin (Flavonoid)

The flavonoid silymarin, which possesses apoptotic and antiproliferative properties, is being assessed in a phase IV trial in patients with mCRC treated with first-line FOLFIRI and bevacizumab (Trial NCT03130634). This study demonstrated that treatment with silymarin may lessen certain treatment-related toxicity [133].

8.10.26 MMC (Hyleneimines)

MMC, an anthralemidomine of the hyleneimine with antiproliferative properties, is currently under study in an active phase II trial (NCT03073694) in conjunction with melphalan in mCRC. Data on the outcomes of the present study are yet to be published, and therefore, it is still uncertain whether MMC is effective in CRC [134].

8.11 CLINICAL APPLICATIONS OF IMMUNOTHERAPY IN COLON CANCER

8.11.1 FDA-Approved Immunotherapy

8.11.1.1 Pembrolizumab

Pembrolizumab (Keytruda) is an engineered monoclonal antibody designed to target PD-1 on immune-acting cells and its ligands, PDL-1 and PDL-2, on tumor cells to unmask its capability of destroying cancer cells. Currently available in the United States, the FDA has approved its use primarily for MSI-H or dMMR metastatic CRC. For unresectable or metastatic CRC, pembrolizumab is effectively prescribed in first-line therapy, and the treatment is most responsive to anti-PD-1 checkpoint inhibitors in high TMB tumor neoantigens that stimulate immune activation [135,136].

The drug has been found to enhance PFS in addition to enhancing ORR when compared to conventional chemotherapy treatments. Just like any other immune checkpoint inhibitor, durvalumab targets a protein called PD-1 on the surface of T-cells. Thus, pembrolizumab suppresses this interaction, thereby ‘waking up’ cytotoxic T-cells to kill tumor cells, which otherwise would not be identified by the immune system. Pembrolizumab is recommended for use in adults and children with MSI-H or dMMR CRC that cannot be resected surgically or have metastasized, especially after standard treatments with fluoropyrimidine and oxaliplatin [100,136]. It has already been recommended for first-line treatment because it has demonstrated superior efficacy in increasing the PFS of patients with MSI-H/dMMR metastatic CRC compared to conventional chemotherapy. In particular, the phase III KEYNOTE-177 trial compared pembrolizumab with chemotherapy in MSI-H/dMMR metastatic CRC and proved the advantage of the former. Pembrolizumab has a favorable safety profile; however, the program’s irAEs are because of enhanced immune response. They comprise pneumonitis, colitis, hepatitis, hypothyroidism, dermatitis, etc. [136].

8.11.1.2 Combination Therapy

Opdivo (nivolumab) and Yervoy (ipilimumab) are both immune checkpoint inhibitors indicated for treating many cancers, including MSI-H/dMMR advanced CRC. They have already been approved to be used by the FDA for individuals with metastatic CRC that is MSI-H/dMMR and has relapsed after standard therapies. Opdivo binds to PD-1, tying up cancer cells that would have otherwise disabled T-cells, while Yervoy suppresses CTLA-4, boosting T-cell activity [93,94,137].

Together with Ipilimumab, a CTLA-4 inhibitor, the therapy introduced two checkpoints leading to increased T-cell stimulation. Ipilimumab binds CTLA-4, which restrains early immune shutdown services, making the combination stronger than nivolumab PD-1 pathway blockade. This combination appears to have enhanced anti-tumor activity, longer-lasting responses, and better survival in CRC patients, particularly those insensitive to conventional cytotoxic therapy [94]. In the CheckMate 142 trial, nivolumab with ipilimumab increased PFS and OS of MSI-H/dMMR CRC with ORRs of approximately 49% with combination therapy and 31% with nivolumab only. This combination therapy gives long-lasting clinical responses and may last for months, if not years. The combination of nivolumab and ipilimumab stems from their mechanisms. Nivolumab enhances immune recognition of tumors by blocking PD-1, a tonically active inhibitory receptor that causes T-cell exhaustion. Ipilimumab augments the initial activation of T-cells by binding to CTLA-4, a natural suppressor of proliferation and the function of T-colon. In combination, these two agents optimize T-cell activation, leading to more effective destruction of tumor cells by the immune system compared to when either is administered alone [93,138].

Ipilimumab and nivolumab, when used together, have higher rates of irAEs from mild to severe, such as fatigue, rash, colitis, hepatitis, and pneumonitis. Most of these adverse events can be controlled with immunosuppressive medications, such as corticosteroids, and require continuous monitoring during treatment. Patient safety depends on long-term management of irAEs [138,139].

8.11.2 Clinical Trial Outcomes

8.11.2.1 Second-Line Clinical Trial

Pembrolizumab and nivolumab, with or without Ipilimumab, were approved in 2017 as second-line treatments for patients with mCRC with dMMR/MSI-H. In dMMR/MSI-H patients in the KEYNOTE 016 trial, pembrolizumab resulted in a 16% ORR and an 89% disease control rate (DCR), while MSI-low patients had no response [136]. In the CheckMate 142 study, nivolumab with or without ipilimumab produced an ORR of 55% and a DCR of 80% in patients with MSI-H tumors, with PFS and OS rates at 12 months of 71% and 85%, respectively [83,136].

8.11.2.2 First-Line Clinical Trial

Immunotherapy is a growing area of interest in first-line treatment of mCRC disease patients with dMMR or MSI-H, after excellent second-line results. This has been explored in several clinical trials, including the KeyNOTE-177 trial (NIH-ClinicalTrials.gov: **NCT02563002; NCT02060188)**, comparing pembrolizumab to standard chemotherapy in first-line mCRC with mSSI-H [140]. Of 852 patients screened, 307 were randomized, and 60% switched to immunotherapy for disease progression. Pembrolizumab versus chemotherapy showed that the median PFS was 16 months versus 8 months, respectively. At the time of analysis, though there weren't statistically significant OS differences, pembrolizumab is now a standard first-line treatment in MSI-H mCRC [140].

The MSI-H mCRC patients in the CheckMate-142 trial were treated with nivolumab in combination with low-dose ipilimumab. The ORR was 60% with a median follow-up of 13.8 months, the DCR was 84%, and the CR rate was 7%. Efficacy was demonstrated by the ORR at 29 months of 69% and CR of 13%, exceeding pembrolizumab monotherapy [139].

The COMMIT (NIHClinicalTrials.gov: **NCT02997228)** trial is also enrolling treatment-naive mCRC patients with dMMR/MSI-H on mFOLFOX6, bevacizumab, and atezolizumab, with PFS as the primary endpoint. More data to refine first-line immunotherapy approaches for MSI-H/dMMR CRC will be provided in ongoing research [141].

8.11.2.3 Postoperative Adjuvant Therapy

Patients with Stage III CRC must undergo postoperative adjuvant therapy. In the ATOMIC trial (NCT02912559), 700 Stage III dMMR/MSI-H colon cancer patients were treated with adjuvant immunotherapy to assess the effectiveness of immunotherapy as an adjuvant treatment [142]. In the experimental group, the patients were administered FOLFOX (along with atezolizumab for 6 months and then for 6 months without the addition of FOLFOX). DFS and OS were primary endpoints, and the frequency of adverse events was a secondary endpoint of the study. Promising outcomes have also been observed in early-stage CRC after neoadjuvant immunotherapy. Of the 40 colon cancer patients in the NICHE trial (NCT03026140; phase II), 21 had dMMR tumors, and 20 had proficient MMR tumors (pMMR). All 21 patients with dMMR tumors achieved pathological responses after ipilimumab and nivolumab post-surgery, suggesting positive safety and survival outcomes [142].

Additionally, the NRG-GI002 study (NCT02921256) evaluated veliparib or pembrolizumab in combination with chemotherapy and RT in patients with LARC. The neoadjuvant rectal cancer score reduction was the primary endpoint, but secondary endpoints included sphincter-sparing surgery rate, pCR, cCR, DFS, toxicity, and OS [142,143].

In another study using capecitabine prior to the radiation course and with sequential neoadjuvant immunotherapy for locally advanced resectable rectal cancer, patients were enrolled. Of three patients with dMMN/MSI-H tumors, three achieved successful outcomes with pathologic complete response and major pathologic response (NCT02948348) [144]. Together, these results suggest neoadjuvant immunotherapy may serve as a replacement for standard therapy for dMMR/MSI-H patients.

8.12 CONCLUSION

The chapter concludes by highlighting the future of the therapeutics landscape for CRC, emphasizing its potential to improve patient outcomes and survival rates. With further research, more targeted therapies, immunotherapies, and customized medicine methods are paving the path for more efficient and individualized therapies. Specific molecular and genetic markers will be identified, allowing healthcare providers to individualize treatment based not only on efficacy but also on minimizing adverse effects. Also important is that ongoing clinical trials help to explain more about CRC and uncover new therapeutic approaches. These advancements will allow us to harness collaboration between researchers, clinicians, and patients to revolutionize the treatment of CRC toward a better quality of life and prognosis for our patients throughout the globe.

REFERENCES

1. Ewing I, Hurley JJ, Josephides E, Millar A. The molecular genetics of colorectal cancer. *Frontline Gastroenterol*. 2014;5:26–30. https://doi.org/10.1136/flgastro-2013-100329.
2. Vogelstein B, Fearon ER, Hamilton SR, Kern SE, Preisinger AC, Leppert M, et al. Genetic alterations during colorectal-tumor development. *N Engl J Med*. 1988;319:525–532. https://doi.org/10.1056/NEJM198809013190901.
3. Bray F, Laversanne M, Sung H, Ferlay J, Siegel RL, Soerjomataram I, et al. Global cancer statistics 2022: GLOBOCAN estimates of incidence and mortality worldwide for 36 cancers in 185 countries. *CA Cancer J Clin*. 2024 May-Jun;74(3):229–263. https://doi.org/10.3322/caac.21834.
4. Ferlay J, Ervik M, Lam F, Laversanne M, Colombet M, Mery L, et al. *Global Cancer Observatory: Cancer Today*. Lyon, France: International Agency for Research on Cancer; Available from: https://gco.iarc.fr/today, Accessed 02 November 2018.
5. Surveillance, Epidemiology, and End Results (SEER). Program (www.seer.cancer.gov) SEER*Stat Database: Incidence – SEER 18 Regs Research Data+Hurricane Katrina Impacted Louisiana Cases, Nov 2015 Sub (1973–2013 varying) – Linked To County Attributes – Total U.S., 1969–2014 Counties, National Cancer Institute, DCCPS, Surveillance Research Program, Surveillance Systems Branch, released April 2016, based on the November 2015 submission.
6. Kuipers EJ, Grady WM, Lieberman D, Seufferlein T, Sung JJ, Boelens PG, et al. Colorectal cancer. *Nat Rev Dis Primers*. 2015;1:15065.
7. Recio-Boiles A, Cagir B. *Colon Cancer*. Treasure Island, FL: StatPearls, 2022.
8. Sandhu J, Lavingia V, Fakih M. Systemic treatment for metastatic colorectal cancer in the era of precision medicine. *J Surg Oncol*. 2019;119(5):564–582.
9. Häfner MF, Debus J. Radiotherapy for colorectal cancer: Current standards and future perspectives. *Visc Med*. 2016 Jun;32(3):172–177. https://doi.org/10.1159/000446486.
10. Golshani G, Zhang Y. Advances in immunotherapy for colorectal cancer: A review. *Therap Adv Gastroenterol*. 2020 Jun 1;13:1756284820917527. https://doi.org/10.1177/1756284820917527.
11. Tumeh PC, Harview CL, Yearley JH, Peter Shintaku I, Taylor EJM, Robert L, et al. PD-1 blockade induces responses by inhibiting adaptive immune resistance. *Nature*. 2014;515:568–571.
12. Stein A, Moehler M, Trojan J, Goekkurt E, Vogel A. Immuno-oncology in GI tumours: Clinical evidence and emerging trials of PD-1/PD-L1 antagonists. *Crit Rev Oncol Hematol*. 2018;130:13–26.
13. Morse MA, Overman MJ, Hartman L, Khoukaz T, Brutcher E, Lenz HJ, et al. Safety of nivolumab plus low-dose ipilimumab in previously treated microsatellite instability-high/mismatch repair-deficient metastatic colorectal cancer. *Oncologist*. 2019 Nov;24(11):1453–1461. https://doi.org/10.1634/theoncologist.2019-0129.
14. Overman MJ, Linardi S, Leone F, McDermott RS, Morse MA, Wong KYM, et al. Nivolumab in patients with DNA mismatch repair deficient/microsatellite instability high metastatic colorectal cancer: Update from CheckMate 142. *J Clin Oncol*. 2017;35(4):519. https://doi.org/10.1200/JCO.2017.35.4_suppl.519.
15. Latchman Y, Wood CR, Chernova T, Chaudhary D, Borde M, Chernova I, et al. PD-L2 is a second ligand for PD-1 and inhibits T cell activation. *Nat Immunol*. 2001;2:261–268. https://doi.org/10.1038/85330.
16. Le DT, Uram JN, Wang H, Bartlett BR, Kemberling H, Eyring AD, et al. PD-1 blockade in tumors with mismatch-repair deficiency. *N Engl J Med*. 2015;372:2509–2520. https://doi.org/10.1056/NEJMoa1500596.

17. Feig C, Jones JO, Kraman M, Wells RJ, Deonarine A, Chan DS, et al. Targeting CXCL12 from FAP-expressing carcinomaassociated fibroblasts synergizes with anti-PD-L1 immunotherapy in pancreatic cancer. *Proc Natl Acad Sci U S A*. 2013;110:20212–20217. https://doi.org/10.1073/pnas.1320318110.
18. Freeman GJ, Long AJ, Iwai Y, Bourque K, Chernova T, Nishimura H, et al. Engagement of the PD-1 immunoinhibitory receptor by a novel B7 family member leads to negative regulation of lymphocyte activation. *J Exp Med*. 2000;192:1027–1034. https://doi.org/10.1084/jem.192.7.1027.
19. Parry RV, Chemnitz JM, Frauwirth KA, Lanfranco AR, Braunstein I, Kobayashi SV, et al. CTLA-4 and PD-1 receptors inhibit T-cell activation by distinct mechanisms. *Mol Cell Biol*. 2005;25(21):9543–9553. https://doi.org/10.1128/MCB.25.21.9543-9553.2005.
20. Omura Y, Toiyama Y, Okugawa Y, Yin C, Shigemori T, Kusunoki K, et al. Prognostic impacts of tumoral expression and serum levels of PD-L1 and CTLA-4 in colorectal cancer patients. *Cancer Immunol Immunother*. 2020;69:2533–2546. https://doi.org/10.1007/s00262-020-02645-1.
21. Zhang H, Dutta P, Liu J, Sabri N, Song Y, Li WX, et al. Tumour cell-intrinsic CTLA 4 regulates PD-L1 expression in non-small cell lung cancer. *J Cell Mol Med*. 2019;23:535–542. https://doi.org/10.1111/jcmm.13956.
22. Liu F, Huang J, Liu X, Cheng Q, Luo C, Liu Z. CTLA-4 correlates with immune and clinical characteristics of glioma. *Cancer Cell Int*. 2020;20:1–10. https://doi.org/10.1186/s12935-019-1085-6.
23. Derakhshani A, Hashemzadeh S, Asadzadeh Z, Shadbad MA, Rasibonab F, Safarpour H, et al. Cytotoxic T-lymphocyte antigen-4 in colorectal cancer: Another therapeutic side of capecitabine. *Cancers (Basel)*. 2021 May 17;13(10):2414. https://doi.org/10.3390/cancers13102414.
24. Nicol AJ, Tokuyama H, Mattarollo SR, Hagi T, Suzuki K, Yokokawa K, et al. Clinical evaluation of autologous gamma delta T cell-based immunotherapy for metastatic solid tumours. *Br J Cancer*. 2011;105(6):778–786.
25. Parkhurst MR, Yang JC, Langan RC, Dudley ME, Nathan D-AN, Feldman SA, et al. T cells targeting carcinoembryonic antigen can mediate regression of metastatic colorectal cancer but induce severe transient colitis. *Mol Ther*. 2011;19(3):620–626.
26. Nicol AJ, Tazbirkova A, Nieda M. Comparison of clinical and immunological effects of intravenous and intradermal administration of α-galactosylceramide (KRN7000)-pulsed dendritic cells. *Clin Cancer Res*. 2011;17(15):5140–5151.
27. Ghanbari Sevari F, Mehdizadeh A, Abbasi K, Hejazian SS, Raeisi, M. Cytokine-induced killer cells: New insights for therapy of hematologic malignancies. *Stem Cell Res Ther*. 2024;15(1):254,. https://doi.org/10.1186/s13287-024-03869-z
28. Ghazi B, El Ghanmi A, Kandoussi S, Ghouzlani A, Badou A. CAR T-cells for colorectal cancer immunotherapy: Ready to go? *Front Immunol*. 2022 Nov 15;13:978195. https://doi.org/10.3389/fimmu.2022.978195.
29. Xiao L, Chen N, Pu C, Zhao L, Li N, Wang C, Huang Y, Luo S, Li X, Yang Z, Bie J, Zhu R, Liang T, Wang Y, Jia B, Chen D, Wu Z, Song Y, Lu V, Cui J.. A phase 1 dose escalation study of GCC19CART: A novel coupled CAR therapy for patients with metastatic colorectal cancer. *J Clin Oncol*. 2023;41 (16_suppl): 3547–3547. https://doi.org/10.1200/JCO.2023.41.16_suppl.3547.
30. Wang Y, Chen M, Wu Z, Tong C, Dai H., Guo Y, Liu Y, Huang J, Lv H, Luo C, Feng K, Yang Q, Li X, Han W. CD133-directed CAR T cells for advanced metastasis malignancies: A phase I trial. *Oncoimmunology*. 2018;7(7):e1440169. https://doi.org/10.1080/2162402X.2018.1440169.
31. Hege KM, Bergsland EK, Fisher GA, Nemunaitis JJ, Warren RS, McArthur JG, et al. Safety, tumor trafficking and immunogenicity of chimeric antigen receptor (CAR)-T cells specific for TAG-72 in colorectal cancer. *J Immuno Ther Cancer*. 2017;5(1):22. https://doi.org/10.1186/s40425-017-0222-9.
32. Zhang C, Wang Z, Yang Z, Wang M, Li S, Li Y, et al. Phase I escalating-dose trial of CAR-T therapy targeting CEA+metastatic colorectal cancers. *Mol Ther*. 2017;25(5):1248–1258. https://doi.org/10.1016/j.ymthe.2017.03.010.
33. Martinis E, Ricci C, Trevisan C, Tomadini G, Tonon S. Cancer vaccines: From the state of the art to the most promising frontiers in the treatment of colorectal cancer. *Pharmaceutics*. 2023 Jul 17;15(7):1969. https://doi.org/10.3390/pharmaceutics15071969.
34. Shahnazari M, Samadi P, Pourjafar M, Jalali A. Therapeutic vaccines for colorectal cancer: The progress and future prospect. *Int Immunopharmacol*. 2020;88:106944. https://doi.org/10.1016/j.intimp.2020.106944.
35. Jia W, Zhang T, Huang H, Feng H, Wang S, Guo Z, et al. Colorectal cancer vaccines: The current scenario and future prospects. *Front Immunol*. 2022;13:942235. https://doi.org/10.3389/fimmu.2022.942235.
36. Moulton HM, Yoshihara PM, Mason DH, Iversen PL, Triozzi PL. Active specific immunotherapy with a β-human chorionic gonadotropin peptide vaccine in patients with metastatic colorectal cancer|Clinical Cancer Research|American Association for Cancer Research. *Clin Cancer Res*. 2002;8:2044–2051.

37. Tsuruma T, Hata F, Torigoe T, Furuhata T, Idenoue S, Kurotaki T, et al. Phase I clinical study of anti-apoptosis protein, survivin-derived peptide vaccine therapy for patients with advanced or recurrent colorectal cancer. *J Transl Med*. 2004;2:19. https://doi.org/10.1186/1479-5876-2-19.
38. Fricke I, Gabrilovich DI. Dendritic cells and tumor microenvironment: A dangerous liaison. *Immunol Invest*. 2006;35(3–4):459–483. https://doi.org/10.1080/08820130600803429.
39. Lewis CE, Pollard JW. Distinct role of macrophages in different tumor microenvironments. *Cancer Res*. 2006;66(2):605–612. https://doi.org/10.1158/0008-5472.CAN-05-4005.
40. Greenhough A, Smartt HJ, Moore AE, Roberts HR, Williams AC, Paraskeva C, et al. The COX-2/PGE2 pathway: Key roles in the hallmarks of cancer and adaptation to the tumour microenvironment. *Carcinogenesis*. 2009;30(3):377–386. https://doi.org/10.1093/carcin/bgp014.
41. Gout S, Huot J. Role of cancer microenvironment in metastasis: Focus on colon cancer. *Cancer Microenviron*. 2008;1(1):69–83. https://doi.org/10.1007/s12307-008-0007-2.
42. Huang B, Lei Z, Zhang GM, Li D, Song C, Li B, et al. SCF-mediated mast cell infiltration and activation exacerbate the inflammation and immunosuppression in tumor microenvironment. *Blood*. 2008;112(4):1269–1279. https://doi.org/10.1182/blood-2008-03-147033.
43. Mueller L, Goumas FA, Affeldt M, Sandtner S, Gehling UM, Brilloff S, et al. Stromal fibroblasts in colorectal liver metastases originate from resident fibroblasts and generate an inflammatory microenvironment. *Am J Pathol*. 2007;171(5):1608–1618. https://doi.org/10.2353/ajpath.2007.060661.
44. Onogawa S, Kitadai Y, Tanaka S, Kuwai T, Kuroda T, Chayama K. Regulation of vascular endothelial growth factor (VEGF)-C and VEGF-D expression by the organ microenvironment in human colon carcinoma. *Eur J Cancer*. 2004;40(10):1604–1609. https://doi.org/10.1016/j.ejca.2004.02.026.
45. Meng H, Nan M, Li Y, Ding Y, Yin Y, Zhang M. Application of CRISPR-Cas9 gene editing technology in basic research, diagnosis and treatment of colon cancer. *Front Endocrinol (Lausanne)*. 2023 Mar 20;14:1148412. https://doi.org/10.3389/fendo.2023.1148412.
46. Wooster AL, Girgis LH, Brazeale H, Anderson TS, Wood LM, Lowe DB. Dendritic cell vaccine therapy for colorectal cancer. *Pharmacol Res*. 2021 Feb;164:105374. https://doi.org/10.1016/j.phrs.2020.105374.
47. Kajihara M, Takakura K, Kanai T, Ito Z, Saito K, Takami S, Shimodaira S, Okamoto M, Ohkusa T, Koido S. Dendritic cell-based cancer immunotherapy for colorectal cancer. *World J Gastroenterol*. 2016 May 7;22(17):4275–86. https://doi.org/10.3748/wjg.v22.i17.4275. PMID: 27158196; PMCID: PMC4853685.
48. Mulet-Margalef N, Linares J, Badia-Ramentol J, Jimeno M, Monte CS, Mozo JLM, et al. Challenges and therapeutic opportunities in the dMMR/MSI-H colorectal cancer landscape. *Cancers (Basel)*. 2023;15(4):1022. https://doi.org/10.3390/cancers15041022.
49. Colle R, Lonardi S, Cachanado M, Overman MJ, Elez E, Fakih M, et al. BRAF V600E/RAS mutations and Lynch syndrome in patients with MSI-H/DMMR metastatic colorectal cancer treated with immune checkpoint inhibitors. *Oncologist*. 2023;28(9):771–779. https://doi.org/10.1093/oncolo/oyad082.
50. Lenz HJ, Lonardi S, Elez E, Van Cutsem E, Jensen LH, Bennouna J, Mendez G, Schenker M, De La Fouchardiere C, Limon ML, Yoshino T, Li J, Manzano JL, Tortora G, Garcia-Carbonero R, Joshi R, Cela E, Chen T, Jin L, Andre T. Nivolumab (NIVO) plus ipilimumab (IPI) vs chemotherapy (chemo) as first-line (1L) treatment for microsatellite instability-high/mismatch repair-deficient (MSI-H/dMMR) metastatic colorectal cancer (mCRC): Expanded efficacy analysis from CheckMate 8HW. *J Clin Oncol*. 2024;42(16_suppl):3503–3503. https://doi.org/10.1200/JCO.2024.42.16_suppl.3503.
51. Pećina-Šlaus N, Kafka A, Salamon I, Bukovac A. Mismatch repair pathway, genome stability and cancer. *Front Mol Biosci*. 2020 Jun 26;7:122. https://doi.org/10.3389/fmolb.2020.00122.
52. Fucà G, Cohen R, Lonardi S, Shitara K, Elez ME, Fakih M, et al. Ascites and resistance to immune checkpoint inhibition in dMMR/MSI-H metastatic colorectal and gastric cancers. *J Immunother Cancer*. 2022;10:e004001.
53. Bonneville R, Krook MA, Chen HZ, Smith A, Samorodnitsky E, Wing MR, Reeser JW, Roychowdhury S. Detection of microsatellite instability biomarkers via next-generation sequencing. *Methods Mol Biol*. 2020;2055:119–132. https://doi.org/10.1007/978-1-4939-9773-2_5.
54. Marques AC, Ferraro-Peyret C, Michaud F, Song L, Smith E, Fabre G, Willig A, Wong MML, Xing X, Chong C, Brayer M, Fenouil T, Hervieu V, Bancel B, Devouassoux M, Balme B, Meyronet D, Menu P, Lopez J, Xu Z. Improved NGS-based detection of microsatellite instability using tumor-only data. *Front Oncol*. 2022 Nov 17;12:969238. https://doi.org/10.3389/fonc.2022.969238.
55. Friker LL, Perwein T, Waha A, Dörner E, Klein R et al. MSH2, MSH6, MLH1, and PMS2 immunohistochemistry as highly sensitive screening method for DNA mismatch repair deficiency syndromes in pediatric high-grade glioma. *Acta Neuropathol*. 2025 Feb 2;149(1):11. https://doi.org/10.1007/s00401-025-02846-x.

56. Newton K, Jorgensen NM, Wallace AJ, Buchanan DD, Lalloo F, McMahon RF, Hill J, Evans DG. Tumour MLH1 promoter region methylation testing is an effective prescreen for Lynch Syndrome (HNPCC). *J Med Genet.* 2014 Dec;51(12):789–96. https://doi.org/10.1136/jmedgenet-2014-102552. Epub 2014 Oct 3.
57. Rico-Méndez MA, Trujillo-Rojas MA, Ayala-Madrigal Mdl L, Hernández-Sandoval JA, González-Mercado A, Gutiérrez-Angulo M, Romero-Quintana JG, Valenzuela-Pérez JA, Ramírez-Ramírez R, Flores-López BA, et al. MLH1 methylation status and microsatellite instability in patients with colorectal cancer. *Genes.* 2025; 16(2):182. https://doi.org/10.3390/genes16020182.
58. Nandakumar V, Mills JR. The now and beyond of tumor mutational burden as a predictor of response to immune checkpoint inhibitors. *Clin Chem.* 2019;65:357. https://doi.org/10.1373/clinchem.2018.295097.
59. Kang K, Xie F, Mao J, Bai Y, Wang X. Significance of tumor mutation burden in immune infiltration and prognosis in cutaneous melanoma. *Front Oncol.* 2020;10:573141. https://doi.org/10.3389/fonc.2020.573141.
60. Sha D, Jin Z, Budczies J, Kluck K, Stenzinger A, Sinicrope FA. Tumor mutational burden as a predictive biomarker in solid tumors. *Cancer Discov.* 2020;10(12):1808–1825. https://doi.org/10.1158/2159-8290.CD-20-0522.
61. Riviere P, Goodman AM, Okamura R, Barkauskas DA, Whitchurch TJ, Lee S, et al. High tumor mutational burden correlates with longer survival in immunotherapy-naïve patients with diverse cancers. *Mol Cancer Ther.* 2020;19(10):2139–2145. https://doi.org/10.1158/1535-7163.MCT-20-0161.
62. Chan TA, Yarchoan M, Jaffee E, Swanton C, Quezada SA, Stenzinger A, et al. Development of tumor mutation burden as an immunotherapy biomarker: Utility for the oncology clinic. *Ann Oncol.* 2019;30(1):44–56. https://doi.org/10.1093/annonc/mdy495.
63. Chung BS, Liao IC, Lin PC, Wu SY, Kang JW, Lin BW, et al. PD-L1 expression in high-risk early-stage colorectal cancer-its clinical and biological significance in immune microenvironment. *Int J Mol Sci.* 2022 Oct 31;23(21):13277. https://doi.org/10.3390/ijms232113277.
64. Azcue P, Encío I, Guerrero Setas D, Suarez Alecha J, Galbete A, Mercado M, et al. PD-L1 as a prognostic factor in early-stage colon carcinoma within the immunohistochemical molecular subtype classification. *Cancers.* 2021;13:1943. https://doi.org/10.3390/cancers13081943.
65. Roach C, Zhang N, Corigliano E, Jansson M, Toland G, Ponto G, et al. Development of a companion diagnostic PD-L1 immunohistochemistry assay for pembrolizumab therapy in non-small-cell lung cancer. *Appl Immunohistochem Mol Morphol.* 2016;24:392–397. https://doi.org/10.1097/PAI.0000000000000408.
66. Li Y, He M, Zhou Y, Yang C, Wei S, Bian X, et al. The prognostic and clinicopathological roles of PD-L1 expression in colorectal cancer: A systematic review and meta-analysis. *Front Pharmacol.* 2019 Feb 28;10:139. https://doi.org/10.3389/fphar.2019.00139.
67. Oldenhuis CN, Oosting SF, Gietema JA, de Vries EG. Prognostic versus predictive value of biomarkers in oncology. *Eur J Cancer.* 2008 May;44(7):946–953. https://doi.org/10.1016/j.ejca.2008.03.006.
68. Chauhan S, Sharma S. Recent approaches on molecular markers, treatment and novel drug delivery system used for the management of colorectal cancer: A comprehensive review. *Curr Pharm Biotechnol.* 2024;25(15):1969–1985. https://doi.org/10.2174/0113892010270975231208113157.
69. Siena S, Sartore-Bianchi A, Di Nicolantonio F, Balfour J, Bardelli A. Biomarkers predicting clinical outcome of epidermal growth factor receptor-targeted therapy in metastatic colorectal cancer. *J Natl Cancer Inst.* 2009 Oct 7;101(19):1308–1324. https://doi.org/10.1093/jnci/djp280.
70. Tortora SC, Bodiwala VM, Quinn A, Martello LA, Vignesh S. Microbiome and colorectal carcinogenesis: Linked mechanisms and racial differences. *World J Gastrointest Oncol.* 2022 Feb 15;14(2):375–395. https://doi.org/10.4251/wjgo.v14.i2.375.
71. Quaglio AEV, Grillo TG, De Oliveira ECS, Di Stasi LC, Sassaki LY. Gut microbiota, inflammatory bowel disease and colorectal cancer. *World J Gastroenterol.* 2022 Aug 14;28(30):4053–4060. https://doi.org/10.3748/wjg.v28.i30.4053.
72. Kwong TNY, Wang X, Nakatsu G, Chow TC, Tipoe T, Dai RZW, et al. Association between bacteremia from specific microbes and subsequent diagnosis of colorectal cancer. *Gastroenterology.* 2018 Aug;155(2):383–390.e8. https://doi.org/10.1053/j.gastro.2018.04.028.
73. Shariati A, Razavi S, Ghaznavi-Rad E, Jahanbin B, Akbari A, Norzaee S, et al. Association between colorectal cancer and Fusobacterium nucleatum and Bacteroides fragilis bacteria in Iranian patients: A preliminary study. *Infect Agent Cancer.* 2021 Jun 9;16(1):41. https://doi.org/10.1186/s13027-021-00381-4.
74. Bokemeyer C, Van Cutsem E, Rougier P, Ciardiello F, Heeger S, Schlichting M, et al. Addition of cetuximab to chemotherapy as first-line treatment for KRAS wild-type metastatic colorectal cancer: Pooled analysis of the CRYSTAL and OPUS randomised clinical trials. *Eur J Cancer.* 2012;48:1466–1475. https://doi.org/10.1016/j.ejca.2012.02.057.

75. Douillard JY, Siena S, Cassidy J, Tabernero J, Burkes R, Barugel M, et al. Final results from PRIME: Randomized phase III study of panitumumab with FOLFOX4 for first-line treatment of metastatic colorectal cancer. *Ann Oncol.* 2014;25:1346–1355. https://doi.org/10.1093/annonc/mdu141.
76. Douillard JY, Siena S, Cassidy J, Tabernero J, Burkes R, Barugel M, et al. Randomized, phase III trial of panitumumab with infusional fluorouracil, leucovorin, and oxaliplatin (FOLFOX4) versus FOLFOX4 alone as first-line treatment in patients with previously untreated metastatic colorectal cancer: The PRIME study. *J Clin Oncol.* 2010;28:4697–4705. https://doi.org/10.1200/JCO.2009.27.4860.
77. Giusti RM, Shastri KA, Cohen MH, Keegan P, Pazdur R. FDA drug approval summary: Panitumumab (Vectibix). *Oncologist.* 2007;12:577–583. https://doi.org/10.1634/theoncologist.12-5-577.
78. Bai S, Zhang R, Chen W, Dong H, Wang G, Li X, Wang W. Clinical efficacy and safety of nimotuzumab plus chemotherapy in patients with advanced colorectal cancer: A retrospective analysis. *J Int Med Res.* 2020;48(1):0300060519895858. https://doi.org/10.1177/0300060519895858.
79. Bagchi A, Haidar JN, Eastman SW, Vieth M, Topper M, Iacolina MD, Walker JM, Forest A, Shen Y, Novosiadly RD, Ferguson KM. Molecular basis for necitumumab inhibition of EGFR variants associated with acquired cetuximab resistance. *Mol Cancer Ther.* 2017;17(2):521. https://doi.org/10.1158/1535-7163.MCT-17-0575.
80. Hurwitz H, Fehrenbacher L, Novotny W, Cartwright T, Hainsworth J, Heim W, et al. Bevacizumab plus irinotecan, fluorouracil, and leucovorin for metastatic colorectal cancer. *N Engl J Med.* 2004;350:2335–2342. https://doi.org/10.1056/NEJMoa032691.
81. Dinu IM, Mihăilă M, Diculescu MM, Croitoru VM, Turcu-Stiolica A, Bogdan D, Miron MI, Lungulescu CV, Alexandrescu ST, Dumitraşcu T, Buică F, Luca IN, Lungulescu C, Negulescu MC, Gramaticu IM, Cazacu IM, Croitoru AE. Bevacizumab treatment for metastatic colorectal cancer in real-world clinical practice. Medicina. 2023;59(2):350. https://doi.org/10.3390/medicina59020350.
82. Yoshino T, Portnoy DC, Obermannová R, Bodoky G, Prausová J, Garcia-Carbonero R, et al. Biomarker analysis beyond angiogenesis: RAS/RAF mutation status, tumour sidedness, and second-line ramucirumab efficacy in patients with metastatic colorectal carcinoma from RAISE-a global phase III study. *Ann Oncol.* 2019;30:124–131. https://doi.org/10.1093/annonc/mdy461.
83. Overman MJ, McDermott R, Leach JL, Lonardi S, Lenz HJ, Morse MA, et al. Nivolumab in patients with metastatic DNA mismatch repair-deficient or microsatellite instability-high colorectal cancer (CheckMate 142): An open-label, multicentre, phase 2 study. *Lancet Oncol.* 2017;18:1182–1191. https://doi.org/10.1016/S1470-2045(17)30422-9.
84. Lenz HJ, Van Cutsem E, Luisa Limon M, Wong KYM, Hendlisz A, Aglietta M, et al. First-line nivolumab plus low-dose ipilimumab for microsatellite instability-high/mismatch repair-deficient metastatic colorectal cancer: The phase II CheckMate 142 study. *J Clin Oncol.* 2022;40:161–170. https://doi.org/10.1200/JCO.21.01015.
85. *The Phoenix Trial: Cemiplimab for Patients with Localized dMMR Colon Cancer : Treatment.* (n.d.). The Phoenix Trial: Cemiplimab for Patients with Localized dMMR Colon Cancer : Treatment. Retrieved October 30, 2024, from https://www.facingourrisk.org/research-clinical-trials/study/299/the-phoenix-trial-cemiplimab-for-patients-with-localized-dmmr-colon-cancer
86. Tapia Rico G, Price TJ. Atezolizumab for the treatment of colorectal cancer: The latest evidence and clinical potential. *Expert Opin Biol Ther.* 2018;18(4):449–457. https://doi.org/10.1080/14712598.2018.1444024.
87. Cousin S, Cantarel C, Guegan JP, Gomez-Roca C, Metges JP, Adenis A, Pernot S, Bellera C, Kind M, Auzanneau C, Le Loarer F, Soubeyran I, Bessede A, Italiano A. Regorafenib-Avelumab Combination in Patients with Microsatellite Stable Colorectal Cancer (REGOMUNE): A single-arm, open-label, phase II trial. *Clin Cancer Res.* 2021;27(8):2139–2147. https://doi.org/10.1158/1078-0432.CCR-20-3416.
88. Taïeb J, Bouche O, André T, Le Malicot K, Laurent-Puig P, Bez J, Toullec C, Borg C, Randrian V, Evesque L, Corbinais S, Perrier H, Buecher B, Di Fiore F, Gallois C, Emile JF, Lepage C, Elhajbi F, Tougeron D, SAMCO-PRODIGE 54 Investigators. Avelumab vs standard second-line chemotherapy in patients with metastatic colorectal cancer and microsatellite instability: A randomized clinical trial. *JAMA Oncol.* 2023;9(10):1356–1363. https://doi.org/10.1001/jamaoncol.2023.2761.
89. Kumar A, Gautam V, Sandhu A, Rawat K, Sharma A, Saha L. Current and emerging therapeutic approaches for colorectal cancer: A comprehensive review. *World J Gastrointest Surg.* 2023 Apr 27;15(4):495–519. https://doi.org/10.4240/wjgs.v15.i4.495. PMID: 37206081; PMCID: PMC10190721.
90. Andre T, Amonkar M, Norquist JM, Shiu KK, Kim TW, Jensen BV, et al. Health-related quality of life in patients with microsatellite instability-high or mismatch repair deficient metastatic colorectal cancer treated with first-line pembrolizumab versus chemotherapy (KEYNOTE-177): An open-label, randomised, phase 3 trial. *Lancet Oncol.* 2021;22:665–677. https://doi.org/10.1016/S1470-2045(21)00064-4.

91. Moretto R, Rossini D, Catteau A, Antoniotti C, Giordano M, Boccaccino A, et al. Dissecting tumor lymphocyte infiltration to predict benefit from immune-checkpoint inhibitors in metastatic colorectal cancer: Lessons from the AtezoT RIBE study. *J ImmunoTher Cancer.* 2023;11(4):e006633. https://doi.org/10.1136/jitc-2022-006633.
92. Cercek A, Lumish M, Sinopoli J, Weiss J, Shia J, Lamendola-Essel M, et al. PD-1 blockade in mismatch repair-deficient, locally advanced rectal cancer. *N Engl J Med.* 2022;386(25):2363–2376. https://doi.org/10.1056/NEJMoa2201445.
93. Chalabi M, Fanchi LF, Dijkstra KK, Van den Berg JG, Aalbers AG, Sikorska K, et al. Neoadjuvant immunotherapy leads to pathological responses in MMRproficient and MMR-deficient early-stage colon cancers. *Nat Med.* 2020;26(4):566–576. https://doi.org/10.1038/s41591-020-0805-8.
94. Le DT, Uram JN, Wang H, Bartlett BR, Kemberling H, Eyring AD, et al. PD-1 blockade in tumors with mismatch-repair deficiency. *N Engl J Med.* 2015;372(26):2509–2520. https://doi.org/10.1056/NEJMoa1500596.
95. André T, Shiu KK, Kim TW, Jensen BV, Jensen LH, Punt C, et al. KEYNOTE-177 investigators: Pembrolizumab in microsatellite-instability-high advanced colorectal cancer. *N Engl J Med.* 2020;383(23):2207–2218. https://doi.org/10.1056/NEJMoa2017699.
96. Andre T, Elez E, Cutsem EV, Jensen LH, Bennouna J, Mendez G, et al. Nivolumab (NIVO) plus ipilimumab (IPI) vs chemotherapy (chemo) as first-line (1L) treatment for microsatellite instability-high/mismatch repair-deficient (MSI-H/dMMR) metastatic colorectal cancer (mCRC): First results of the CheckMate 8HW study. *J Clin Oncol.* 2024;42(suppl 3; abstr LBA768). https://doi.org/10.1200/JCO.2024.42.3_suppl.LBA768.
97. Lenz HJ, Van Cutsem E, Luisa Limon M, Wong KYM, Hendlisz A, Aglietta M, et al. First-line nivolumab plus low-dose ipilimumab for microsatellite instability-high/mismatch repair-deficient metastatic colorectal cancer: The phase II CheckMate 142 study. *J Clin Oncol.* 2022;40(2):161–170. https://doi.org/10.1200/JCO.21.01015.
98. Rousseau B, Foote MB, Maron SB, Diplas BH, Lu S, Argilés G, et al. The spectrum of benefit from checkpoint blockade in hypermutated tumors. *N Engl J Med.* 2021;384(12):1168–1170. https://doi.org/10.1056/NEJMc2031965.
99. Bikhchandani M, Amersi F, Hendifar A, Gangi A, Osipov A, Zaghiyan K, et al. POLE-mutant colon cancer treated with PD-1 blockade showing clearance of circulating tumor DNA and prolonged disease-free interval. *Genes (Basel).* 2023;14(5):1054. https://doi.org/10.3390/genes14051054.
100. Chen J, Lou H. Complete response to pembrolizumab in advanced colon cancer harboring somatic POLE F367S mutation with microsatellite stability status: A case study. *Onco Targets Ther.* 2021 Mar 9;14:1791–1796. https://doi.org/10.2147/OTT.S300987.
101. Antoniotti, C., Rossini, D., Pietrantonio, F., Salvatore, L., Lonardi, S. et al. Upfront Fluorouracil, Leucovorin, Oxaliplatin, and irinotecan plus bevacizumab with or without Atezolizumab for patients with metastatic colorectal cancer: Updated and overall survival results of the ATEZOTRIBE study. *J Clin Oncol.* 2024:42(22):2637–2644. https://doi.org/10.1200/JCO.23.02728.
102. Maring JG, Eijsink JFH, Tichelaar FD, Veluwenkamp-Worawutputtapong P, Postma MJ, Touw DJ, et al. Role of patient-reported outcomes in clinical trials in metastatic colorectal cancer: A scoping review. *Cancers (Basel).* 2023 Feb 10;15(4):1135. https://doi.org/10.3390/cancers15041135.
103. Conroy T, Hebbar M, Bennouna J, Ducreux M, Ychou M, Llédo G, et al. Quality-of-life findings from a randomised phase-III study of XELOX vs FOLFOX-6 in metastatic colorectal cancer. *Br J Cancer.* 2010 Jan 5;102(1):59–67. https://doi.org/10.1038/sj.bjc.6605442.
104. Seymour MT, Thompson LC, Wasan HS, Middleton G, Brewster AE, Shepherd SF, et al. Chemotherapy options in elderly and frail patients with metastatic colorectal cancer (MRC FOCUS2): An open-label, randomised factorial trial. *Lancet.* 2011 May 21;377(9779):1749–1759. https://doi.org/10.1016/S0140-6736(11)60399-1.
105. Braun DP, Gupta D, Grutsch JF, Staren ED. Can changes in health related quality of life scores predict survival in stages III and IV colorectal cancer? *Health Qual Life Outcomes.* 2011 Aug 3;9:62. https://doi.org/10.1186/1477-7525-9-62.
106. Vaseghi Maghvan P, Jeibouei S, Akbari ME, Niazi V, Karami F, Rezvani A, et al. Personalized medicine in colorectal cancer. *Gastroenterol Hepatol Bed Bench.* 2020 Winter;13(Suppl1):S18–S28.
107. Włodarczyk M, Włodarczyk J, Siwiński P, Sobolewska-Włodarczyk A, Fichna J. Genetic molecular subtypes in optimizing personalized therapy for metastatic colorectal cancer. *Curr Drug Targets.* 2018;19(15):1731–1737. https://doi.org/10.2174/1389450119666180803122744.
108. Rebersek M. Consensus molecular subtypes (CMS) in metastatic colorectal cancer: Personalized medicine decision. *Radiol Oncol.* 2020 May 28;54(3):272–277. https://doi.org/10.2478/raon-2020-0031.

109. Le Tourneau C, Raymond E, Faivre S. Sunitinib: A novel tyrosine kinase inhibitor. A brief review of its therapeutic potential in the treatment of renal carcinoma and gastrointestinal stromal tumors (GIST). *Ther Clin Risk Manag.* 2007 Jun;3(2):341–348. https://doi.org/10.2147/tcrm.2007.3.2.341.
110. Infante JR, Reid TR, Cohn AL, Edenfield WJ, Cescon TP, Hamm JT, et al. Axitinib and/or bevacizumab with modified FOLFOX-6 as first-line therapy for metastatic colorectal cancer: A randomized phase 2 study. *Cancer.* 2013 Jul 15;119(14):2555–2563. https://doi.org/10.1002/cncr.28112.
111. Do K, Cao L, Kang Z, Turkbey B, Lindenberg ML, Larkins E, et al. A Phase II study of sorafenib combined with cetuximab in EGFR-expressing, KRAS-mutated metastatic colorectal cancer. *Clin Colorectal Cancer.* 2015 Sep;14(3):154–161. https://doi.org/10.1016/j.clcc.2015.02.007.
112. Chen D, Wei L, Yu J, Zhang L. Regorafenib inhibits colorectal tumor growth through PUMA-mediated apoptosis. *Clin Cancer Res.* 2014 Jul 1;20(13):3472–3484. https://doi.org/10.1158/1078-0432.CCR-13-2944.
113. Tabernero J, Grothey A, Van Cutsem E, Yaeger R, Wasan H, Yoshino T, et al. Encorafenib plus cetuximab as a new standard of care for previously treated BRAF V600E-mutant metastatic colorectal cancer: Updated survival results and subgroup analyses from the BEACON study. *J Clin Oncol.* 2021;39:273–284. https://doi.org/10.1200/JCO.20.02088.
114. Hecht JR, Benson AB 3rd, Vyushkov D, Yang Y, Bendell J, Verma U. A phase II, randomized, double-blind, placebo-controlled study of simtuzumab in combination with FOLFIRI for the second-line treatment of metastatic KRAS mutant colorectal adenocarcinoma. *Oncologist.* 2017 Mar;22(3):243–223. https://doi.org/10.1634/theoncologist.2016-0479.
115. Zschäbitz S, Grüllich C. Lenvantinib: A tyrosine kinase inhibitor of VEGFR 1–3, FGFR 1–4, PDGFRα, KIT and RET. *Recent Results Cancer Res.* 2018;211:187–198. https://doi.org/10.1007/978-3-319-91442-8_13.
116. Niwakawa M, Yamaguchi R, Onozawa Y, Yasui H, Taku K, Naito T, et al. Phase I study of highly selective inhibitor of VEGFR tyrosine kinase, tivozanib, in Japanese patients with solid tumors. *Cancer Sci.* 2013 Aug;104(8):1039–1044. https://doi.org/10.1111/cas.12197.
117. Lancet JE, Gojo I, Gotlib J, Feldman EJ, Greer J, Liesveld JL, et al. A phase 2 study of the farnesyltransferase inhibitor tipifarnib in poor-risk and elderly patients with previously untreated acute myelogenous leukemia. *Blood.* 2007 Feb 15;109(4):1387–1394. https://doi.org/10.1182/blood-2006-04-014357.
118. Miyashita H, Hong DS. Combining EGFR and KRAS G12C inhibitors for KRAS G12C mutated advanced colorectal cancer. *J Cancer Immunol (Wilmington).* 2024;6(2):62–69. https://doi.org/10.33696/cancerimmunol.6.086.
119. Ciombor KK, Berlin J, Chan E. Aflibercept. Clin Cancer Res. 2013 Apr 15;19(8):1920–1925. https://doi.org/10.1158/1078-0432.CCR-12-2911.
120. Samad MA, Saiman MZ, Abdul Majid N, Karsani SA, Yaacob JS. Berberine inhibits telomerase activity and induces cell cycle arrest and telomere erosion in colorectal cancer cell line, HCT 116. *Molecules.* 2021 Jan 13;26(2):376. https://doi.org/10.3390/molecules26020376.
121. Wu R, Zhao J, Wei P, Tang M, Ma Z, Zhao Y, et al. Piper nigrum extract inhibits the growth of human colorectal cancer HT-29 cells by inducing p53-mediated apoptosis. *Pharmaceuticals.* 2023;16(9):1325. https://doi.org/10.3390/ph16091325.
122. Huang CW, Chen YC, Yin TC, Chen PJ, Chang TK, Su WC, et al. Low-molecular-weight fucoidan as complementary therapy of fluoropyrimidine-based chemotherapy in colorectal cancer. *Int J Mol Sci.* 2021 Jul 27;22(15):8041. https://doi.org/10.3390/ijms22158041.
123. Howells LM, Iwuji COO, Irving GRB, Barber S, Walter H, Sidat Z, et al. Curcumin combined with FOLFOX chemotherapy is safe and tolerable in patients with metastatic colorectal cancer in a randomized phase IIa trial. *J Nutr.* 2019 Jul 1;149(7):1133–1139. https://doi.org/10.1093/jn/nxz029.
124. Lee SH, Cekanova M, Baek SJ. Multiple mechanisms are involved in 6-gingerol-induced cell growth arrest and apoptosis in human colorectal cancer cells. *Mol Carcinog.* 2008 Mar;47(3):197–208. https://doi.org/10.1002/mc.20374.
125. Moloudizargari M, Mortaz E, Asghari MH, Adcock IM, Redegeld FA, Garssen J. Effects of the polyunsaturated fatty acids, EPA and DHA, on hematological malignancies: A systematic review. *Oncotarget.* 2018 Feb 5;9(14):11858–11875. https://doi.org/10.18632/oncotarget.24405.
126. Kciuk M, Alam M, Ali N, Rashid S, Głowacka P, Sundaraj R, et al. Epigallocatechin-3-gallate therapeutic potential in cancer: Mechanism of action and clinical implications. *Molecules.* 2023 Jul 6;28(13):5246. https://doi.org/10.3390/molecules28135246.
127. Ohwada S, Ikeya T, Yokomori T, Kusaba T, Roppongi T, Takahashi T, et al. Adjuvant immunochemotherapy with oral Tegafur/Uracil plus PSK in patients with stage II or III colorectal cancer: A randomised controlled study. *Br J Cancer.* 2004 Mar 8;90(5):1003–1010. https://doi.org/10.1038/sj.bjc.6601619.

128. Brockmueller A, Sajeev A, Koklesova L, Samuel SM, Kubatka P, Büsselberg D, et al. Resveratrol as sensitizer in colorectal cancer plasticity. *Cancer Metastasis Rev.* 2024 Mar;43(1):55–85. https://doi.org/10.1007/s10555-023-10126-x.
129. Creemers GJ, Wanders J, Gamucci T, Vallentin S, Dirix LY, Schöffski P, et al. Topotecan in colorectal cancer: A phase II study of the EORTC early clinical trials group. *Ann Oncol.* 1995 Oct;6(8):844–846. https://doi.org/10.1093/oxfordjournals.annonc.a059328.
130. Kamarudin MNA, Sarker MMR, Zhou JR, Parhar I. Metformin in colorectal cancer: Molecular mechanism, preclinical and clinical aspects. *J Exp Clin Cancer Res.* 2019;38:491. https://doi.org/10.1186/s13046-019-1495-2.
131. Wolpin BM, Ng K, Zhu AX, Abrams T, Enzinger PC, McCleary NJ, et al. Multicenter phase II study of tivozanib (AV-951) and everolimus (RAD001) for patients with refractory, metastatic colorectal cancer. *Oncologist.* 2013;18(4):377–378. https://doi.org/10.1634/theoncologist.2012-0378.
132. Tundis R, Patra JK, Bonesi M, Das S, Nath R, Das Talukdar A et al. Anti-cancer agent: The labdane diterpenoid-andrographolide. *Plants.* 2023;12(10):1969. https://doi.org/10.3390/plants12101969.
133. Chang TK, Yin TC, Su WC, Tsai HL, Huang CW, Chen YC, et al. A pilot study of silymarin as supplementation to reduce toxicities in metastatic colorectal cancer patients treated with first-line FOLFIRI plus bevacizumab. *Oncol Res.* 2021 Sep 7;28(7):801–809. https://doi.org/10.3727/096504021X16218531628569.
134. Tang YL, Li DD, Duan JY, Sheng LM, Wang X. Resistance to targeted therapy in metastatic colorectal cancer: Current status and new developments. *World J Gastroenterol.* 2023 Feb 14;29(6):926–948. https://doi.org/10.3748/wjg.v29.i6.926.
135. Le DT, Kavan P, Kim TW, Burge ME, Cutsem EV, Hara H, et al. KEYNOTE-164: Pembrolizumab for patients with advanced microsatellite instability high (MSI-H) colorectal cancer. *J Clin Oncol.* 2018;36:3514a.
136. Andre T, Lonardi S, Wong M, Lenz H-J, Gelsomino F, Aglietta M, et al. Nivolumab+ipilimumab combination in patients with DNA mismatch repair-deficient/microsatellite instability-high (dMMR/MSIH) metastatic colorectal cancer (mCRC): First report of the full cohort from CheckMate-142. *J Clin Oncol.* 2018;36:553. https://doi.org/10.1200/JCO.2018.36.4_suppl.553.
137. Overman MJ, Lonardi S, Wong KYM, Lenz HJ, Gelsomino F, Aglietta M, et al. Durable clinical benefit with nivolumab plus ipilimumab in DNA mismatch repair-deficient/microsatellite instability-high metastatic colorectal cancer. *J Clin Oncol.* 2018;36(8):773–779. https://doi.org/10.1200/JCO.2017.76.9901.
138. Andre T, Lonardi S, Wong M, Lenz H-J, Gelsomino F, Aglietta M, et al. Nivolumab+ipilimumab combination in patients with DNA mismatch repair-deficient/microsatellite instability-high (dMMR/MSIH) metastatic colorectal cancer (mCRC): First report of the full cohort from CheckMate-142. *J Clin Oncol.* 2018;36:553. https://doi.org/10.1200/JCO.2018.36.4_suppl.553.
139. Overman MJ, Lonardi S, Wong KYM, Lenz HJ, Gelsomino F, Aglietta M, et al. Durable clinical benefit with nivolumab plus ipilimumab in DNA mismatch repair-deficient/microsatellite instability-high metastatic colorectal cancer. *J Clin Oncol.* 2018;36(8):773–779. https://doi.org/10.1200/JCO.2017.76.9901.
140. Andre T, Shiu KK, Kim TW, Jensen BV, Jensen LH, Punt CJA, et al. Pembrolizumab versus chemotherapy for microsatellite instability-high/mismatch repair deficient metastatic colorectal cancer: The phase 3 KEYNOTE-177 study. *J Clin Oncol.* 2020;38:LBA4–LBA. https://doi.org/10.1200/ JCO.2020.38.18_suppl.LBA4.
141. Sinicrope FA, Ou FS, Shi Q, Nixon AB, Mody K, Levasseur A, et al. Randomized trial of FOLFOX alone or combined with atezolizumab as adjuvant therapy for patients with stage III colon cancer and deficient DNA mismatch repair or microsatellite instability (ATOMIC, Alliance A021502). *J Clin Oncol.* 2017;35:TPS3630. https://doi.org/10.1200/JCO.2017.35.15_suppl.TPS3630.
142. Rahma OE, Yothers G, Hong TS, Russell MM, You YN, Parker W, et al. NRG-GI002: A phase II clinical trial platform using total neoadjuvant therapy (TNT) in locally advanced rectal cancer (LARC)-Pembrolizumab experimental arm (EA) primary results. *J Clin Oncol.* 2021;39:8.
143. George TJ, Yothers G, Rahma OE, Hong TS, Russell MM, You YN, Parker W, Jacobs SA, Lucas PC, Colangelo LH, Gollub MJ, Hall WA, Kachnic LA, Bajaj M, Gross HM, Peterson RA, Dorth JA, Vijayvergia N, Wolmark N. Long-term results from NRG-GI002: A phase II clinical trial platform using total neoadjuvant therapy (TNT) in locally advanced rectal cancer (LARC). *J Clin Oncol.* 2023;41(4_suppl):7–7. https://doi.org/10.1200/JCO.2023.41.4_suppl.7.
144. Yuki S, Bando H, Tsukada Y, Inamori K, Komatsu Y, Homma S, et al. Short-term results of VOLTAGE-A: Nivolumab monotherapy and subsequent radical surgery following preoperative chemoradiotherapy in patients with microsatellite stable and microsatellite instabilityhigh locally advanced rectal cancer. *J Clin Oncol.* 2020;38:4100. https://doi.org/10.1200/JCO.2020.38.15_suppl.4100.

9 Colorectal Cancer
Problems with Current Treatments, Side Effects, and Lifestyle

Firdos Alam Khan
Department of Stem Cell Research, Institute for Research and Medical Consultations (IRMC), Imam Abdulrahman Bin Faisal University, Dammam, Saudi Arabia

9.1 LIFESTYLE AND COLORECTAL CANCER

Colorectal cancer (CRC) is one of the most widely occurring cancers in the world. As the number of CRC patient's increases, there is also a rise in the diagnosis and treatment of CRC. There are various benefits of currently applied treatments for CRC, as they not only control the progression of the cancer cells but also improve the life expectancy of the patients. In spite of having many useful benefits, the currently used treatment approaches, such as chemotherapy and radiation therapy, have many side effects. Epidemiologic data obtained from CRC patients suggest that lifestyle factors such as foods, dietary supplements, and physical activity can be associated with the risk of CRC more than any other cancer (American Institute of Cancer Research, 2011) (Figure 9.1). A report prepared by the American Institute for Cancer Research and the World Cancer Research Fund has suggested that these factors, such as foods, dietary supplements, and physical activity, are also linked with the increased risk of cardiovascular and metabolic diseases (Lichtenstein et al., 2006). It has been revealed that patients with CRC have to invest more money in the diagnosis of CRC compared to other types of cancer, including breast cancer and prostate cancer (Edwards et al., 2014). It has been estimated that more than 40% of CRC patients have other types of cancer; whereas 32% of breast cancer patients and 31% of prostate patients have other types of cancer. Moreover, among CRC patients, other types of diseases also coexist; for example, 17% with diabetes and 13% with CRC.

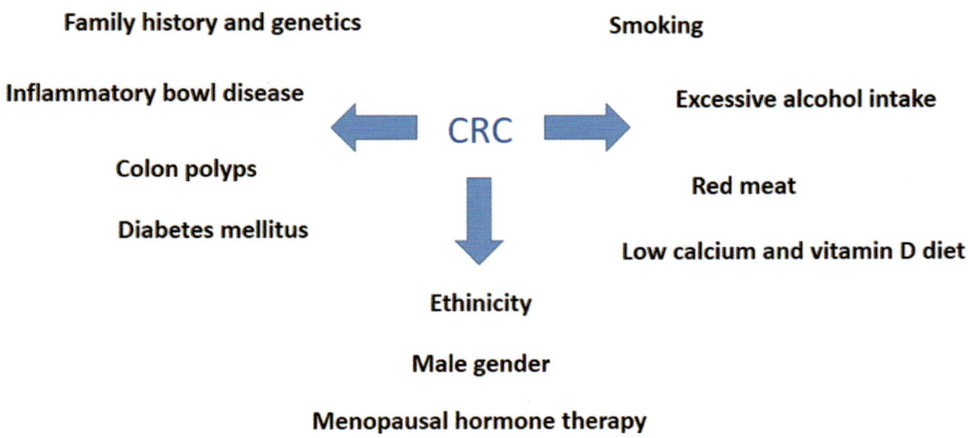

FIGURE 9.1 The figure shows the impact of various factors such as eating, smoking, family history and ethnicity on the development of CRC.

DOI: 10.1201/9781003503842-9

The most common comorbidities at diagnosis are diabetes (17%), chronic obstructive pulmonary disease, and congestive heart failure (12%) (Edwards et al., 2014). Nonetheless, the occurrence of these illnesses among cancer-free people is 14% for diabetes, 9% for chronic obstructive pulmonary disease, and 7% for congestive heart failure, respectively. It has been reported that the risk of cardiovascular disease and diabetes can be reduced by using special dietary supplements and physical exercise activity (Ley et al., 2014; Salas-Salvadó et al., 2014; Schwab et al., 2014; Estruch et al., 2013; Schmid and Leitzmann, 2014). Consequently, the use of dietary food supplements and regular physical exercise can also reduce the risk of CRC in patients.

9.2 PHYSICAL EXERCISE

It has been reported that physical exercise or activities is one of the most promising approaches to lowering the risk of CRC and increasing the life expectancy of CRC patients. One study conducted on CRC patients, in which 7,522 patients were recruited, found that those with low physical exercise had a higher risk of CRC compared to those CRC patients who used to do high physical activities. The patients with higher physical activities showed a lesser risk or occurrence of CRC (Schmid and Leitzmann, 2014) (Figure 9.2). In another study involving 251 CRC patients, the expression of p27 was examined, and it was found that patients with high physical activity had a 67% lower risk of CRC and lower mortality rates (Meyerhardt et al., 2009).

The signaling pathways are involved in the development of CRC; for example, the WNT-CTNNB1 signaling pathway shows a role in the development of CRC, obesity, and other metabolic diseases (Schinner, 2009). It has been found that activation of CTNNBI with accumulation of nuclear signaling causes a decrease in cancer cell proliferation. In another study, patients with CRC tumors that were CTNNB1-negative showed a decreased risk of CRC with high physical activity, as well as reduced mortality, compared to CTNNB1-positive patients (Morikawa et al., 2011).

The question of how physical activities can reduce the risk or occurrence of CRC remains to be explored, as many studies have shown that enhanced physical activities are directly proportional to a decrease in the occurrence of CRC. The CRC patients who have applied physical activities in their daily lives have been found to improve their life expectancy after CRC diagnosis. Another critical factor is the intensity of physical activities, as higher intensity of physical activities leads to a lower occurrence of CRC (Kenfield et al., 2011; Richman et al., 2011). It has been reported that male prostate patients who have done more than 3 hours of physical activities showed a 16% lower risk of prostate cancer-related mortality compared with patients who did less than 1 hour/week of

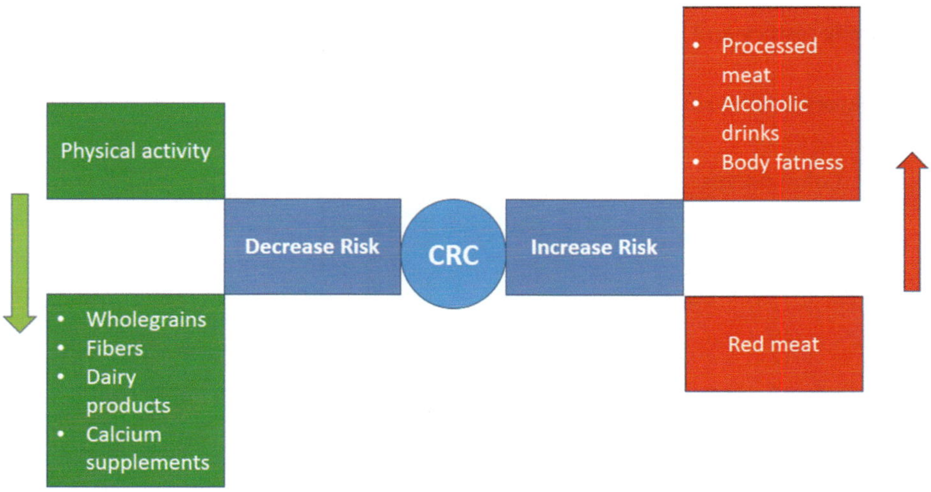

FIGURE 9.2 The figure shows the factors which increases and decreases the development of CRC.

physical activities (Richman et al., 2011). Physical activities are classified as vigorous activities that require ≥ six metabolic equivalents of task, whereas non-vigorous activities require < six metabolic equivalents of the task (Ainsworth et al., 2011). In another study, CRC survival patients showed total metabolic equivalent of task (hour/week) after diagnosis; for example, an individual can achieve 18 metabolic equivalents of task hours/week by walking for 6 hours/week. It has been reported that different metabolic equivalents of task intensities and durations of activity produce specific biological responses (Kraus et al., 2002). In addition to studying total physical activities, future studies should examine vigorous and non-vigorous activities separately in relation to CRC survival patients.

9.3 DIETARY SUPPLEMENT

The role of dietary supplements in reducing CRC occurrence has been studied in the United States, and it has been found that the Cancer and Leukemia Group B (CALGB) 89803 Diet was embedded in a randomized phase of a clinical III trial of using adjuvant chemotherapy. A total of 1,264 patients with stage III CRC were recruited in the study and were asked to fill out the food-related food frequency questionnaires (FFQs) who have received the chemotherapy on different time intervals from midway through chemotherapy and 6 months after completion of chemotherapy (Saltz et al., 2007). Another group of trials consisted of 2,284 patients as per American Cancer Prevention Study (CPS) II trials, and patients' nutritional data FFQ from the 1992 to 1993 period were collected immediately after CRC diagnosis (Flagg et al., 2000), and all participants submitted the dietary data in 1999 and 2003. It has been reported that CRC patients who were on Western dietary supplements showed an increased occurrence of CRC and a 2.3-fold increase in CRC-related mortality compared to CRC patients who were not on Western dietary supplements. It has been found that Western dietary supplements mainly contain refined grains, condiments, red, high- and low-fat dairy, processed meats, desserts, and potatoes. There was no link found between prudent dietary patterns, which mainly contain vegetable, legume, and fruit intake, and after diagnosis of CRC and CRC-related mortality. In addition, similar reports were found in the patients with stage I to III CRC, where 529 patients participated in the investigation in Newfoundland, Canada. In this study, patients were asked to submit their dietary report 1 year before going to diagnosis. The patients in the fourth quartile who were using processed meat dietary products were found to have an 82% increased risk of death and CRC occurrence compared to patients who were in the lowest quartile. Interestingly, no link was found between a prudent or high-sugar dietary pattern in the year before diagnosis and disease-free survival.

9.4 RED AND PROCESSED MEATS

The consumption of red and processed meats is reported to increase the risk of CRC occurrence, and the rate of occurrence of CRC may vary from patient to patient, as post-diagnostic consumption of red and processed meat intake was not associated with CRC-specific mortality in the Cancer Prevention Study II (CPS-II) Nutrition Cohort. Nonetheless, individuals who consumed more than the median amount of red and processed meats before and after CRC diagnosis had a higher risk (79%) of CRC-specific mortality compared with individuals with intakes smaller than the median at both time points (McCullough et al., 2013). Moreover, the California Teachers Study, in which 704 women participated, showed that consumption of total meat and red meat intake was not associated with mortality among women with CRC (Zell et al., 2010).

9.5 MILK AND CALCIUM INTAKES

Cancer Prevention Study II was conducted to examine the link between milk and calcium supplement consumption and the occurrence of CRC. It found milk consumption after a CRC diagnosis was not associated with CRC-related risk and mortality (Yang et al., 2014). However, individuals who took seven servings/week of milk after CRC diagnosis showed a 28% decline in CRC-related mortality compared to those who have taken one serving/week. Moreover, the consumption of calcium also showed a similar pattern in relation to CRC mortality and CRC risk, with the highest and lowest quartiles of total calcium intake after CRC diagnosis. Interestingly, there was no link associated with pre-diagnostic intake of total dairy, milk, vitamin D, or calcium and CRC-related mortality or death. Correspondingly, Dik et al. (2014) observed no link between total dairy, milk, yogurt, cheese, or dietary calcium intake and CRC-related mortality in 3,859 CRC survivors.

9.6 CARBOHYDRATES, GLYCEMIC LOAD, AND SUGAR-SWEETENED BEVERAGES

The consumption of carbohydrates has been reported to be associated with CRC development and survival. It has been reported that high-glycemic and high-carbohydrate intake can increase the risk of CRC occurrence and mortality, thereby decreasing the survival of CRC patients (Meyerhardt et al., 2012). The people with the highest level of glycemic load, compared to those with the lowest level after a CRC diagnosis, showed an increased (97%) risk of CRC occurrence and an increased (74%) risk of mortality. It has been reported that glycemic load may influence the glycemic response to carbohydrates in the dietary supplement. As the relationship observed in CALGB 89803, the impact of consumed food is based on the amount of carbohydrates consumed after CRC diagnosis rather than the quality of food. Moreover, people with the highest carbohydrate intake after CRC diagnosis had a twofold increased risk of disease recurrence and all-cause mortality compared to those with the lowest level of carbohydrate intake. Moreover, a damaging association was detected with the consumption of sugar-sweetened beverages. It has been found that patients who have consumed more than two sugar-sweetened beverages each day after the diagnosis of CRC showed a 75% increase in CRC occurrence compared with patients who consumed less than two sugar-sweetened beverages per month (Fuchs et al., 2014).

9.7 VITAMIN D

Vitamin D is synthesized in the skin when exposed to sunlight, and it is obtained through some food supplements. Vitamin D is present in fortified foods such as milk, fish, and eggs. The association between vitamin D and CRC occurrence has been studied in the 2,330 patients who have CRC, and it has been found that higher levels of circulating vitamin D decrease CRC occurrence and mortality by 29% (Dray et al., 2003). In another study, it was reported that higher circulating vitamin D in 1,598 patients with stage I to III CRC showed a decreased occurrence of CRC risk and mortality compared to low levels of vitamin D. Additionally, the difference in the vitamin D receptor gene considerably altered these associations, suggesting a physiological impact of vitamin D on the CRC growth and progression. Nevertheless, no association between vitamin D intake (from foods or supplements) and the risk of CRC-related mortality was observed. Further studies are required to examine whether elevated levels of vitamin D in the patients after diagnosis of CRC are linked with CRC survivors (Yang et al., 2014).

9.8 ALL-INCLUSIVE LIFESTYLE INTERVENTIONS

The change in lifestyle can produce a significant impact on the lives of CRC patients. The change in lifestyle includes eating food and dietary supplements after the diagnosis of the CRC. Studies have been done where surviving CRC patients were given special food supplements and were asked to do physical exercise. The impact of behavioral changes, eating special food, and doing cardiovascular exercises is measured, and their impact on the occurrence of CRC and CRC mortality has been studied (Bourke et al., 2011; Gillis et al., 2014; Morey et al., 2009; Hawkes et al., 2013; Van Blarigan and Meyerhardt, 2015). One study was conducted in Australia in which 410 patients who had been diagnosed with CRC in the preceding 12 months were arbitrarily allocated to health coaching. The study was conducted with many sessions for almost 6 months in doing activities such as workbooks, postcard prompts, and pedometers (Van Blarigan and Meyerhardt, 2015). Each patient was counseled for 150 min/week and was provided two servings of fruit and five portions of vegetables each day. Their body weight was monitored, and after 12 months of comprehensive training and dietary control, the patients were evaluated. It was found that the patients had lost weight and reduced their saturated fat intake. Additionally, in CRC patients, vegetable consumption was found to be increased when compared to non-CRC groups for 6 months but could not be sustained through 12 months. Overall, the study suggests that such comprehensive training and food management did not produce a significant positive impact on the lives of CRC survivors.

9.9 SIDE EFFECTS OF CHEMOTHERAPY

The anticancer agents or medicines are very effective in controlling the growth and progression of CRC, and they have been widely used in cancer therapy, but these anticancer agents have also caused serious side effects on CRC patients and their day-to-day life activities. It has been reported that the anticancer drug oxaliplatin is effective in the treatment of CRC (Ferlay et al., 2015; NCCN Org, 2017), and after a few days of treatment, nearly 72%–95.2% of patients have experienced acute and chronic peripheral neuropathy (Pietrangeli et al., 2006; Argyriou et al., 2013). It has been reported that the development of acute peripheral neuropathic symptoms in patients developed after receiving the first course of the anticancer drug (Argyriou et al., 2013; Dault et al., 2016). The total cumulative dose of oxaliplatin is about 540–850 mg/m^2. Because of this high dose of the drug, the symptoms persist for 1 month, or in a few cases, they may last for 1 year also (Pietrangeli et al., 2006; Dault et al., 2016; Gupta and Bhaskar, 2016). The side effects of oxaliplatin in CRC patients are chemotherapy-induced peripheral neuropathy with constraint of daily activity, which was a common consequence for CRC patients. The main side effects of oxaliplatin are chronic peripheral neuropathic symptoms, persistent paresthesia, dysesthesia, altered proprioception, sensory ataxia, progressive sensory loss, and decreased vibratory and temperature sensations (Pietrangeli et al., 2006; Dault et al., 2016; Gupta and Bhaskar, 2016; Lu et al., 2018). In addition, patients may get induced peripheral neuropathy in limbs; the disorder often coincides with the restriction of daily activity (Dault et al., 2016; Kanda et al., 2017; Tofthagen et al., 2013; Lu et al., 2019).

REFERENCES

Ainsworth BE, Haskell WL, Herrmann SD, et al. 2011 Compendium of physical activities: a second update of codes and MET values. *Med Sci Sports Exerc.* 2011;43:1575–1581. doi: 10.1249/MSS.0b013e31821ece12.

American Institute of Cancer Research, World Cancer Research Fund. *Food, Nutrition, Physical Activity, and the Prevention of Cancer: A Global Perspective,* London, UK (ed 2), 2011. https://www.wcrf.org/int/research-we-fund/continuous-update-project-cup/second-expert-report.

Argyriou AA, Cavaletti G, Briani C, et al. Clinical pattern and associations of oxaliplatin acute neurotoxicity: a prospective study in 170 patients with colorectal cancer. *Cancer.* 2013;119(2):438-444. doi: 10.1002/cncr.27732.

Bourke L, Thompson G, Gibson DJ, et al. Pragmatic lifestyle intervention in patients recovering from colon cancer: a randomized controlled pilot study. *Arch Phys Med Rehabil.* 2011;92:749–755. doi: 10.1016/j.apmr.2010.12.020.

Dault R, Rousseau MP, Beaudoin A, Frenette MA, Lemay F, Beauchesne MF. Impact of oxaliplatin-induced neuropathy in patients with colorectal cancer: a prospective evaluation at a single institution. *Curr Oncol.* 2016;23(1):e65–e69. doi: 10.3747/co.23.2780.

Dik VK, Murphy N, Siersema PD, et al. Prediagnostic intake of dairy products and dietary calcium and colorectal cancer survival: results from the EPIC cohort study. *Cancer Epidemiol Biomarkers Prev.* 2014;23:1813–1823. doi: 10.1158/1055-9965.EPI-14-0172.

Dray X, Boutron-Ruault MC, Bertrais S, et al. Influence of dietary factors on colorectal cancer survival. *Gut.* 2003;52:868–873. doi: 10.1136/gut.52.6.868.

Edwards BK, Noone AM, Mariotto AB, et al. Annual report to the nation on the status of cancer, 1975–2010, featuring prevalence of comorbidity and impact on survival among persons with lung, colorectal, breast, or prostate cancer. *Cancer*. 2014;120:1290–1314. doi: 10.1002/cncr.28509.

Estruch R, Ros E, Salas-Salvadó J, et al. Primary prevention of cardiovascular disease with a Mediterranean diet. *N Engl J Med*. 2013;368:1279–1290. doi: 10.1056/NEJMc1806491.

Flagg EW, Coates RJ, Calle EE, Potischman N, Thun MJ. Validation of the American Cancer Society Cancer Prevention Study II Nutrition Survey Cohort Food Frequency Questionnaire. *Epidemiology*. 2000 Jul;11(4):462–468. doi: 10.1097/00001648-200007000-00017. PMID: 10874556.

Ferlay J, Soerjomataram I, Dikshit R, et al. Cancer incidence and mortality worldwide: sources, methods and major patterns in GLOBOCAN 2012. *Int J Cancer*. 2015;136(5):E359–E386. doi: 10.1002/ijc.29210.

Fuchs MA, Sato K, Niedzwiecki D, et al. Sugar-sweetened beverage intake and cancer recurrence and survival in CALGB 89803 (Alliance). *PLoS ONE*. 2014;9:e99816. doi: 10.1371/journal.pone.0099816.

Gillis C, Li C, Lee L, et al. Prehabilitation versus rehabilitation: a randomized control trial in patients undergoing colorectal resection for cancer. *Anesthesiology*. 2014;121:937–947. doi: 10.1097/ALN.0000000000000393.

Gupta R, Bhaskar A. Chemotherapy-induced peripheral neuropathic pain. *BJA Educ*. 2016;16(4):115–119. doi: 10.1093/bjaed/mkv044.

Hawkes AL, Chambers SK, Pakenham KI, et al. Effects of a telephone-delivered multiple health behavior change intervention (CanChange) on health and behavioral outcomes in survivors of colorectal cancer: a randomized controlled trial. *J Clin Oncol*. 2013;31:2313–2321. doi: 10.1200/JCO.2012.45.5873.

Kanda K, Fujimoto K, Kyota A. Emotional responses to persistent chemotherapy-induced peripheral neuropathy experienced by patients with colorectal cancer in Japan. *Asia Pac J Oncol Nurs*. 2017;4(3):233–240. doi: 10.4103/apjon.apjon_12_17.

Kenfield SA, Stampfer MJ, Giovannucci E, et al. Physical activity and survival after prostate cancer diagnosis in the health professionals follow-up study. *J Clin Oncol*. 2011;29:726–732. doi: 10.1200/JCO.2010.31.5226.

Kraus WE, Houmard JA, Duscha BD, et al. Effects of the amount and intensity of exercise on plasma lipoproteins. *N Engl J Med*. 2002;347:1483–1492. doi: 10.1056/NEJMoa020194.

Ley SH, Hamdy O, Mohan V, et al. Prevention and management of type 2 diabetes: dietary components and nutritional strategies. *Lancet*. 2014;383:1999–2007. doi: 10.1016/S0140-6736(14)60613-9.

Lichtenstein AH, Appel LJ, Brands M, et al. Diet and lifestyle recommendations revision 2006: a scientific statement from the American Heart Association Nutrition Committee. *Circulation*. 2006;114:82–96. doi: 10.1161/CIRCULATIONAHA.106.176158.

Lu LC, Chang SY, Liu CY, Tsay SL. Reliability and validity of the Chinese version neuropathic pain symptom inventory in patients with colorectal cancer. *J Formos Med Assoc*. 2018;117:1019–1026. doi: 10.10.1016/j.jfma.2017.11.010.

Lu LC, Tsay SL, Chang Y, Chen CM, Liu CY. Daily activity, mood, and quality of life in colorectal cancer patients with chemotherapy-induced peripheral neuropathy: a mediation effect analysis. *Cancer Med*. 2019;8(3), 963–971. doi: 10.1002/cam4.1976.

McCullough ML, Gapstur SM, Shah R, et al. Association between red and processed meat intake and mortality among colorectal cancer survivors. *J Clin Oncol*. 2013;31:2773–2782. doi: 10.1200/JCO.2013.49.1126.

Meyerhardt JA, Ogino S, Kirkner GJ, et al. Interaction of molecular markers and physical activity on mortality in patients with colon cancer. *Clin Cancer Res*. 2009;15:5931–5936. doi: 10.1158/1078-0432.CCR-09-0496.

Meyerhardt JA, Sato K, Niedzwiecki D, et al. Dietary glycemic load and cancer recurrence and survival in patients with stage III colon cancer: findings from CALGB 89803. *J Natl Cancer Inst*. 2012;104:1702–1711. doi: 10.1093/jnci/djs399.

Morey MC, Snyder DC, Sloane R, et al. Effects of home-based diet and exercise on functional outcomes among older, overweight long-term cancer survivors: RENEW—a randomized controlled trial. *JAMA*. 2009;301:1883–1891. doi: 10.1001/jama.2009.643.

Morikawa T, Kuchiba A, Yamauchi M, et al. Association of CTNNB1 (beta-catenin) alterations, body mass index, and physical activity with survival in patients with colorectal cancer. *JAMA*. 2011;305:1685–1694. doi: 10.1001/jama.2011.513.

NCCN. NCCN Clinical Guidelines in Oncology: Colon Cancer & Rectal Cancer, 2017. https://www.nccn.org/. Accessed March 13, 2017.

Pietrangeli A, Leandri M, Terzoli E, Jandolo B, Garufi C. Persistence of high-dose oxaliplatin-induced neuropathy at long-term follow-up. *Eur Neurol*. 2006;56(1):13–16. doi: 10.1159/000094376.

Richman EL, Kenfield SA, Stampfer MJ, et al. Physical activity after diagnosis and risk of prostate cancer progression: data from the cancer of the prostate strategic urologic research endeavor. *Cancer Res.* 2011;71:3889–3895. doi: 10.1158/0008-5472.CAN-10-3932.

Salas -Salvadó J, Bulló M, Estruch R, et al. Prevention of diabetes with Mediterranean diets: a subgroup analysis of a randomized trial. *Ann Intern Med.* 2014;160:1–10. doi: 10.7326/M13-1725.

Saltz LB, Niedzwiecki D, Hollis D, et al. Irinotecan fluorouracil plus leucovorin is not superior to fluorouracil plus leucovorin alone as adjuvant treatment for stage III colon cancer: results of CALGB 89803. *J Clin Oncol.* 2007;25:3456–3461. doi: 10.1200/JCO.2007.11.2144.

Schinner S. Wnt-signalling and the metabolic syndrome. *Horm Metab Res.* 2009;41:159–163. doi: 10.1055/s-0028-1119408.

Schmid D, Leitzmann MF. Association between physical activity and mortality among breast cancer and colorectal cancer survivors: a systematic review and meta-analysis. *Ann Oncol.* 2014;25:1293–1311. doi: 10.1093/annonc/mdu012.

Schwab U, Lauritzen L, Tholstrup T, et al. Effect of the amount and type of dietary fat on cardiometabolic risk factors and risk of developing type 2 diabetes, cardiovascular diseases, and cancer: a systematic review. *Food Nutr Res.* 2014;10:58. doi: 10.3402/fnr.v58.25145.

Tofthagen C , Visovsky CM, Hopgood R. Chemotherapy-induced peripheral neuropathy: an algorithm to guide nursing management. *Clin J Oncol Nurs.* 2013;17(2):138–144. doi: 10.1188/13.CJON.138-144.

Van Blarigan EL, Meyerhardt JA. Role of physical activity and diet after colorectal cancer diagnosis. *J Clin Oncol.* 2015;33(16):1825–1834. doi: 10.1200/JCO.2014.59.7799.

Yang B, McCullough ML, Gapstur SM, et al. Calcium, vitamin D, dairy products, and mortality among colorectal cancer survivors: the Cancer Prevention Study-II Nutrition Cohort. *J Clin Oncol.* 2014;32:2335–2343. doi: 10.1200/JCO.2014.55.3024.

Zell JA, Ziogas A, Bernstein L, et al. Meat consumption, nonsteroidal anti-inflammatory drug use, and mortality among colorectal cancer patients in the California Teachers Study. *Cancer Prev Res (Phila).* 2010;3:865–875. doi: 10.1158/1940-6207.CAPR-09-0262.

10 Future Trends and Innovation in Colorectal Cancer Diagnosis and Treatment

The Role of Machine Learning, Artificial Intelligence, Stem Cells, and Nanodrug Delivery

Priya Chauhan, Preeti Chauhan, Naimish Pandey, Marilyn Tanushka Gomes, and Gayatri Sharma
Centre for Medical Biotechnology, Amity Institute of Biotechnology, Amity University Uttar Pradesh, Noida, Uttar Pradesh, India

10.1 INTRODUCTION

In recent decades, rapid advancements in information technology have pushed significant progress in the medical applications of artificial intelligence (AI). Colorectal cancer (CRC), the third most diagnosed cancer globally, has reported incidence and mortality rates that continue to rise worldwide. This chapter highlights the latest advancements in AI for diagnosing and treating CRC.

10.2 TRADITIONAL METHODS OF DIAGNOSIS AND TREATMENT OF COLORECTAL CANCER

Worldwide, CRC is the third most common type of cancer. Environmental variables, dietary habits, and lifestyle choices can all have an impact on its occurrence. Conventional methods for the diagnosis and treatment of CRC include physical exams, radiography, computed tomography (CT) scans, and surgery in addition to chemotherapy and radiation.

10.2.1 Diagnosis

10.2.1.1 Symptoms and Physical Examination

The first step in diagnosing any medical problem is to determine any accompanying symptoms. Rectal bleeding resulting in a change in stool color (red or black), a drop in red blood cell count, discomfort in the abdomen, ongoing constipation or diarrhea, and other associated symptoms are typical signs of CRC. After that, a medical practitioner would usually perform a physical examination, such as the digital rectal exam, to check for any abnormalities in the organs or the rectal area.

10.2.1.2 Fecal Tests

To find blood in fecal matter samples, immunochemical procedures using antibodies and stool DNA tests are used. For this reason, the guaiac-based FOBT is commonly used. It uses guaiac,

DOI: 10.1201/9781003503842-10

a plant-based chemical that changes color when exposed to traces of blood, making it possible to find hidden blood in the stool.

10.2.1.3 Blood Tests

Blood testing has a limited function in the detection of indicators for CRC. However, blood tests—which are FDA-approved procedures—can identify some cancer indicators, like the altered SEPT9 gene. Important CRC indicators include APC, KRAS, BRAF, PI3K, PTEN, and CA19-9. For detecting recurrences, carcinoembryonic antigen (CEA) may be utilized. Furthermore, assays based on blood, such as nutritional panels and complete blood counts, can help determine how the disease is progressing.

10.2.1.4 Colonoscopy and Sigmoidoscopy

A sigmoidoscopy is a medical technique used to look for and evaluate the existence of polyps, abnormal regions, or possible cancerous symptoms in the rectum and the lower part of the colon, known as the sigmoid colon. A thin, tube-like device known as a sigmoidoscope is inserted into the sigmoid colon via the rectum during this surgery. In addition to having a light and lens for visual inspection, the sigmoidoscope may also have a tool for removing polyps or tissue samples, which are then inspected under a microscope to look for any signs of cancer.

A colonoscopy offers a thorough evaluation of the entire large bowel, including the terminal ileum, whereas a flexible sigmoidoscopy is a quick assessment of the rectum and sigmoid colon. Using this technique, the rectum and colon are visually inspected for the presence of polyps, abnormalities, or signs of cancer. A colonoscope is a thin, tube-shaped device that is inserted into the colon through the rectum. It has lights and a viewing lens. It might also have a system in place for removing polyps or tissue samples, which are then examined under a microscope to look for any indications of malignancy (Triantafillidis et al. 2015).

10.2.1.5 Imaging/Radiology

When it comes to colon cancer staging, morphological characterization, and prognostic monitoring, imaging modalities are essential. For colon cancer, CT is the most used imaging modality; for rectal exams, magnetic resonance imaging (MRI) is the preferred modality. CT is a type of X-ray that is used in virtual colonoscopy to provide a series of images of the colon. The photos are combined by a computer to produce detailed views that could reveal polyps and other anomalies on the colons inside surface. Another name for this examination is computed tomography colonography, or CTC. Furthermore, fluoroscopy methods like barium enemas are also applied. Studies show that arterial angiography—an X-ray-based imaging modality that visualizes the heart's blood vessels—can help in staging as well as diagnosis.

10.2.2 Treatment

10.2.2.1 Surgery

There are several surgical treatments available for CRC patients in the early stages of treatment. These consist of endoscopic mucosal resection (EMR), polypectomy, and local excision. A polypectomy is the removal of a colorectal polyp, whereas a local excision is the use of a colonoscope to remove malignant cells along with some surrounding tissue. As local excision surgery is more involved than polypectomy surgery, recuperation from a local excision may take longer. Using a colonoscope, stage 1 or stage 2 CRC is removed during EMR. When it comes to colon cancer, complete mesocolic excision entails carefully excising the impacted colon segment and its mesocolon to preserve the visceral peritoneum's ability to enclose the tumor and any lymph nodes that may be implicated.

A colectomy is required when colon cancer has progressed to an advanced stage and has spread to the upper parts of the colon. A colectomy entails removing the whole colon or a part of it. Laparoscopic colectomy is a minimally invasive form of this technique; open surgery with a large abdominal incision may be necessary. A hemicolectomy (removal of part of the colon) or a total colectomy (removal of the entire colon) may be performed during open surgery in cases when the colon has expanded widely due to cancer or other disorders such as inflammation or polyps. Depending on the amount of resection needed, either a laparoscopic or open colectomy is the preferred method. A colostomy, which involves removing the damaged colon and creating an artificial opening called a stoma to join the remaining colon to the abdominal wall, may be done after a colectomy.

10.2.2.2 Chemotherapy

KRAS or NRAS gene mutations are important biomarkers for CRC and can be used as a target for anti-EGFR therapy such as cetuximab and panitumumab (Petrelli et al. 2011). Resistance to EGFR-targeted treatments is common in individuals with KRAS/NRAS mutations, highlighting the significance of mutation testing in directing treatment and guaranteeing that these medicines are only administered to receptive patients. Although the BRAFV600E mutation in CRC is linked to a poor prognosis, it also provides a target for beneficial treatments like encorafenib. For BRAFV600E-mutant CRC, the standardization of targeted combination therapy with medications like cetuximab or panitumumab is becoming more advantageous, as individualized care leads to better results (Donada et al. 2013). In CRC, the existence of mismatch repair deficit (dMMR) or microsatellite instability (MSI) is a crucial biomarker. Pembrolizumab and anti-CTLA4 antibodies are examples of immune checkpoint inhibitors that can be very beneficial for patients with dMMR or MSI-high (MSI-H) malignancies. Accordingly, MSI-H/dMMR testing is crucial for choosing an appropriate course of treatment for advanced CRC. Additionally, the VEGF signaling pathway is linked to the expression of vascular endothelial growth factor D (Gockel et al. 2008) and mutations in the kinase insert domain receptor. Anti-angiogenic drugs such as bevacizumab, ramucirumab, fruquintinib, and regorafenib target this pathway when treating metastatic CRC (Mohammad Rezaei et al. 2019).

The use of certain biomarkers to direct the administration of these drugs is still being studied, even though their overall survival rates have improved. Furthermore, medications that target these receptors, such as trastuzumab, pertuzumab, and tucatinib, are used in cases of CRC where there is overexpression of HER2 proteins. Furthermore, a small percentage of CRCs exhibit abnormal TRK proteins because of NTRK gene alterations (H. Wang et al. 2022). For advanced cancers that continue to exhibit NTRK gene alterations despite previous treatments, larotrectinib and entrectinib must be used to target these proteins (Suh et al. 2022).

10.2.2.3 Radiotherapy

Radiation therapy is rarely used to treat colon cancer, but when it is, it can be useful in certain situations. For example, it can be used in conjunction with chemotherapy before surgery to shrink tumors, during surgery (intraoperative radiation therapy) to remove any remaining cells, or after surgery to treat metastasized cancer that has spread to organs or the abdominal lining. However, rectal cancer patients are more frequently treated with radiation therapy. To reduce the risk of recurrence and facilitate tumor excision, it is frequently administered before, during, or after surgery in conjunction with chemotherapy (neoadjuvant treatment). Radiation therapy, known as external-beam radiation therapy, is commonly used to target cancer by directing radiation from outside the body. The procedure known as internal radiation therapy, or brachytherapy, entails putting a radioactive source inside the tumor or near it (Feeney et al. 2019).

The present challenges with CRC detection and therapy include false positives and negatives, which reduce the efficacy of current approaches. Moreover, there are issues with procedure expenses and invasiveness. Probes may break, physical examinations can be uncomfortable and invasive, and anesthesia has risks like cardiac issues. Biomarker-based CRC diagnosis also faces several difficulties, including the requirement for robust biomarker validation, the intrinsic heterogeneity of CRC that exacerbates diagnostic reliability, and the requirement for biomarkers to be broadly accessible and reasonably priced. Additionally crucial to ensuring accurate and dependable biomarker testing is assay standardization. Significant obstacles also arise from ethical concerns about the management of private genetic data and the incorporation of biomarkers into standard therapeutic procedures (Kamel et al. 2022).

10.3 DEVELOPMENT OF ARTIFICIAL INTELLIGENCE AND APPLICATION IN MEDICINE

The last 50 years have seen a significant improvement in AI. AI has increased precision, decreased costs, maximized time efficiency, and minimized errors in diagnosis and treatment options. These days, predictive models are essential for diagnosis, the prediction of treatment response, and possibly even preventive medicine. The effectiveness of provider workflow, disease monitoring, procedure accuracy, diagnostic accuracy, and overall patient outcomes have all been greatly improved by AI (Amisha et al. 2019). In addition, AI can help with personalized treatment by analyzing large datasets for outcome prediction and treatment regimen optimization, including dose optimization and adverse drug event prediction. It can also monitor emerging disease threats, such as COVID-19, and possible outbreaks (Schork 2019). In population health management, predictive analytics—which analyzes historical and present data using machine learning (ML) and AI to identify at-risk populations and direct preventive interventions for chronic diseases like endocrine and cardiac conditions—is becoming more and more crucial.

To lower the risk of CRC development, AI has been implemented in several avenues of diagnosis and treatment—from robotic surgery to computer-aided detection (CADe) systems that improve polyp and adenoma detection rates (ADRs) during early colonoscopy screenings. Furthermore, bioinformatics and ML can identify CRC biomarkers for non-invasive screening, and convolutional neural networks (CNNs) help evaluate histopathologic pictures, reducing differences in medical professionals' knowledge (Nduma et al. 2024). The impact of AI and ML on the diagnosis and treatment of CRC is represented in Figure 10.1.

10.3.1 Machine Learning

A field of computer science called ML is devoted to creating algorithms that can evaluate input data, forecast the future with a given level of accuracy, identify patterns and trends, and iteratively improve models based on historical data. ML functions at the nexus of computer science and statistics. Applications for it may be found in many different fields, but in the field of healthcare research, it provides tools that are automated, flexible, and computationally demanding to extract complex patterns from data, such as subgroups, underlying dimensions, and nonlinear correlations.

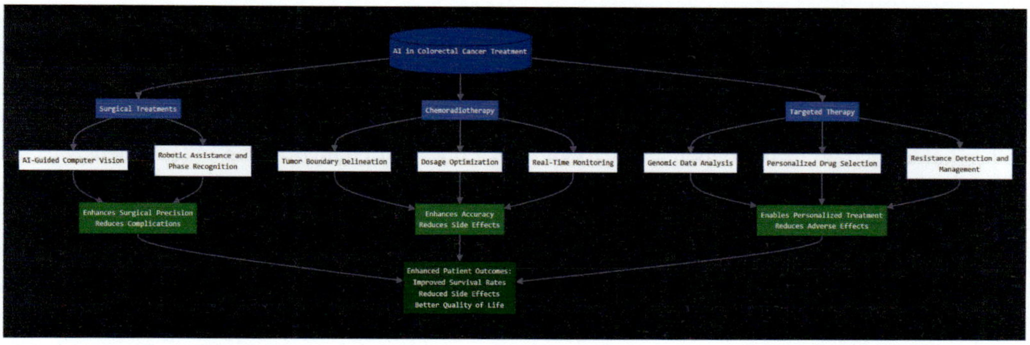

FIGURE 10.1 Role of artificial intelligence (AI) in colorectal cancer development, treatment, and diagnosis

Fundamentally, ML relies on a semi-automated process in which computers receive data so they can build analytical models inside a learning framework, ultimately improving prediction accuracy. This does not lessen its integration with conventional statistics since ML frequently incorporates statistical underpinnings into its functions. Predictive accuracy optimization is a key component of ML, and it performs very well when working with high-dimensional inputs—even in the absence of explicit knowledge about the combined probability distribution of these variables. ML differs from traditional programming in that it generates algorithms using subsets of data, potentially producing novel feature and weight combinations not possible with first-principles algorithms. Four main learning approaches—supervised, unsupervised, semi-supervised, and reinforcement learning—are at the center of ML. Each is designed for a particular purpose. The emphasis of ML is mostly on the mathematical aspects of prediction performance and generalization, with interpretability taking on a secondary role.

10.3.2 Deep Learning

Deep learning (DL), a subset of ML, employs artificial neural networks (ANNs) to enable computers to process data in a manner like the human brain. It is versatile and can be applied to various problems, such as recognizing complex patterns in images, recognizing speech, processing natural language, and transcribing audio files into text. DL is making significant strides in solving problems that have long challenged the AI community. Its strength lies in uncovering complex structures within high-dimensional data, making it applicable across various fields, including science, business, and government. Beyond breaking records in image and speech recognition, DL has outperformed other ML techniques in predicting drug molecule activity, analyzing particle accelerator data, reconstructing brain circuits, and forecasting the effects of mutations in non-coding DNA on gene expression and disease. Additionally, DL has shown remarkable promise in natural language understanding tasks such as topic classification, sentiment analysis, question answering, and language translation. When it comes to the diagnosis and management of CRC, DL has demonstrated significant potential in biomarker and polyp identification, for example, and in diagnosis by classifying lesions using radiological data. For example, using grayscale photos, Hsu et al. employed a DL network to identify colorectal polyps (Hsu et al. 2021).

10.3.3 Supervised Learning, Semi-Supervised Learning, Support Vector Machine, Random Forest, and Convolutional Neural Network

10.3.3.1 Supervised Learning

Using labeled training data, supervised ML trains algorithms to make predictions on unlabeled test data. This includes regression and classification, which are the two main forms of learning (Mohamed et al. 2022). The process of classifying data involves grouping it into predetermined groups according to labeled examples. It analyzes the properties of new instances and compares them to the training data using methods like support vector machines (SVMs) and random forests (RFs) to place them in the appropriate categories. However, regression creates a mathematical model that explains the link between independent and dependent variables to predict continuous values for a target variable. The predicted change in the dependent variable for a one-unit change in the independent variable is shown by the regression coefficient. Regression analysis involves the use of several algorithms, such as SVMs, RFs, neural networks (NNs), polynomial regression, linear regression, and Gaussian processes. Logistic regression has displayed a high accuracy in CRC diagnosis using several tumor markers such as blood counts and biomarkers like CEA and AFP (Li et al. 2021).

By combining several decision trees, Leo Breiman and Adele Cutler's ensemble learning technique known as RF improves accuracy and reduces overfitting. A varied "forest" of trees is produced by building each tree using randomly chosen subsets of data and predictor variables (Breiman

2001). In comparison to a single decision tree, overall performance is improved because the final prediction is generated from the average of all trees' predictions. The successful application of RF algorithms in microbial and genetic biomarker detection for CRC diagnosis has been proved by the research of scientists like Kolisnik et al. (2023).

10.3.3.2 Support Vector Machine

SVMs are used to place a hyperplane as far away from the closest data points of each class as feasible to increase the distinction between the two classes. To identify the best-separating hyperplane, this method includes elevating input vectors into a higher-dimensional space using a kernel function. SVMs provide for precise prediction by clearly delineating the boundaries between classes (Wang 2005). To find novel treatments for colon cancer, Sengupta et al. have created a quantitative structure-activity connection model with SVM as the underlying algorithm (Sengupta et al. 2024).

10.3.3.3 Semi-supervised Learning

By using a combination of labeled and unlabeled data in a training dataset, semi-supervised learning (SSL) is an affordable method for enhancing natural picture categorization, among other uses. In SSL, the knowledge from unlabeled data is used to improve the classifier, which is first built only from labeled data, using two different sets of samples: labeled and unlabeled data. When both labeled and unlabeled data are present, this approach bridges the gap between supervised and unsupervised learning and is very helpful as seen in the SSL model created by Yu et al., which uses whole slide images to effectively diagnose CRC (Padmanabha Reddy et al.,2018; Yu et al., 2021).

10.3.3.4 Convolutional Neural Network

DL's core component, NNs, uses ANNs to process complex, nonlinear data. NNs are artificial neurons composed of several layers that are coupled and have nonlinear activation functions. They are very flexible and may be used to simulate intricate patterns with few assumptions. Their proficiency in handling high-dimensional data, including epigenomic data, stems from the abundance of large-scale training data. However, NNs are more complex to train and comprehend than conventional algorithms, requiring large amounts of data. NN topologies are typified by interconnected layers that allow nodes to optimize prediction accuracy by adjusting weights through forward propagation. NNs mimic the structure and operations of the human brain by using nonlinear functions to learn complex representations directly from labeled input, eliminating the need for carefully designed features (Choi et al. 2020). CNNs are a particular type of NN designed to process grid-like input, such as images and movies. CNNs identify features such as edges and textures by using convolutional layers with tiny filters (kernels), which capture hierarchical patterns (Alzubaidi et al. 2021). CNNs can now efficiently process and identify intricate patterns in images, which is useful when diagnosing CRC based on radiological or histological scans. A prime example is the Color-CADx system, which uses several CNNs to provide an accurate diagnosis (Sharkas and Attallah 2024).

10.4 APPLICATIONS OF AI IN THE DIAGNOSIS OF COLORECTAL CANCER

AI is emerging as a transformative tool in CRC diagnosis, particularly through the use of CADe systems in endoscopy. AI enhances detection rates, reduces human error, and standardizes diagnostic procedures, thereby improving outcomes in CRC screening.

10.4.1 Endoscopic Diagnosis

AI is increasingly transforming the diagnosis of CRC, particularly through endoscopic procedures like colonoscopy, which remains the gold standard for detecting precancerous polyps and lesions. Traditional colonoscopy outcomes depend on operator expertise and are prone to variability, which can result in missed adenomas. AI-driven CADe systems, utilizing advanced DL algorithms like CNNs, have been developed to enhance real-time polyp detection. These systems analyze video footage during colonoscopy, highlighting suspicious lesions that may otherwise go unnoticed. Studies have demonstrated that AI significantly improves ADRs by identifying subtle lesions, even in challenging cases such as small, flat polyps (Avram et al. 2023; Yin et al. 2023).

Beyond enhancing detection, AI contributes to standardizing the diagnostic process. Traditionally, diagnostic accuracy in colonoscopy can vary depending on the experience level of the endoscopist. AI eliminates much of this human variability by providing consistent detection capabilities based on extensive datasets of annotated medical images. These AI models can accurately identify and classify lesions, offering real-time diagnostic insights that improve decision-making during procedures. This standardization is particularly valuable in resource-limited settings where access to highly skilled endoscopists may be limited, ensuring that patients receive reliable diagnostic care regardless of location or clinician expertise (Avram et al. 2023; Yin et al. 2023).

In addition to detection and standardization, AI plays a pivotal role in reducing the adenoma miss rate, a critical factor in CRC screening efficacy. CADe systems equipped with AI provide continuous monitoring of the endoscopic field, alerting clinicians to regions that require further inspection and reducing the likelihood of missed lesions. Moreover, AI-driven bioinformatics tools are increasingly being used to classify lesions based on their characteristics, distinguishing between benign and malignant tissue in real time. This capability accelerates clinical decision-making and facilitates more personalized treatment strategies. As AI continues to advance, its integration into CRC diagnostics will enhance early detection, improve patient outcomes, and significantly contribute to the precision medicine landscape (Avram et al. 2023; Yin et al. 2023).

10.4.2 Non-invasive Screening

Non-invasive screening primarily involves identifying specific tumor markers from samples such as ascites, stool, and blood. Unlike colonoscopy, non-invasive methods present lower risks and require significantly less preparation time. Conventional non-invasive screening techniques, like the FOBT and CEA, have limited sensitivity and specificity. Recently, ML has been increasingly applied in medical data analysis, offering the potential to enhance the accuracy of current biomarkers and identify additional marker genes (Yin et al. 2023).

Bioinformatics tools are being utilized more frequently to study cancer pathogenesis. By effectively integrating bioinformatics with ML, early biomarkers for CRC can be identified and screened, which helps assess the prognosis of CRC patients (Zhou et al. 2020). CEA is one of the most comprehensively studied tumor markers for CRC; however, serum CEA screening shows limited sensitivity in asymptomatic patients. BRAF and KRAS are the commonly analyzed genes for detecting DNA mutations in CRC, though these tumor markers lack sufficient sensitivity and specificity to indicate CRC reliably. In 2019, Zhang et al. employed near-infrared (NIR) spectroscopy combined with a counter-propagation artificial neural network to differentiate between BRAF V600E mutant and wild-type samples in CRC tissues. This approach achieved a sensitivity of 100% and a specificity of 87.5% in detecting the BRAF V600E mutation, showing promising results for use in the auxiliary diagnosis of BRAF mutations in CRC (Zhang et al. 2019).

Detecting abnormal DNA methylation markers in plasma or stool presents a promising method for the non-invasive early detection of CRC. The SEPT9 gene methylation test is already commercially available as an alternative for CRC screening. However, the specificity of detection techniques that rely on single DNA methylation sites remains limited. The collective detection of abnormal DNA methylation markers can significantly improve the early detection of CRC (Ferrari et al. 2021).

There are numerous non-coding RNAs, including messenger RNAs (mRNAs), microRNAs (miRNAs), and long non-coding RNAs. These non-coding RNAs are quite stable and easy to preserve and are therefore extensively used for research. miRNAs, a class of endogenous ncRNAs approximately 22 nucleotides in length, play crucial regulatory roles within cells and are linked to cancer proliferation and metastasis. Growing evidence suggests that miRNAs hold promise as potential biomarkers for CRC. In 2019, researchers investigated the potential of miR-31 as a biomarker for diagnosing CRC lymph node metastasis (LNM) using various bioinformatics analysis techniques. These included KEGG enrichment analysis, correlation analysis, survival analysis, validation of expression levels, protein–protein interaction network construction, and gene ontology analysis. They discovered that miR-31 levels were significantly elevated in the plasma and tissue of CRC patients with LNM. Additionally, Wang and his team found that the expression of miR-1–3p was decreased in CRC according to their bioinformatics analysis, suggesting that miR-1–3p could serve as a potential diagnostic tool for CRC and may help inhibit tumor progression (Wang et al. 2018; Yin et al. 2023).

A recent study utilized bioinformatics analysis of gene expression microarray data from the Gene Expression Omnibus database, which resulted in the identification of 105 differentially expressed genes and ten hub genes. Researchers employed tools such as SVM, receiver operating characteristic curve analysis, and survival analyses to assess the potential of these hub genes as diagnostic biomarkers for CRC.

These biomarkers can therefore be used for the non-invasive screening of CRC and thus help in the early prognosis of CRC.

10.4.3 Histopathological Diagnosis

Research has demonstrated that the integration of AI technology can help pathologists enhance diagnostic efficiency, reduce their workload, and improve overall working conditions. These improvements result in higher diagnostic accuracy while reducing misdiagnosis rates (Acs et al. 2020). Digital pathology plays a crucial role in AI applications, including image retrieval, pattern recognition, ML, and DL. By extracting relevant quantitative features and identifying specific regions of interest, digital pathology supports the development of AI-driven diagnostic systems that feature automatic recognition capabilities (Acs et al. 2020).

In a study conducted by Kasahara et al., AI was applied to a dataset of 146 T1 CRC cases. The AI analyzed nuclear morphological features from hematoxylin and eosin-stained slides. The findings suggested that AI could improve the accuracy of predicting LNM before surgery, although further clinical validation is necessary (Kasahara et al. 2022). Similarly, Rathore et al. developed a CRC detection system using an SVM radial basis function algorithm. This system was able to classify normal and malignant colon biopsy images, and it also automatically determined the grade of malignancy. When compared to earlier methods, this detection system exhibited superior accuracy in both cancer detection (95.40%) and malignancy grading (93.47%) (Wang et al. 2020). Another approach by Yang et al. utilized a combination of sub-patch weight color histograms and least squares SVM, which allowed for a novel AI application in CRC pathology. This method not only captured the color and spatial details of tumor images but also highlighted heterogeneous data, achieving a tumor classification accuracy of 96.78% (Wang et al. 2020).

At present, most AI-based pathology recognition technologies rely on supervised learning (SL). To overcome the challenge of large datasets required for SL, researchers have introduced SSL based on an average teacher model. This method has been shown to reduce the demand for extensive labeled data while achieving results comparable to SL. Interestingly, a monogram model that integrates ML-pathomics, immunoscore, radiomics, and clinical factors has been proven effective in predicting the postoperative outcomes of patients with CRC metastasized to the lungs. These developments indicate significant progress in AI-assisted pathology, with the potential to move beyond early-stage screening and provide more comprehensive guidance for the treatment and prognosis of CRC patients (Yu et al. 2021).

10.4.4 Radiologic Diagnosis

Radiomics is a technology that transforms medical imaging into high-dimensional data, providing valuable insights for cancer diagnosis and prognosis (Li et al. 2021). In CRC, common imaging modalities include MRI, ultrasonography, and CT. However, traditional imaging methods have limitations, such as restricted tumor evaluation, low accuracy in tumor staging, and heavy reliance on the expertise of imaging physicians.. While functional magnetic resonance techniques, including diffusion-weighted imaging and T2-weighted imaging, have shown promising predictive capabilities, there is room for improvement (Horvat et al. 2022).

AI is increasingly being integrated with radiomics to enhance diagnostic accuracy. By processing and analyzing large volumes of imaging data, AI can facilitate tumor segmentation, feature extraction, and the development of predictive models. This combination is becoming essential in the precision diagnosis and treatment of CRC. For instance, in 2018, Liu et al. introduced a label assignment generative adversarial network, which successfully segmented regions of interest in CRC CT scans. This AI-driven segmentation technique has proven to save time and reduce labor costs while also demonstrating robustness in various medical imaging tasks (Liu et al. 2019). While MRI radiomics has been extensively studied for CRC, ultrasound remains a valuable tool for early detection. Song et al. developed a deep multi-view fusion network using endorectal ultrasonography (ERUS) to differentiate between benign and malignant colorectal tumors, significantly reducing both the workload of ultrasound specialists and the risk of misdiagnosis (Song et al. 2022).

Another critical application of radiomics is in predicting the response of locally advanced rectal cancer (LARC) following treatment. While ERUS is useful for early detection, its ability to assess metastasis in advanced CRC is limited (Wong et al. 2022). Post-neoadjuvant MRI may also fall short in accurately identifying pathological complete response (PCR). However, combining multiparametric MRI data with clinicopathological factors in a nomogram allows for more accurate risk stratification and treatment planning for patient's post-neoadjuvant chemoradiotherapy (post-NCRT) (Liu et al. 2021). In this context, Farri et al. created an AI model using MRI texture features to predict PCR in 55 LARC patients post-NCRT (Yin et al. 2023).

Radiomics also plays a role in detecting distant metastases. Rocca et al. recently employed a formal method combined with CT imaging to monitor liver metastases in CRC patients, achieving an overall accuracy of 93.3% (Rocca et al. 2021). In addition, combined models integrating radiomics with immunomics and pathomics have shown promise in predicting lung metastasis in CRC patients (Liu et al. 2021).

10.5 APPLICATIONS OF AI IN COLORECTAL CANCER TREATMENT

AI has shown great promise in various fields of medicine, particularly in improving treatment outcomes for CRC. The integration of AI-based techniques in surgical procedures, chemoradiotherapy, and targeted therapies is revolutionizing how clinicians approach this complex disease. This chapter delves into the role AI plays in enhancing treatment efficacy and patient outcomes for CRC, particularly focusing on three key treatment modalities: surgical treatments, chemoradiotherapy, and targeted therapy (Figure 10.1).

10.5.1 Surgical Treatments

AI has increasingly impacted the surgical treatment of CRC, particularly through computer vision (CV) applications. In colorectal surgeries, AI technologies assist in identifying critical anatomical structures, such as vessels and surgical planes, improving intraoperative decision-making. AI-based systems have been shown to recognize surgical phases and actions with high accuracy, supporting real-time adjustments during procedures like total mesorectal excision. This enhances both surgical safety and efficiency, reducing intraoperative adverse events such as hemorrhage and nerve injuries.

AI-driven CV has also demonstrated potential in guiding surgeons during complex procedures by automatically interpreting fluorescent signals, facilitating precise dissection, and minimizing complications (Quero et al. 2022).

One of the most notable applications of AI is in robotic surgery, where systems like the da Vinci Surgical System are now used widely. Robotic surgery offers several advantages, including higher precision, improved visualization, and less invasiveness, which translate to quicker recovery times and lower complication rates (Avram et al. 2023; Yin et al. 2023).

Moreover, AI is being utilized to assist surgeons during procedures by offering real-time image analysis, identifying tumor margins, and even recognizing surgical phases. For instance, phase recognition technology uses DL to identify specific phases in colorectal surgery, helping surgeons navigate complex anatomical structures. Additionally, autonomous robotic systems, such as the Smart Tissue Autonomous Robot, have shown promise in performing critical procedures like anastomosis, where AI surpasses human performance in terms of precision. AI also plays a role in real-time microcirculation analysis during surgery, using fluorescence imaging to predict anastomotic complications (Avram et al. 2023).

10.5.2 Chemoradiotherapy

In chemoradiotherapy, AI applications are primarily focused on improving precision in treatment delivery and optimizing radiation dosing. AI-driven algorithms assist in delineating tumor boundaries in radiotherapy, ensuring that radiation targets the cancer cells while sparing surrounding healthy tissues. This not only improves treatment efficacy but also minimizes side effects. Furthermore, AI facilitates real-time treatment monitoring, allowing clinicians to adjust treatment plans dynamically based on a patient's response. AI's ability to analyze large datasets enables personalized treatment approaches, improving patient outcomes in both chemotherapy and radiotherapy (Weerarathna et al. 2023).

Moreover, AI has been instrumental in treatment planning by automating radiation therapy workflows. ML algorithms analyze medical imaging data to identify cancerous tissues and suggest optimal radiation dosages. This level of precision reduces the likelihood of radiation exposure to surrounding healthy tissues, improving treatment safety (Yin et al. 2023).

The role of AI in chemoradiotherapy also extends to post-treatment monitoring, where it can predict recurrence by analyzing patterns in imaging and clinical data. This allows for timely intervention and tailored follow-up strategies (Yin et al. 2023).

10.5.3 Targeted Therapy

The integration of AI has markedly advanced targeted therapies for CRC by facilitating the identification of molecular targets and optimizing treatment selection. AI's capacity to analyze large-scale, complex datasets allows for the precise tailoring of therapeutic interventions based on the specific genetic mutations identified within a patient's tumor, thereby enhancing treatment efficacy while minimizing adverse effects. AI-driven models, applied to high-throughput sequencing data, have been instrumental in predicting the effectiveness of targeted therapies such as monoclonal antibodies and kinase inhibitors. This enables a more personalized approach to managing advanced CRC by identifying patients most likely to respond to specific treatments. Furthermore, AI plays a critical role in drug discovery and development, streamlining the optimization of therapeutic strategies and potentially introducing novel drugs specifically tailored to the molecular profile of individual tumors (Weerarathna et al. 2023).

AI is significantly influencing targeted therapy for CRC through its advanced application in genomic data analysis. In the era of precision medicine, AI algorithms are employed to identify critical genetic mutations and biomarkers, such as EGFR or BRAF, that drive CRC, enabling the selection of the most effective targeted therapies. AI-driven bioinformatics further facilitates the analysis of large genomic datasets, accelerating the discovery of novel biomarkers and therapeutic targets. This capability supports the development of personalized treatment regimens tailored to the unique genetic profile of each patient's cancer. AI's ability to process and interpret vast amounts of genomic data far surpasses human capacity, expediting the identification of actionable mutations. Beyond drug selection, AI is crucial for monitoring the ongoing efficacy of targeted therapies, allowing for real-time adjustments based on evolving genetic landscapes. Importantly, AI can detect emerging resistance mechanisms earlier, enabling clinicians to proactively adapt treatment strategies to prevent disease progression (Yin et al. 2023).

10.6 ROLE OF CANCER STEM CELLS IN COLORECTAL CANCER

Recent findings suggest that a small group of cancer cells, known as tumor-initiating cells with stem cell-like characteristics, or cancer stem cells (CSCs), exist within tumors and play a crucial role in the initiation, development, recurrence, metastasis, and proliferation of the tumor (Ma et al. 2020). Colorectal cancer stem cells (CR-CSCs) show tumor heterogeneity and have different molecular and phenotypic characteristics compared to the other tumor cells. These cells have self-renewal properties and show resistance to the standard chemotherapy and radiotherapy treatments, due to which they play a crucial role in tumor recurrence. Various factors, including alterations in the tumor microenvironment and genetic and epigenetic mutations, can induce the transformation of non-cancer stem cells (non-CSCs) into CSCs (Parizadeh et al. 2019).

CRC carcinogenesis could be explained by two proposed models: the stochastic model, in which each cell has an equal ability to initiate and promote CRC. This model suggests that cancer cells arise from a series of mutations, leading to a population of increasingly diverse cells. This cellular heterogeneity enables various clones to adapt, survive, and proliferate under certain conditions, contributing to tumor relapse, metastasis, and resistance to radiochemotherapy. In contrast, the new CSC model suggests that only the pluripotent CR-CSCs can initiate and promote tumor growth (Cherciu et al. 2014). Traditional treatment methods target the rapidly dividing tumor cells, failing to affect the CSCs. Therefore, targeting the primary pathogenic factors of CRC has led to a major breakthrough in treatment, while the development of CSC theory has unveiled a new approach to combating cancer (Figure 10.2).

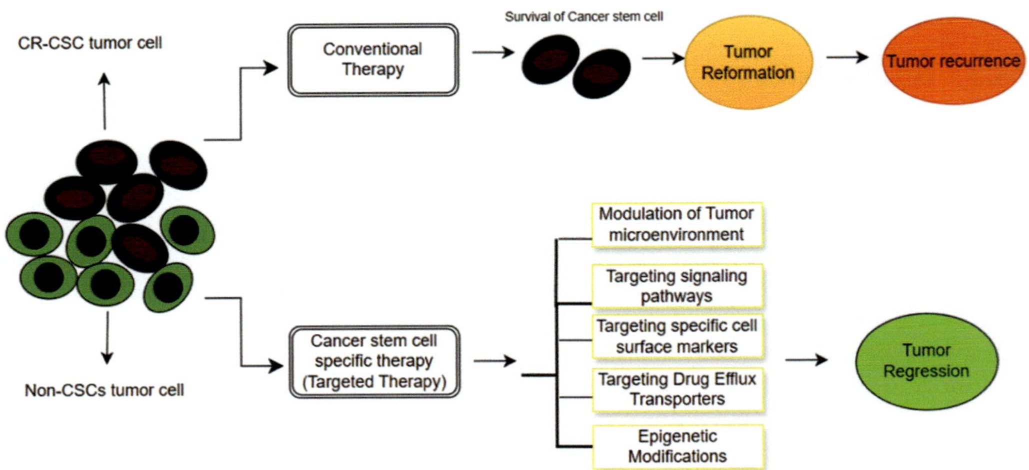

FIGURE 10.2 The effect of conventional therapy and targeted therapy on colorectal cancer stem cells (CR-CSCs).

10.6.1 Surface Markers

CR-CSC isolation can be accomplished by identifying and targeting a subset of tumor cells based on the expression of one or more cell surface markers linked to cancer self-renewal. These markers include CD133, CD166, CD44, CD24, betal integrin (CD29), Lgr5, EpCAM (ESA), ALDH-1, Msi-1, DCAMLK1, and EphB receptors (Cherciu et al. 2014). Three cell surface markers, namely CD133, CD166, and CD44, were found to be associated with CR-CSCs. The comparative expression and diagnostic significance of these three markers was indicative of low risk, intermediate risk, and high risk of CRC in patients (Horst et al. 2009).

CD133 Surface Marker: CD133, having a molecular weight of 120kDa, is a transmembrane cell surface protein. CD133 was first discovered in 1997 on normal human hematopoietic stem cells. Several immunohistochemical studies were performed to study the expression of the CD133 surface marker. CD133 antigen could be found both at the luminal surface of the cell membrane of cancer cells as well as in the cytoplasm of CRC. The expression of this marker at different locations is linked to different clinical relevance. Membranous expression of CD133 is linked to patient survival, recurrence-free survival, and resistance to chemotherapy, whereas cytoplasmic expression does not serve as an independent marker for patient outcomes (Cherciu et al. 2014). The shift in CD133 localization from the cytoplasm to the membrane is associated with the transition of epithelial cells to a more invasive phenotype (Chen et al. 2013). In addition to immunohistochemical staining, CD133 expression can also be assessed using reverse transcriptase polymerase chain reaction, which analyzes the expression of the CD133 gene or its mRNA (Reggiani et al. 2012). This technique suggests that elevated levels of the CD133 gene may indicate a higher likelihood of recurrence and an overall poor prognosis for CRC patients (Choi et al. 2009).

In 2007, O'Brien et al. showed that all CSCs expressed $CD133^+$ protein, while $CD133^-$ cells were incapable of tumor formation. This finding functionally validated the use of CD133 as a surface marker for CSCs. It was found that $CD133^+$ cells showed more resistance to chemoradiotherapy than the $CD133^-$ cells in CRC (Moossavi and Ansari 2013).

CD166 Surface Marker: Also known as activated leukocyte cell adhesion molecule, CD166 is a type-1 glycoprotein and a member of the immunoglobulin superfamily. It is associated with tumor initiation in xenografts, the formation of tumor colonies, and is linked to the prognosis and survival of CRC patients (Cherciu et al. 2014). The relationship between the heterogeneous marker expression and poor prognosis and survival has led to the use of CD166 as a valuable prognostic surface marker. It holds significant prognostic value, as positive cytoplasmic expression is linked to poor survival rates. Typically, CD166 expression signifies an advanced T category and N-positive status in CRC (Weichert et al. 2004).

CD44 Surface Marker: It is a cell adhesion molecule that is used as a CRC marker. The CD44 marker plays an important role in cancer metastasis. CD44-positive cells have CSC properties, which means they can self-renew and differentiate from a new tumor. $CD44^+$ cells are associated with high neoplastic potential, while the $CD44^-$ cells are unable to form a tumor (Vaiopoulos et al. 2012).

$CD133^+CD44^+$ cells were found to be tumorigenic, while the $CD133^+CD44^-$ and $CD133^-CD44^+$ cells were non-tumorigenic (Haraguchi et al. 2008). Another study on immunodeficient mice revealed that $CD166^+CD44^+$ cells have a higher rate of forming tumors than the $CD166^-CD44^+$ and $CD166^+CD44^-$ cancer cells. These inter-relationships between different tumor markers help in the early prognosis of CRC (Cherciu et al. 2014).

Lgr5, a leucine-rich G-protein receptor 5, is another surface marker for the identification of CRC. Lgr5-positive CRC cells are known to be involved in tumor growth, proliferation, and metastasis (Parizadeh et al. 2019). Acetaldehyde dehydrogenase (ALDH) is a more specific marker than CD133 and CD44 markers. $CD26^+$ cells exhibited much more drug resistance and invasiveness than the $CD26^-$ cancer cells. Similarly, epithelial cell adhesion factor molecule (EpCAM) is also used as a surface marker. $CD44^+$/EpCAM (high) cells exhibit the properties of CSCs (Liu et al. 2014) (Table 10.1).

TABLE 10.1
Surface Markers for Colorectal Cancer (CRC) (Langan et al. 2013; Ma et al. 2020)

Marker	Function
CD133	• Chemoresistance, tumorigenesis, invasiveness, proliferation, angiogenesis, resistance to apoptosis • $CD133^{+}$ tumor cells are enriched 200-fold for CSCs
CD166	• Tumorigenesis, proliferation, chemoresistance • Role in neuronal growth, embryonic blood vessel formation, and embryonic hematopoiesis
CD44	• Involved in malignant transformation of cells from adenoma to carcinoma • Metastasis • Cell adhesion
CD24	• Metastasis • Invasiveness • Strong expression associated with poor survival rates • Chemoresistance
CD29	• Involved in tumor growth, proliferation, metastasis, and death
EpCAM	• Decreased expression linked to aggressive rectal cancer • Involved in the cadherin–catenin pathway and Wnt pathway • Enhanced expression linked to poor survival rate
ALDH1	• Higher expression in colorectal cancer • Resistance to chemotherapy • Detoxifying enzyme that facilitates the oxidation of intracellular aldehydes
Lgr5	• Negatively regulates the Wnt pathway • Higher expression in CRC
Lamin A	• Chemoresistance • Tumorigenic
Aurora-A	• EMT (epithelial to mesenchymal transition) • Metastasis

10.6.2 Resistance to Therapy

CSCs play a pivotal role in driving therapy resistance and relapse in CRC. As therapy resistance remains a major challenge in cancer treatment, numerous underlying mechanisms have been suggested over the past several decades. These include mutations in key signaling pathway molecules, elevated expression of anti-apoptotic proteins, the presence of quiescent or resistant tumor cells, and the over-activation of drug efflux pumps, all of which contribute to therapy failure (Patsalias & Kozovska, 2021).

Two models explain the origin of CSC-driven multidrug resistance (MDR) in CRCs that show heightened resistance to standard chemotherapeutic treatments. The first model suggests that following exposure to anticancer drugs, only CRC stem cells expressing ATP-binding cassette transporters are capable of regenerating the tumor through asymmetrical cell division, producing both new CRC-SCs and partially differentiated precursor cells. The second model posits that after chemotherapy, only CSCs persist, and those that gain chemoresistance evolve into more aggressive and drug-resistant cell phenotypes due to accumulating mutations (Patsalias & Kozovska, 2021).

While non-CSCs or differentiated cancer cells typically undergo apoptosis in response to chemotherapy, CSCs demonstrate resistance to these treatments, evading programmed cell death. The remaining CSCs are then capable of restoring their population, ultimately leading to cancer recurrence. CSCs have been observed to withstand therapeutic interventions across various cancers, including CRC. The selective activation of checkpoints of DNA damage in these CSCs exhibits resistance to radiotherapy. Therapy resistance in CRC is attributed to several mechanisms involving

CSCs. These include the activation of key growth signaling pathways like Hedgehog, TGF-β, Wnt/β-catenin, Hippo, Notch, and PI3K/AKT, as well as the dysregulation of miRNAs. For example, the dysregulation of various miRNAs plays a key role in the development of a chemo-resistant phenotype in CR-CSCs. Additionally, the ability to evade drug-induced DNA damage, activate a quiescent state, and undergo a metabolic shift to a more stable phenotype collectively strengthens CSCs in CRC, making them highly resistant to treatment (Das et al. 2020).

The idea of CSCs being responsible for resistance to therapy originated from the observations of cell cultures and animal models with CRC. In colon carcinoma cell cultures, direct analysis of apoptosis revealed that chemotherapy led to the death of differentiated cancer cells, or non-CSCs. In contrast, CSCs were able to withstand the treatment and survived.

In vitro studies have provided substantial evidence of CSCs evading chemotherapy in colon carcinoma models. Specifically, in xenograft models, an increase in $CD133^+$ tumor cells, a marker of CSCs, was observed following oxaliplatin treatment. These $CD133^+$ CSCs demonstrated higher resistance to oxaliplatin compared to $CD133^-$ differentiated non-CSCs in vivo. Moreover, colon cancer cells expressing stem cell surface markers such as ESA^+, $CD44^+$, and $CD166^+$—which were derived from patients and xenografted into mice—showed heightened resistance to irinotecan, further underscoring the therapeutic challenges posed by CSCs in CRC (Das et al. 2020).

10.6.3 Targeting Signaling Pathways in Colorectal Cancer Stem Cells

Numerous studies have identified several developmental pathways that play a crucial role in regulating CSCs. These signaling pathways are mainly involved in the self-renewal of these CSCs, including Wnt, Hedgehog, and Notch pathways (Ma et al. 2020) (Figure 10.3).

The activation of the **Wnt pathway** in CRC has made it a key target for molecular therapies. Nearly all intestinal tumors exhibit mutations in the Wnt signaling pathway, involving alterations either in the tumor suppressor genes APC and Axin2 or the oncogene β-catenin. In vitro studies have demonstrated that using small molecule inhibitors to either suppress β-catenin activity in the nucleus or prevent its aggregation is effective (Li et al. 2018). Additionally, treatment with β-catenin antisense primers has been shown to reduce β-catenin expression and cancer cell stemness, leading to significant inhibition of tumor growth in CRC animal models (Krishnamurthy & Kurzrock, 2018).

Research has identified the **Notch pathway** as crucial for preserving CSC traits in CR-CSCs. In a CRC animal model, blocking DLL4 in the Notch pathway with the polyclonal antibody 21M18 resulted in a notable reduction in tumor stem cells and significantly suppressed tumor growth. Additionally, in a CRC metastasis model, although imatinib treatment effectively slowed tumor progression, it resulted in a marked increase in the proportion of $ESA^+CD44^+CD166^+$ CCSCs.

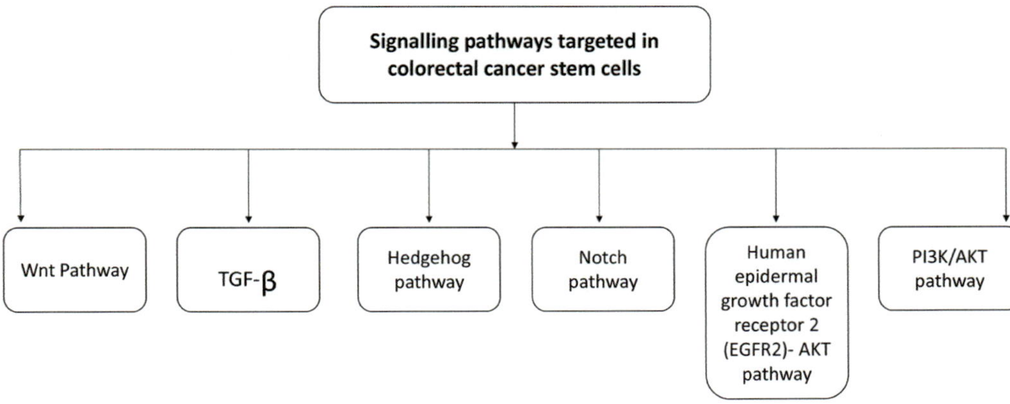

FIGURE 10.3 Commonly targeted signaling pathways in the colorectal cancer stem cells.

However, combining imatinib with hDLL4 antibodies proved more effective, as it significantly reduced both tumor growth and the CSC population compared to using the antibody alone (Wei et al. 2019).

Hedgehog binds to its receptor and activates the **Hedgehog signaling pathway**. $CD133^+$ CRC cells have high levels of the transcription factor Gli1, which triggers Hedgehog activation, whereas this activation is absent in $CD133^-$ CRC cells. The SMO gene can be knocked out, which will lead to the inhibition of the Hedgehog pathway and apoptosis of the CR-CSCs (Ma et al. 2020).

Several other signaling pathways, including the human epidermal growth factor receptor 2 (EGFR2)-AKT pathway and BMI1/signal transducer and activator of transcription 3 (STAT3) pathway, are involved in regulating CSCs. The knockdown of BMI1 in cancer cells has been shown to inhibit the self-renewal of CSCs. BMI1 functions as an epigenetic regulator of several genes and is overexpressed in certain cancers, including CRC. This overexpression drives metastasis, increases tumor grade, and supports the self-renewal of CSCs.

10.7 ROLE OF NANOTECHNOLOGY IN COLORECTAL CANCER

Nanotechnology employs nanoparticles (NPs) for the targeted detection of tumors and cancer biomarkers, advanced drug delivery systems, biologically targeted contrast agents, and new therapeutic approaches (Viswanath et al. 2016).

Nanotechnology plays a pivotal role in cancer treatment by enabling the precise delivery of chemotherapies directly to cancer cells, significantly improving the therapeutic outcome. Furthermore, nanotechnology increases patient survival rates compared to conventional chemotherapy methods. A notable feature of NPs is their dual functionality, allowing them to be used for both diagnosis and treatment. These promising attributes have captivated researchers globally, driving efforts to explore their application in the early detection of CRC and as a potent treatment option (Gogoi et al. 2022). NPs were initially recognized as drug carriers in 1986 when it was discovered that they naturally accumulate in tumor tissues. This passive accumulation of NPs, referred to as the "enhanced permeability and retention" effect, represented a major advancement in cancer treatment using NPs. NPs are considered a promising tool in the field of cancer therapeutics due to their high specificity, rapid response, effective targeting of tumor sites, and extended blood circulation time (Carvalho et al. 2019).

10.7.1 Role of Nanotechnology in Diagnosis of Colorectal Cancer: Image Enhancing

Recent advancements in nanoscience have enabled the development of various NP-based systems for therapeutic and diagnostic purposes. While the clinical applications of nano-based theranostics remain somewhat limited—likely due to their complex pharmacokinetics—NPs have the potential to enhance our understanding of the biochemical and physiological principles underlying several diseases and their treatments (Siddique & Chow, 2020).

NPs hold great potential in facilitating early diagnosis and monitoring the effectiveness of treatments. Early detection of CRC is crucial for its prevention and can significantly influence the long-term survival rates of patients diagnosed with the disease (Viswanath et al. 2016).

Various nanostructures are also being employed in in vitro diagnostic assays targeting protein markers or nucleic acids. In vivo imaging with nanostructures, similar to in vitro diagnostics, presents a promising avenue for colon cancer diagnosis. However, there are still limited examples of nanotechnology applications in CRC imaging. Notably, novel MRI contrast agents represent some of the most promising technologies in CRC detection. The introduction of nanostructures that modify conventional contrast agents, like gadolinium, or imaging agents such as iron oxide, can enhance the detection capabilities of medical imaging (Siddique & Chow, 2020). These nanostructures not only improve the functionality of traditional MRI imaging but also open up new avenues for the early diagnosis and treatment of CRC.

Incorporating diverse contrast agents—such as radioactive materials, superparamagnetic particles, or fluorescent compounds—along with targeting moieties and biocompatible coatings represents a viable approach to NP design (Naseri et al. 2018). Gadolinium and metal chelate-based contrast agents offer several limitations, such as low tissue specificity, rapid clearance, and nonspecific distribution. Therefore, nanotechnology can enhance these agents to improve the sensitivity and specificity of CRC diagnostics (Cisterna et al. 2016). Nanotechnology-based contrast agents significantly improve the effectiveness of cancer diagnostic imaging methods, including positron emission tomography, MRI, and various other optical imaging techniques. The main goal of this technology is to identify the smallest possible clusters of cancer cells. Positron emission tomography stands out as a powerful cancer diagnostic tool due to its high sensitivity, employing a range of biomarkers for molecular and metabolic imaging, such as fluorodeoxyglucose, 9mTc-labeled polyethylene glycol, and [^{18}F–DCFPyL] (Shahbazi-Gahrouei et al. 2019).

Recently, NPs like 56Fe and 14C linked with superparamagnetic iron oxide nanoparticles (SPIONs), a 99mTc-radiolabeled nanosilica system conjugated with a trastuzumab half-chain, and MUC1 aptamer-conjugated mesoporous silica NPs have been utilized as radiolabeled tracers for positron emission tomography (Gogoi et al. 2022).

Since positron emission tomography does not provide anatomical details, it is often combined with other imaging modalities, such as CT or MRI. Common contrast agents employed in MRI include SPIONs, gadolinium-encapsulated silicon microparticles, and gadolinium-ion-doped upconversion NPs. Meanwhile, 2-deoxy-D-glucose-labeled gold NPs and iodinated gold nanoclusters (AuNCs-BSA-I) are employed in CT imaging. The emergence of nanotechnology has led to significant advancements in imaging techniques, enabling earlier detection of cancer and facilitating prompt treatment (Shahbazi-Gahrouei et al. 2019).

Apart from various imaging methods, nanocrystals are often employed in the early detection and diagnosis of CRC. Quantum dots (QDs) are semiconductor nanocrystals that emit fluorescence upon exposure to light. They possess exceptional optical properties, including high brightness, resistance to photobleaching, and the capability to fluoresce across a wide range of wavelengths (Wagner et al. 2019). QDs-based nanotechnology is an emerging platform in cancer research, particularly in the study of CRC, due to its advantageous optical and chemical properties (Al-Joufi et al. 2022).

Park et al. introduced a multiplexed detection technique for the swift and precise diagnosis of CRC by integrating an enzyme-sensitive fluorescent dye with antibody-QD conjugates. The researchers utilized a cresyl violet–glutamic acid derivative as the fluorescent molecular probe, which displayed a two-color fluorescence response depending on the enzymatic activity of k-glutamyl transpeptidase (GGT). They developed a biocompatible $AgInS_2/ZnS$ (core/shell) QD probe that emitted fluorescence at 690 nm by conjugating it with an anti-matrix metalloproteinase-14 antibody. This probe yielded significant results in both ex vivo and in vitro experiments, utilizing human CRC cell lines, patient tumor samples, and tissues from experimental animal models. Through two-photon microscopy, the team demonstrated that the combined application of both probes allowed for quick tissue penetration at depths of 10–20 mm. The findings indicated that employing this dual-probe approach facilitated rapid (within 5 minutes) and accurate imaging of tumor lesions, which traditional colonoscopy methods often overlook. Furthermore, it effectively identified hyperplastic polyps and adenomas in the early stages of CRC (Park et al. 2018).

Carbary-Ganz et al. introduced a dual-modality imaging approach that integrates optical coherence tomography with laser-induced fluorescence, allowing for non-invasive endoscopic CRC visualization while minimizing patient discomfort. This technique enables concurrent monitoring of both biochemical and morphological changes as CRC progresses over time. The method achieved high sensitivity and specificity by employing QDot655-VEGFR2, which selectively targets the colon in animals exposed to carcinogens, thereby differentiating between diseased and healthy tissues in ex vivo analyses (Carbary-Ganz et al. 2015).

The development of ultra-sensitive nanobiosensors has gained considerable attention in the field of metal nanostructures (Wang et al. 2017). It has been established that noble metal NPs, such as platinum (Pt), gold (Au), and silver (Ag), are essential in creating biosensors that cater to the increasing need for precise and highly sensitive diagnostic platforms (Sugumaran et al. 2018). Over the past few decades, gold NPs have emerged as particularly promising candidates for use in biological applications, especially in cancer detection, due to their unique properties. One significant advantage of gold NPs is their ability to utilize surface plasmon resonance, which makes them ideal for incorporation into NIR imaging systems (Elahi et al. 2019).

10.7.2 Role of Nanotechnology in the Treatment of Colorectal Cancer

NP-based medicines are being developed as therapeutic options for cancer treatment, resulting in significant advancements in pharmaceutical research. These innovations have led to a reduction in the adverse effects of cytotoxic drugs while simultaneously enhancing their efficacy and solubility. Utilizing the enhanced permeability and retention effect afforded by tumor vascularization and lymphatic drainage, first-generation anticancer NPs can passively infiltrate tumor tissue. This allows for the extravasation of NPs and their accumulation within cancer cells, thereby improving therapeutic efficacy (Al-Joufi et al. 2022).

The main functional application of nanotechnology in oncology is in drug delivery. Utilizing NPs to formulate drug delivery systems has addressed several challenges, including MDR, stability, efficacy, and biocompatibility. By enhancing these factors, nanotechnology facilitates more effective and targeted treatment options for cancer patients (Gogoi et al. 2022). Nanometer-sized particles of various designs and configurations have proven to be effective and promising new strategies for diagnosing, developing, and treating various malignancies (Al-Joufi et al. 2022). The different NPs that have been currently demonstrated to play an important role in the diagnosis and prognosis of CRC are discussed below and depicted in Figure 10.4.

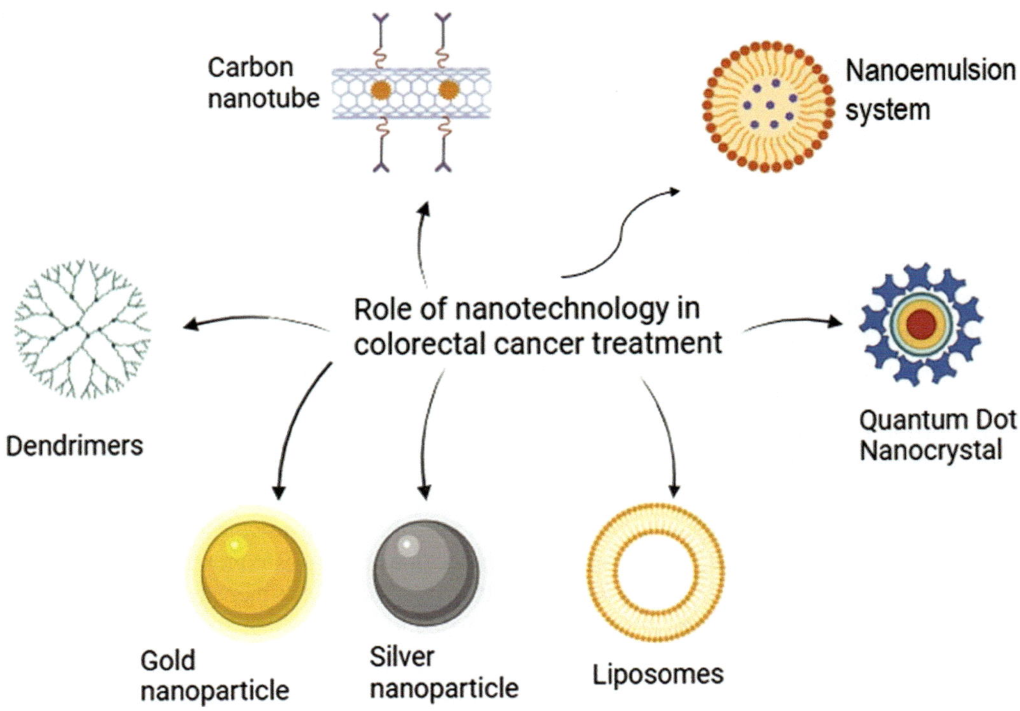

FIGURE 10.4 Different nanoparticles employed in the diagnosis and treatment of colorectal cancer.

10.7.2.1 Dendrimers

Dendrimers are intricately branched, spherical molecules with three-dimensional chemical structures that have proven to be exceptionally beneficial in nanopharmaceuticals due to their biodegradable backbones. Their adaptability in the delivery of anticancer drugs and their role in theranostic applications in cancer therapy have been well established. Anticancer-conjugated dendrimers can efficiently deliver drugs into cells, bypassing efflux transporters and significantly enhancing the bioavailability of the therapeutic compounds they carry. They are also employed to deliver diagnostic agents specifically designed for imaging tumors (Li et al. 2016). Several techniques have been employed to detect colorectal circulating tumor cells using dendrimer–antibody conjugates. One such method involves PAMAM dendrimers linked to Sialyl Lewis X antibodies, which are used to capture colon cancer HT29 cells (Gogoi et al. 2022).

In addition to their diagnostic applications, dendrimers have also been explored for use in anticancer therapy in vitro. For instance, G4 PAMAM dendrimers conjugated with capecitabine have been shown to reduce tumor size while minimizing the side effects associated with capecitabine treatment. Gold NPs encapsulated within PAMAM dendrimers conjugated with curcumin exhibited enhanced cellular uptake, internalization, and cytotoxicity in C26 and HT29 CRC cells. Furthermore, camptothecin-loaded PEGylated PAMAM dendrimers functionalized with AS1411, an anti-nucleolin aptamer, have demonstrated site-specific targeting of CRC cells. Similarly, L-lysine dendrimers with polyoxazoline conjugated to SN-38 (the active metabolite of irinotecan) improved therapeutic efficacy and minimized side effects. PAMAM G4 dendrimers with oxaliplatin also showed enhanced targeting efficiency toward folic acid receptor-expressing CRC cells in vitro. Additionally, gemcitabine-loaded YIGSR-CMCht/PAMAM dendrimer NPs induced targeted cell death in HCT-116 cancer cells (Gogoi et al. 2022).

10.7.2.2 Liposomes

Liposomes are man-made, non-toxic vesicles made from lipids and are commonly used as drug delivery carriers. They were first introduced as nanocarriers in 1961 and have since become an FDA-approved system for delivering drugs. Liposomes consist of a small, spherical aqueous core surrounded by a phospholipid bilayer. One of their major advantages is their small size, which enables efficient and targeted delivery of drugs for both treatment and diagnosis while reducing side effects. Among NPs, liposomes are particularly well-suited for delivering nucleic acids, proteins, and peptides. Depending on their properties, liposomes can be classified into three types: (i) stealth or long-circulating liposomes, (ii) active targeting liposomes, and (iii) thermosensitive, pH-sensitive, or magnetic liposomes. Examples of FDA-approved liposomal drugs include Doxil (doxorubicin), DaunoXome, and Marqibo®, while Thermodox® is a thermosensitive liposome used in CRC treatment (Allen & Cullis 2013).

10.7.2.3 Nanoparticles

Silica and gold NPs are extensively used for precise drug delivery, playing a vital role in CRC treatment. Silica NPs are particularly advantageous due to their highly porous structure, low toxicity, pH sensitivity, biocompatibility, and ease of functionalization, making them highly suitable for anticancer therapies. In CRC treatment, silica NPs are often combined with photosensitizer chemicals. When exposed to light, these photosensitizers become activated, releasing reactive oxygen species that target and destroy cancer cells (Bharti et al. 2015). However, gold NPs are considered safer and more efficient for drug delivery, with a high specificity for targeting cancer cells. They are the most stable among noble metal NPs, making them particularly suitable for therapeutic applications. Gold NPs have been shown to enhance the delivery efficacy of cisplatin, a common chemotherapy drug, and have demonstrated the ability to effectively decompress CRC blood vessels, improving treatment outcomes (Zhao et al. 2018). Nanodrug delivery systems enclose therapeutic agents and enable targeted delivery specifically to the tumor environment, which helps to reduce side effects (Gogoi et al. 2022).

10.7.2.4 Carbon Nanotubes

Carbon nanotubes are nanoscale tubular structures made up of carbon atoms arranged in a honeycomb lattice, exhibiting remarkable physicochemical properties. These carbon nanomaterials have diverse applications in areas such as diagnostics, gene therapy, immunotherapy, and drug delivery systems. For instance, CpG-conjugated carbon nanotubes enhance the uptake of CpG sequences in mouse colon cancer cells and activate nuclear transcription factor-kappa B signaling pathways. The conjugation of CpG with carbon nanotubes has been shown to effectively reduce local xenograft tumor growth and limit liver metastasis (Gogoi et al. 2022).

10.7.2.5 Nanoemulsion System

The nanoemulsion system creates a transparent colloidal solution with several advantageous properties, including low toxicity, pH sensitivity, high stability, thermosensitivity, and enhanced efficacy. Within the tumor microenvironment, vascularized tissue serves as a barrier, which nanoemulsions can effectively bypass to reach cancer cells. These systems are particularly valuable for delivering water-insoluble drugs housed within a hydrophobic core. Nanoemulsions are frequently conjugated with antibodies or polyethylene glycol to achieve effective drug delivery and precise targeting (Silva et al. 2011).

10.8 CONCLUSION

Over the past decade, the field of nanomedicine has seen remarkable progress, with nanotechnology becoming increasingly significant due to its highly effective drug delivery systems. The widespread application of nanomaterials for disease diagnosis and treatment is currently a notable trend. The integration of therapeutic and diagnostic functions, known as theranostics, simplifies and accelerates treatment protocols. CRC is primarily linked to factors such as gender, lifestyle, and race, indicating that specific populations are at a higher risk. Therefore, routine screening for CRC is strongly recommended to aid in prevention.

Early diagnosis plays a critical role in improving treatment outcomes and survival rates for CRC patients. Although the theranostic application of nanomaterials for CRC is still in the developmental phase, numerous researchers have reported successful in vitro treatments in cancer cell lines as well as promising in vivo results in various CRC animal models. However, novel strategies are essential to enhance existing therapeutic options. There is optimism that innovative therapies utilizing nanomaterials for CRC will soon become available for clinical use.

REFERENCES

Acs, B., Rantalainen, M., & Hartman, J. (2020). Artificial intelligence as the next step towards precision pathology. *Journal of Internal Medicine*, 288(1), 62–81.

Al-Joufi, F. A., Setia, A., Salem-Bekhit, M. M., Sahu, R. K., Alqahtani, F. Y., Widyowati, R., & Aleanizy, F. S. (2022). Molecular pathogenesis of colorectal cancer with an emphasis on recent advances in biomarkers, as well as nanotechnology-based diagnostic and therapeutic approaches. *Nanomaterials*, 12(1), 169.

Allen, T. M., & Cullis, P. R. (2013). Liposomal drug delivery systems: from concept to clinical applications. *Advanced Drug Delivery Reviews*, 65(1), 36–48.

Alzubaidi, L., Zhang, J., Humaidi, A. J., Al-Dujaili, A., Duan, Y., Al-Shamma, O., Santamaria, J., Fadhel, M. A., Al-Amidie, M., & Farhan, L. (2021). Review of deep learning: concepts, CNN architectures, challenges, applications, future directions. *Journal of Big Data*, 8(1), 53.

Amisha, P. M., Pathania, M., & Rathaur, V. K. (2019). Overview of artificial intelligence in medicine. *Journal of Family Medicine and Primary Care*, 8(7), 2328–2331.

Avram, M. F., Lazăr, D. C., Mariş, M. I., & Olariu, S. (2023). Artificial intelligence in improving the outcome of surgical treatment in colorectal cancer. In *Frontiers in Oncology* (Vol. 13). Frontiers Media S.A. https://doi.org/10.3389/fonc.2023.1116761.

Bharti, C., Nagaich, U., Pal, A. K., & Gulati, N. (2015). Mesoporous silica nanoparticles in target drug delivery system: a review. *International Journal of Pharmaceutical Investigation*, 5(3), 124.

Breiman, L. (2001). Random forests. *Machine Learning*, 45(1), 5–32.
Cai, Y., Zhou, Y., Li, Z., Xiang, J., & Chen, Z. (2013). Surgical outcome of laparoscopic colectomy for colorectal cancer in obese patients: a comparative study with open colectomy. *Oncology Letters*, 6(4), 1057–1062.
Carbary-Ganz, J. L., Welge, W. A., Barton, J. K., & Utzinger, U. (2015). In vivo molecular imaging of colorectal cancer using quantum dots targeted to vascular endothelial growth factor receptor 2 and optical coherence tomography/laser-induced fluorescence dual-modality imaging. *Journal of Biomedical Optics*, 20(9), 096015–096015.
Carvalho, M. R., Carvalho, C. R., Maia, F. R., Caballero, D., Kundu, S. C., Reis, R. L., & Oliveira, J. M. (2019). Peptide-modified dendrimer nanoparticles for targeted therapy of colorectal cancer. *Advanced Therapeutics*, 2(11), 1900132.
Chen, S., Song, X., Chen, Z., Li, X., Li, M., Liu, H., & Li, J. (2013). CD133 expression and the prognosis of colorectal cancer: a systematic review and meta-analysis. *PloS One*, 8(2), e56380.
Cherciu, I. R. I. N. A., Bărbălan, A., Pirici, D., Mărgăritescu, C., & Săftoiu, A. (2014). Stem cells, colorectal cancer and cancer stem cell markers correlations. *Current Health Sciences Journal*, 40(3), 153.
Choi, D., Lee, H. W., Hur, K. Y., Kim, J. J., Park, G. S., Jang, S. H., & Paik, S. S. (2009). Cancer stem cell markers CD133 and CD24 correlate with invasiveness and differentiation in colorectal adenocarcinoma. *World Journal of Gastroenterology: WJG*, 15(18), 2258.
Choi, R. Y., Coyner, A. S., Kalpathy-Cramer, J., Chiang, M. F., & Peter Campbell, J. (2020). Introduction to machine learning, neural networks, and deep learning. *Translational Vision Science & Technology*, 9(2). https://doi.org/10.1167/tvst.9.2.14.
Cisterna, B. A., Kamaly, N., Choi, W. I., Tavakkoli, A., Farokhzad, O. C., & Vilos, C. (2016). Targeted nanoparticles for colorectal cancer. *Nanomedicine*, 11(18), 2443–2456.
Das, P. K., Islam, F., & Lam, A. K. (2020). The roles of cancer stem cells and therapy resistance in colorectal carcinoma. *Cells*, 9(6), 1392.
Donada, M., Bonin, S., Barbazza, R., Pettirosso, D., & Stanta, G. (2013). Management of stage II colon cancer - the use of molecular biomarkers for adjuvant therapy decision. *BMC Gastroenterology,* 13 (February): 36.
Elahi, N., Kamali, M., Baghersad, M. H., & Amini, B. (2019). A fluorescence nano-biosensors immobilization on iron (MNPs) and gold (AuNPs) nanoparticles for detection of Shigella spp. *Materials Science and Engineering: C,* 105, 110113.
Feeney, G., Sehgal, R., Sheehan, M., Hogan, A., Regan, M., Joyce, M., & Kerin, M. (2019). Neoadjuvant radiotherapy for rectal cancer management. *World Journal of Gastroenterology: WJG*, 25(33): 4850–4869.
Ferrari, A., Neefs, I., Hoeck, S., Peeters, M., & Van Hal, G. (2021). Towards novel non-invasive colorectal cancer screening methods: a comprehensive review. *Cancers*, 13(8), 1820.
Gockel, I., Moehler, M., Frerichs, K., Drescher, D., Trinh, T. T., Duenschede, F., Borschitz, T., Schimanski, K., Biesterfeld, S., Herzer, K., Galle, P. R., Lang, H., Junginger, T., & Schimanski, C. C. (2008). Co-expression of receptor tyrosine kinases in esophageal adenocarcinoma and squamous cell cancer. *Oncology Reports*, 20(4), 845–850.
Gogoi, P., Kaur, G., & Singh, N. K. (2022). Nanotechnology for colorectal cancer detection and treatment. *World Journal of Gastroenterology*, 28(46), 6497.
Haraguchi, N., Ohkuma, M., Sakashita, H., Matsuzaki, S., Tanaka, F., Mimori, K., … & Mori, M. (2008). CD133+ CD44+ population efficiently enriches colon cancer initiating cells. *Annals of Surgical Oncology*, 15, 2927–2933.
Horst, D., Kriegl, L., Engel, J., Kirchner, T., & Jung, A. (2009). Prognostic significance of the cancer stem cell markers CD133, CD44, and CD166 in colorectal cancer. *Cancer Investigation*, 27(8), 844–850.
Horvat, N., Veeraraghavan, H., Nahas, C. S., Bates, D. D., Ferreira, F. R., Zheng, J., … & Petkovska, I. (2022). Combined artificial intelligence and radiologist model for predicting rectal cancer treatment response from magnetic resonance imaging: an external validation study. *Abdominal Radiology,* 47(8), 2770–2782.
Hsu, C.-M., Hsu, C.-C., Hsu, Z.-M., Shih, F.-Y., Chang, M.-L., & Chen, T.-H. (2021). Colorectal polyp image detection and classification through grayscale images and deep learning. *Sensors*, 21(18) https://doi.org/10.3390/s21185995.
Kamel, F., Eltarhoni, K., Nisar, P., & Soloviev, M. (2022). Colorectal cancer diagnosis: the obstacles we face in determining a non-invasive test and current advances in biomarker detection. *Cancers*, 14(8). https://doi.org/10.3390/cancers14081889.
Kasahara, K., Katsumata, K., Saito, A., Ishizaki, T., Enomoto, M., Mazaki, J., … & Tsuchida, A. (2022). Artificial intelligence predicts lymph node metastasis or risk of lymph node metastasis in T1 colorectal cancer. *International Journal of Clinical Oncology*, 27(10), 1570–1579.

Kolisnik, T., Sulit, A. K., Schmeier, S., Frizelle, F., Purcell, R., Smith, A., & Silander. O. 2023. Identifying important microbial and genomic biomarkers for differentiating rightversus left-sided colorectal cancer using random forest models. *BMC Cancer*, 23(1): 647.

Krishnamurthy, N., & Kurzrock, R. (2018). Targeting the Wnt/beta-catenin pathway in cancer: update on effectors and inhibitors. *Cancer Treatment Reviews*, 62, 50–60.

Langan, R. C., Mullinax, J. E., Raiji, M. T., Upham, T., Summers, T., Stojadinovic, A., & Avital, I. (2013). Colorectal cancer biomarkers and the potential role of cancer stem cells. *Journal of Cancer*, 4(3), 241.

Li, H., Lin, J., Xiao, Y., Zheng, W., Zhao, L., Yang, X., Zhong, M., & Liu, H. (2021). Colorectal cancer detected by machine learning models using conventional laboratory test data. *Technology in Cancer Research & Treatment*, 20, 15330338211058352.

Li, P., Song, G., Wu, R., Li, H., Zhang, R., Zuo, P., & Li, A. (2021). Multiparametric MRI-based machine learning models for preoperatively predicting rectal adenoma with canceration. *Magnetic Resonance Materials in Physics, Biology and Medicine*, 34(5), 1–10.

Li, T., Jian, X., He, H., Lai, Q., Li, X., Deng, D., ... & Ding, Y. (2018). MiR-452 promotes an aggressive colorectal cancer phenotype by regulating a Wnt/β-catenin positive feedback loop. *Journal of Experimental & Clinical Cancer Research*, 37, 1–15.

Li, T., Smet, M., Dehaen, W., & Xu, H. (2016). Selenium–platinum coordination dendrimers with controlled anti-cancer activity. *ACS Applied Materials & Interfaces*, 8(6), 3609–3614.

Liu, X., Guo, S., Zhang, H., He, K., Mu, S., Guo, Y., & Li, X. (2019). Accurate colorectal tumor segmentation for CT scans based on the label assignment generative adversarial network. *Medical Physics*, 46(8), 3532–3542.

Liu, X., Zhang, D., Liu, Z., Li, Z., Xie, P., Sun, K., ... & Tian, J. (2021). Deep learning radiomics-based prediction of distant metastasis in patients with locally advanced rectal cancer after neoadjuvant chemoradiotherapy: a multicentre study. *EBioMedicine*, 69, 103442.

Ma, Y. S., Li, W., Liu, Y., Shi, Y., Lin, Q. L., & Fu, D. (2020). Targeting colorectal cancer stem cells as an effective treatment for colorectal cancer. *Technology in Cancer Research & Treatment*, 19, 1533033819892261.

Mohamed, S., Ashraf, R., Ghanem, A., Sakr, M., & Mohamed, R. (2022). Supervised machine learning techniques: a comparison.

Mohammad Rezaei, F., Hashemzadeh, S., Gavgani, R. R., Feizi, M. H., Pouladi, N., Kafil, H. S., Rostamizadeh, L., Oskooei, V. K., Taheri, M., & Sakhinia, E. (2019). Dysregulated KDR and FLT1 gene expression in colorectal cancer patients. *Reports of Biochemistry & Molecular Biology*, 8(3), 244–252.

Moossavi, S., & Ansari, R. (2013). Intestinal stem cell imaging in colorectal cancer screening. *Journal of Stem Cells & Regenerative Medicine*, 9(2), 37.

Naseri, N., Ajorlou, E., Asghari, F., & Pilehvar-Soltanahmadi, Y. (2018). An update on nanoparticle-based contrast agents in medical imaging. Artificial Cells, *Nanomedicine, and Biotechnology*, 46(6), 1111–1121.

Nduma, B. N., Nkeonye, S., Uwawah, T. D., Kaur, D., Ekhator, C., & Ambe, S. (2024). Use of artificial intelligence in the diagnosis of colorectal cancer. *Cureus*, 16(1), e53024.

Padmanabha Reddy, Y. C. A., Viswanath, P., & Reddy, B. E. (2018). Semi-supervised learning: a brief review. *International Journal of Engineering & Technology*, 7(1.8), 81.

Parizadeh, S. M., Jafarzadeh-Esfehani, R., Hassanian, S. M., Parizadeh, S. M. R., Vojdani, S., Ghandehari, M., ... & Avan, A. (2019). Targeting cancer stem cells as therapeutic approach in the treatment of colorectal cancer. *The International Journal of Biochemistry & Cell Biology*, 110, 75–83.

Park, Y., Ryu, Y. M., Wang, T., Jung, Y., Kim, S., Hwang, S., ... & Myung, S. J. (2018). Colorectal cancer diagnosis using enzyme-sensitive ratiometric fluorescence dye and antibody–quantum dot conjugates for multiplexed detection. *Advanced Functional Materials*, 28(4), 1703450.

Patsalias, A., & Kozovska, Z. (2021). Personalized medicine: stem cells in colorectal cancer treatment. *Biomedicine & Pharmacotherapy*, 141, 111821.

Petrelli, F., Borgonovo, K., Cabiddu, M., Ghilardi, M., & Barni, S. (2011). Cetuximab and panitumumab in KRAS wild-type colorectal cancer: a meta-analysis. *International Journal of Colorectal Disease*, 26(7), 823–833.

Quero, G., Mascagni, P., Kolbinger, F. R., Fiorillo, C., De Sio, D., Longo, F., Schena, C. A., Laterza, V., Rosa, F., Menghi, R., Papa, V., Tondolo, V., Cina, C., Distler, M., Weitz, J., Speidel, S., Padoy, N., & Alfieri, S. (2022). Artificial Intelligence in colorectal cancer surgery: present and future perspectives. *Cancers*, 14(15).https://doi.org/10.3390/cancers14153803

Rathore, S., Hussain, M., & Khan, A. (2015). Automated colon cancer detection using hybrid of novel geometric features and some traditional features. *Computers in Biology and Medicine*, 65, 279–296.

Reggiani Bonetti, L., Migaldi, M., Caredda, E., Boninsegna, A., Ponz De Leon, M., Di Gregorio, C., ... & Sgambato, A. (2012). Increased expression of CD133 is a strong predictor of poor outcome in stage I colorectal cancer patients. *Scandinavian Journal of Gastroenterology*, 47(10), 1211–1217.

Rocca, A., Brunese, M. C., Santone, A., Avella, P., Bianco, P., Scacchi, A., ... & Brunese, L. (2021). Early diagnosis of liver metastases from colorectal cancer through CT radiomics and formal methods: a pilot study. *Journal of Clinical Medicine*, 11(1), 31.

Schork, N. J. (2019). Artificial intelligence and personalized medicine. *Cancer Treatment and Research*, 178, 265–283.

Sengupta, A., Singh, S. K., & Kumar, R. (2024). Support vector machine-based prediction models for drug repurposing and designing novel drugs for colorectal cancer. *ACS Omega*, 9(16), 18584–18592.

Shahbazi-Gahrouei, D., Khaniabadi, P. M., Khaniabadi, B. M., & Shahbazi-Gahrouei, S. (2019). Medical imaging modalities using nanoprobes for cancer diagnosis: a literature review on recent findings. *Journal of Research in Medical Sciences*, 24(1), 38.

Sharkas, M., & Attallah, O. (2024). Color-CADx: A deep learning approach for colorectal cancer classification through triple convolutional neural networks and discrete cosine transform. *Scientific Reports*, 14(1), 6914.

Siddique, S., & Chow, J. C. (2020). Gold nanoparticles for drug delivery and cancer therapy. *Applied Sciences*, 10(11), 3824.

Silva, R., Ferreira, H., & Cavaco-Paulo, A. (2011). Sonoproduction of liposomes and protein particles as templates for delivery purposes. *Biomacromolecules*, 12(10), 3353–3368.

Song, D., Zhang, Z., Li, W., Yuan, L., & Zhang, W. (2022). Judgment of benign and early malignant colorectal tumors from ultrasound images with deep multi-view fusion. *Computer Methods and Programs in Biomedicine*, 215, 106634.

Sugumaran, S., Jamlos, M. F., Ahmad, M. N., Bellan, C. S., & Schreurs, D. (2018). Nanostructured materials with plasmonic nanobiosensors for early cancer detection: a past and future prospect. *Biosensors and Bioelectronics*, 100, 361–373.

Suh, K., Carlson, J. J., Xia, F., Williamson, T., & Sullivan, S. D. (2022). Comparative effectiveness of larotrectinib versus entrectinib for the treatment of metastatic NTRK gene fusion cancers. *Journal of Comparative Effectiveness Research*, 11(14), 1011–1019.

Triantafillidis, J. K., Vagianos, C., & Malgarinos, G. (2015). Colonoscopy in colorectal cancer screening: current aspects. *Indian Journal of Surgical Oncology*, 6(3), 237–250.

Vaiopoulos, A. G., Kostakis, I. D., Koutsilieris, M., & Papavassiliou, A. G. (2012). Colorectal cancer stem cells. *Stem Cells*, 30(3), 363–371.

Viswanath, B., Kim, S., & Lee, K. (2016). Recent insights into nanotechnology development for detection and treatment of colorectal cancer. *International Journal of Nanomedicine*, 2491–2504.

Wagner, A. M., Knipe, J. M., Orive, G., & Peppas, N. A. (2019). Quantum dots in biomedical applications. *Acta Biomaterialia*, 94, 44–63.

Wang, H., Li, Z.-W., Ou, Q., Wu, X., Nagasaka, M., Shao, Y., Ou, S.-H. I., & Yang, Y. (2022). NTRK fusion positive colorectal cancer is a unique subset of CRC with high TMB and microsatellite instability. *Cancer Medicine*, 11(13), 2541–2549.

Wang, Y., He, X., Nie, H., Zhou, J., Cao, P., & Ou, C. (2020). Application of artificial intelligence to the diagnosis and therapy of colorectal cancer. *American Journal of Cancer Research*, 10(11), 3575.

Wang, Y., Zhou, J., & Li, J. (2017). Construction of plasmonic nano-biosensor-based devices for point-of-care testing. *Small Methods*, 1(11), 1700197.

Wang, J. Y., Huang, J. C., Chen, G., & Wei, D. M. (2018). Expression level and potential target pathways of miR-1-3p in colorectal carcinoma based on 645 cases from 9 microarray datasets. *Molecular Medicine Reports*, 17(4), 5013-5020.

Wang, L. (2005). Support Vector Machines: Theory and Applications. Springer Science & Business Media.

Weerarathna, I. N., Kamble, A. R., & Luharia, A. (2023). Artificial intelligence applications for biomedical cancer research: a review. *Cureus*. https://doi.org/10.7759/cureus.48307

Wei, F., Zhang, T., Deng, S. C., Wei, J. C., Yang, P., Wang, Q.,... & Cao, J. (2019). PD-L1 promotes colorectal cancer stem cell expansion by activating HMGA1-dependent signaling pathways. *Cancer Letters*, 450, 1–13.

Weichert, W., Knösel, T., Bellach, J., Dietel, M., & Kristiansen, G. (2004). ALCAM/CD166 is overexpressed in colorectal carcinoma and correlates with shortened patient survival. *Journal of Clinical Pathology*, 57(11), 1160–1164.

Wong, C., Fu, Y., Li, M., Mu, S., Chu, X., Fu, J., & Others. (2022). MRI-based artificial intelligence in rectal cancer. *Journal of Magnetic Resonance Imaging*, 57(1), 45–56. https://doi.org/10.1002/jmri.28381

Yin, Z., Yao, C., Zhang, L., & Qi, S. (2023). Application of artificial intelligence in diagnosis and treatment of colorectal cancer: a novel prospect. *Frontiers in Medicine*, 10, 1128084.

Yu, G., Sun, K., Xu, C., Shi, X. H., Wu, C., Xie, T., … & Deng, H. W. (2021). Accurate recognition of colorectal cancer with semi-supervised deep learning on pathological images. *Nature Communications*, 12(1), 6311.

Zhang, X., Yang, Y., Wang, Y., & Fan, Q. (2019). Detection of the BRAF V600E mutation in colorectal cancer by NIR spectroscopy in conjunction with counter propagation artificial neural network. *Molecules*, 24(12), 2238.

Zhao, X., Pan, J., Li, W., Yang, W., Qin, L., & Pan, Y. (2018). Gold nanoparticles enhance cisplatin delivery and potentiate chemotherapy by decompressing colorectal cancer vessels. *International Journal of Nanomedicine*, 13, 6207–6221.

Zhou, H., Yang, Z., Yue, J., Chen, Y., Chen, T., Mu, T., … & Bi, X. (2020). Identification of potential hub genes via bioinformatics analysis combined with experimental verification in colorectal cancer. *Molecular Carcinogenesis*, 59(4), 425–438.

Index

D

E

F

G

H

I

J

K

L

M

Q

R

S

T

U

V

W

X